THE MATHEMATICS OF SECRETS

THE MATHEMATICS OF
SECRETS

CRYPTOGRAPHY FROM
CAESAR CIPHERS TO
DIGITAL ENCRYPTION

■ ■ ■

JOSHUA HOLDEN

■ ■ ■

PRINCETON UNIVERSITY PRESS
PRINCETON AND OXFORD

press.princeton.edu

Jacket image courtesy of Shutterstock; design by Lorraine Betz Doneker

Library of Congress Cataloging-in-Publication Data

Names: Holden, Joshua, 1970– author.
Title: The mathematics of secrets : cryptography from Caesar
ciphers to digital encryption / Joshua Holden.
Description: Princeton : Princeton University Press, [2017] |
Includes bibliographical references and index.
Identifiers: LCCN 2016014840 | ISBN 9780691141756
(hardcover : alk. paper)
Subjects: LCSH: Cryptography—Mathematics. | Ciphers. |
Computer security.
Classification: LCC Z103 .H664 2017 | DDC 005.8/2—dc23 LC
record available at https://lccn.loc.gov/2016014840

British Library Cataloging-in-Publication Data is available

This book has been composed in Linux Libertine
Printed on acid-free paper. ∞

Printed in the United States of America

1   3   5   7   9   10   8   6   4   2

To Lana and Richard for their love and support

▪ ▪ ▪ CONTENTS ▪ ▪ ▪

■ ■ ■ PREFACE ■ ■ ■

This book is about the mathematics behind the modern science of sending secret messages, or cryptography. Modern cryptography *is* a science, and like all modern science, it relies on mathematics. Without the mathematics, you can only go so far in understanding cryptography. I want you to be able to go farther, not only because I think you should know about cryptography, but also because I think the particular kinds of mathematics the cryptographers use are really pretty, and I want to introduce you to them.

In *A Brief History of Time*, Stephen Hawking says that someone told him that each equation he included in the book would halve the sales. I hope that's not true of this book, because there are lots of equations. But I don't think the math is necessarily that hard. I once taught a class on cryptography in which I said that the prerequisite was high school algebra. Probably I should have said that the prerequisite was high school algebra and a willingness to think really hard about it. There's no trigonometry here, no calculus, no differential equations. There are some ideas that don't usually come up in an algebra course, and I'll try to walk you through them. If you want to really understand these ideas, you can do it without any previous college-level math—but you might have to think hard. (The math in some of the sidebars is a little harder, but you can skip those and still understand the rest of the book just fine.)

Mathematics isn't all there is to cryptography. Unlike most sciences, cryptography is about intelligent adversaries who are actively fighting over whether secrets will be revealed. Ian Cassels, who was both a prominent mathematician at Cambridge and a former British cryptanalyst from World War II, had a good perspective on this. He said that "cryptography is a mixture of mathematics and muddle, and without the muddle the mathematics can be used against you." In this book I've removed some of the muddle in order to focus on the mathematics. Some professional cryptographers may take issue with that, because I am not really showing you the most secure systems that I could. In response, I can say only that this book for those interested in learning about a particular part of cryptography, namely, the mathematical foundations. There are many additional books in *Suggestions for Further Reading* and the Bibliography that you should read if you want to become a well-rounded professional.

Here is where I have drawn my personal line: I have tried not to say anything false in this book in the name of simplification, but I have left things out. I have left out some details of how to use the systems most securely, and I have left out some systems that I don't feel contribute to the mathematical story I want to tell. When possible, I have tried to present systems that have actually been used to protect real secrets. However, I have included some that were made up by me or another academic type when I feel that they best illustrate a point.

Computer technology has changed both the types of data with which cryptographers work and the techniques that are feasible. Some of the systems for protecting data that I discuss are either no longer applicable or no longer secure in today's world, even if they were in the past. Likewise, some of the techniques I discuss for breaking these systems are no longer effective in the forms presented here. Despite this, I feel that all the topics in this book illustrate issues that are still important and relevant to modern cryptography. I have tried to indicate how the principles are still used today, even when the actual systems are not. "Looking Forward" at the end of each chapter gives you a preview of how the chapter you just finished relates to the chapters yet to come or to future developments that I think are possible or likely.

A lot of the chapters follow the historical development of their topic, because that development is often a logical progression through the ideas I'm describing. History is also a good way to tell a story, so I like to use it when it fits. There's lots more about the history of cryptography out there, so if you would like to know more, definitely check out *Suggestions for Further Reading.*

I tell my students that I became a math professor because I like math and I like to talk. This book is me talking to you about a particular application of mathematics that I really like. My hope is that by the end of the book, you will really like it too.

■ ■ ■ ACKNOWLEDGMENTS ■ ■ ■

I wish I could individually thank everyone with whom I have ever had a good conversation about math or cryptography, but obviously I can't. I do want to single out some of the people who have particularly helped with my teaching of cryptography: by letting me sit in on their classes, by encouraging me, by teaching with me, or by sharing relevant materials. In roughly chronological order, these include David Hayes, Susan Landau (from whom I learned the "cosmic ray" principle, among many other cryptographic things), Richard Hain, Stephen Greenfield, Gary Sherman (from whom I learned the "shoes and socks" principle), and David Mutchler. I apologize if I've left anyone out.

Thank you to all the attendees of the Algorithmic Number Theory Symposia, particularly Carl Pomerance, Jon Sorenson, Hugh Williams, and all the members of Hugh's "posse" at (or formerly at) the University of Calgary. I'd also like to thank Brian Winkel, Craig Bauer, and the present and past members of the Editorial Board of *Cryptologia*. Without the friendship and encouragement of all of you, I'm sure my cryptography research would never have gotten off of the ground. And thanks go to all my research students at Rose-Hulman and at the Rose-Hulman Summer Research Experience for Undergraduates, who gave me the best reason to keep my research going.

This book has been in progress for a long time and many people have reviewed various drafts of it over the years. Many of you I don't know personally, and I don't even know some of your names, but thank you to all of you. Two people I particularly would like to thank are Jean Donaldson and Jon Sorenson. Jean volunteered to read a very early draft despite my being unable to offer any personal or professional incentive whatsoever. Not being a professional mathematician or cryptographer, she was the perfect audience and everything she said was immensely useful. Jon Sorenson likewise read an early draft and made encouraging and helpful comments. In addition to being a reviewer, Jon has been a colleague and a friend for many years and has helped my career in numerous ways. Paul Nahin, David Kahn, and John MacCormick are also among those who gave me encouraging and helpful reviews.

The staff at Rose-Hulman's Logan Library have been invaluable through this process. Amy Harshbarger has come up with articles and technical reports through Interlibrary Loan that I thought would never be found. And Jan Jerrell let me keep library books far beyond the limits of a reasonable circulation policy. I thank them both, and everyone else at the library, profusely. Speaking of the library, Heather Chenette and Michelle Marincel Payne helped organize the "Shut Up and Write" group that met there and got me through the final revisions.

I could not have done this without the support and tolerance of my wife, Lana, our housemate, Richard, and the cats, who "tolerated" the occasional late dinner. You've put up with a lot through this process. I really appreciate it.

Finally, thank you to everyone at Princeton University Press, especially my editor, Vickie Kearn. Vickie first approached me about writing a cryptography book 12 years ago, and in all that time she never lost faith that it would happen some day. I can't believe it's finally finished. Thanks so much.

THE MATHEMATICS OF SECRETS

1

███████ **1** ████████

Introduction to Ciphers and Substitution

1.1 ALICE AND BOB AND CARL AND JULIUS: TERMINOLOGY AND CAESAR CIPHER

People have been trying to hide the content of written messages almost as long as writing itself has existed and have developed a multitude of different methods of doing it. And almost as soon as people started trying to hide their messages, scholars started trying to classify and describe these methods. Unfortunately, this means that I've got to hit you straight up with a bunch of terminology. Even worse, a lot of words that are used interchangeably in ordinary conversation have specific meanings to experts in the field. It's not really hard to get the hang of what's what, though.

As our first example, people who study secret messages often use the terms **code** and **cipher** to mean two different things. David Kahn, author of perhaps the definitive account of the history of cryptography, said it about as well as anyone could: "A code consists of thousands of words, phrases, letters, and syllables with the codewords or codenumbers . . . that replace these plaintext elements . . . . In ciphers, on the other hand, the basic unit is the letter, sometimes the letter-pair . . . , very rarely larger groups of letters . . . ." A third method of sending secret messages, **steganography**, consists of concealing the very existence of the message, for example, through the use of invisible ink. In this book we will concentrate on ciphers as they are generally the most interesting mathematically, although examples of the other methods may come up from time to time.

A few more terms will be helpful before we get started. The study of how to send secret messages by codes and ciphers is called **cryptography**, whereas the study of how to read such secret messages without permission is called **cryptanalysis**, or **codebreaking**. Together, the two fields make up the field of **cryptology**. (Sometimes cryptography is also used for the two fields combined, but we will try to keep the terms separate.)

It's become customary when talking about cryptology to talk about Alice, who wants to send a message to Bob. We're going to start with Julius, though. That's **Julius Caesar**, who in addition to being *dictator perpetuus* of Rome was also a military genius, a writer, and ... a cryptographer.

Caesar probably wasn't the original inventor of what we now call the **Caesar cipher**, but he certainly made it popular. The Roman historian Suetonius describes the cipher:

> There are also letters of his [Caesar's] to Cicero, as well as to his intimates on private affairs, and in the latter, if he had anything confidential to say, he wrote it in cipher, that is, by so changing the order of the letters of the alphabet, that not a word could be made out. If anyone wishes to decipher these, and get at their meaning, he must substitute the fourth letter of the alphabet, namely D, for A, and so with the others.

In other words, whenever Alice wants to send a message, she first writes out the **plaintext**, or the text of the message in ordinary language. She is going to **encipher** this message, or put it into secret form using a cipher, and the result will be the **ciphertext** of the message. To put it into code would be to **encode** it, and the term **encrypt** can be used for either. For every a in the plaintext, Alice substitutes D in the ciphertext, for every b, she substitutes E, and so on. Each letter is shifted 3 letters down the alphabet. That's perfectly straightforward. The interesting part happens when Alice gets to the end of the alphabet and runs out of letters. The letter w becomes Z, so where does the letter x go? It wraps around, to A! The letter y becomes B and z becomes C. For example, the message "and you too, Brutus" becomes

plaintext:	a	n	d	y	o	u	t	o	o	b	r	u	t	u	s
ciphertext:	D	Q	G	B	R	X	W	R	R	E	U	X	W	X	V

This would be the message Alice sends to Bob.

You have actually used this "wraparound" idea in daily life since you were a child. What's 3 hours after 1:00? It's 4:00. Three hours after 2:00 is 5:00. What's 3 hours after 10:00? It's 1:00. You wrapped around.

It was around 1800 CE when **Carl Friedrich Gauss** codified this wraparound idea formally. It's now called **modular arithmetic**, and the wraparound number is called the **modulus**. A mathematician would write our wraparound clock example like this:

$$10 + 3 \equiv 1 \quad (\text{mod } 12)$$

and read it as "ten plus three is one modulo twelve."

But what about Caesar cipher? We can represent it using modular arithmetic if we are willing to change our letters into numbers. Instead of a think of the number 1, instead of b think of the number 2, and so on. This changing of letters to numbers is not really considered part of the secret cipher. It's a pretty obvious idea to those of us in the digital age, and Alice shouldn't really expect to keep it a secret. Only the operations that we do on the numbers are considered secret.

Now our modulus is 26 and our Caesar cipher looks like this.

plaintext	number	plus 3	ciphertext
a	1	4	D
b	2	5	E
⋮	⋮	⋮	⋮
x	24	1	A
y	25	2	B
z	26	3	C

Remember that the "plus 3" wraps around at 26.

To **decipher** the message, or take it from ciphertext to plaintext, Bob shifts three letters in the opposite direction, left. This time, he has to wrap around when he goes past a, or in terms of numbers, when he goes past 1. 0 wraps to 26, -1 wraps to 25, and so on. In the form we used earlier, that looks like the following.

ciphertext	number	minus 3	plaintext
A	1	24	x
B	2	25	y
C	3	26	z
⋮	⋮	⋮	⋮
Y	25	22	v
Z	26	23	w

1.2 THE KEY TO THE MATTER: GENERALIZING THE CAESAR CIPHER

From Caesar's point of view, he had a pretty secure cipher. After all, most of the people who might intercept one of his messages couldn't even read, much less analyze a cipher. However, from a modern cryptologic point of view it has a major drawback—once you have figured out that someone is using Caesar cipher, you know everything about the system. There's no **key**, or extra piece of information, that lets you vary the cipher. This is considered to be a very bad thing.

Stop to think about that a moment. What's the big deal? Your cipher is either secret or it isn't, right? That was the view in Caesar's time and for many centuries afterward. But in 1883, **Auguste Kerckhoffs** published a revolutionary essay, in which he stated, "The system must not require secrecy and can be stolen by the enemy without causing trouble." Amazing! How can having your system stolen not cause trouble?

Kerckhoffs went on to point out that it is just too easy for Eve the Eavesdropper to find out what system Alice and Bob are using. In Kerckhoffs' time, like Caesar's, cryptography was used mostly by militaries and governments, so Kerckhoffs was thinking about the information that an enemy might get through bribing or capturing a member of Alice or Bob's staff. These are still valid concerns in many situations today, and we can add to them the possibilities of Eve tapping phone lines, installing spyware on computers, and plain lucky guessing.

On the other hand, if Alice and Bob have a system that requires a key to encipher and decipher, then things aren't so bad. If Eve finds out what general system is being used, she still can't easily read any messages. Attempting to read a message without the key and/or determining the key used for a message is called **cryptanalyzing** the message or cipher or, more colloquially, **breaking** it. And even if she manages to find out Alice and Bob's key, all is not lost. If Alice and Bob are smart, they are changing the key regularly. Since the basic system is the same, this isn't too hard, and then even if Eve gets the key to some of the messages, she won't be able to read all of them.

So we need to find a way to make small changes to Caesar cipher, depending on the value of some key. A logical place to start would be to ask why Alice is shifting her plaintext 3 places and not some other number? There is no particular reason; perhaps Caesar was just fond of the number 3. His successor, Augustus, used a similar system but shifted each letter only 1 place to the right. The "rot13" ("rot" stands for rotate) cipher shifts each letter forward by 13 places, wrapping around when you get to the end. This cipher is often used on the Internet to hide the punchlines of jokes or things that some people might find offensive. The general idea of shifting by k letters (or adding k modulo 26) is called a **shift cipher**, or **additive cipher**, with a key of k. For example, consider a shift cipher with a key of 21. Then Caesar's message would be:

plaintext:	a	n	d	y	o	u	t	o	o	b	r	u	t	u	s
numbers:	1	14	4	25	15	21	20	15	15	2	18	21	20	21	19
plus 21:	22	9	25	20	10	16	15	10	10	23	13	16	15	16	14
ciphertext:	V	I	Y	T	J	P	O	J	J	W	M	P	O	P	N

How many different keys are there? Shifting by 0 letters is probably not a good idea, but you could do it. Shifting by 26 letters is the same as shifting by 0 letters—or, in other words, 26 is the same as 0 modulo 26. Shifting by 27 letters is the same as shifting by 1 letter, and so on. So there are 26 ways of shifting that actually give you different results, or 26 keys. Note that this includes 0, the "stupid key," which doesn't do anything to the message. The technical term for when a cipher doesn't do anything is the **trivial cipher**. Suppose Alice sends Bob a message using a shift cipher and Eve intercepts it. Even if Eve has somehow learned that Alice and Bob are using a shift cipher, she still has to try 26 different keys in order to decipher the message. That's not a large number, but it's better than Caesar cipher.

Can we add some more keys? What about shifting our letters left instead of right? Unfortunately, that doesn't help. Suppose we shift our plaintext 1 letter to the left and wrap around the other direction.

plaintext:	a	n	d	y	o	u	t	o	o	b	r	u	t	u	s
numbers:	1	14	4	25	15	21	20	15	15	2	18	21	20	21	19
minus 1:	0	13	3	24	14	20	19	14	14	1	17	20	19	20	18
ciphertext:	Z	M	C	X	N	T	S	N	N	A	Q	T	S	T	R

Note that since 0 is the same as 26 modulo 26, we can assign them both
to the ciphertext letter "Z" interchangeably. If you think about it, you'll
see that shifting 1 letter to the left is the same as shifting 25 letters to the
right. Or in terms of modular arithmetic, you can think of left shifts as
negative, so we are saying −1 is the same as 25 modulo 26. So left shifts
don't help.

1.3 MULTIPLICATIVE CIPHERS

Let's look at a different type of cipher for some inspiration. This is called
the **decimation method** of constructing a cipher. We need to pick a key,
say 3. We start by writing out our plaintext alphabet.

plaintext: a b c d e f g h i j k l m n o p q r s t u v w x y z

Then we count off every third letter, crossing each out (or "decimating"
it) and writing each such letter as our ciphertext alphabet.

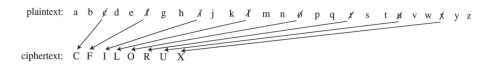

When you get to the end, "wrap around" to the beginning. In this case,
cross out the "a" and keep going.

Finally, wrap around to the "b" and finish up:

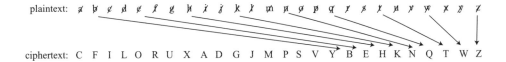

So our final translation of plaintext to ciphertext is

plaintext:	a	b	c	d	e	f	g	h	i	j	k	l	m
ciphertext:	C	F	I	L	O	R	U	X	A	D	G	J	M

plaintext:	n	o	p	q	r	s	t	u	v	w	x	y	z
ciphertext:	P	S	V	Y	B	E	H	K	N	Q	T	W	Z

Okay, now let's try to look at this like a mathematician. How can we describe the decimation method in terms of modular arithmetic? Well, we should translate our numbers into letters, of course.

plaintext:	a	b	c	d	e	f	g	h	i	j	⋯	y	z
numbers:	1	2	3	4	5	6	7	8	9	10	⋯	25	26
some operation?:	3	6	9	12	15	18	21	24	1	4	⋯	23	26
ciphertext:	C	F	I	L	O	R	U	X	A	D	⋯	W	Z

Very interesting! For the first eight letters, all we have to do is multiply the number corresponding to the plaintext by 3 (the key) and we get the ciphertext. For the letter i this doesn't quite work, because 9 times 3 is 27—but 27 is the same as 1 modulo 26, which corresponds correctly to our ciphertext letter A.

Apparently there was nothing much special about the addition part of our additive cipher. Instead of adding 3 to each plaintext number, we can multiply by 3 instead, wrapping around when we get to 26. This makes sense from the "clock arithmetic" point of view also: Start at midnight. Three times 3 hours later is 9:00. Three times 4 hours later is 12:00. Three times 5 hours later is 3:00, and so on. Our new **multiplicative cipher** with key 3 looks like this:

plaintext	number	times 3	ciphertext
a	1	3	C
b	2	6	F
⋮	⋮	⋮	⋮
y	25	23	W
z	26	26	Z

If we want to encipher the message "be fruitful and multiply," it would look like this:

plaintext:	b	e	f	r	u	i	t	f	u	l	a	n	d
numbers:	2	5	6	18	21	9	20	6	21	12	1	14	4
times 3:	6	15	18	2	11	1	8	18	11	10	3	16	12
ciphertext:	F	O	R	B	K	A	H	R	K	J	C	P	L

plaintext:	m	u	l	t	i	p	l	y
numbers:	13	21	12	20	9	16	12	25
times 3:	13	11	10	8	1	22	10	23
ciphertext:	M	K	J	H	A	V	J	W

Incidentally, it's often useful to have a faster way of dealing with the wraparound than subtracting 26 over and over again. Luckily, you already know one—it's division with remainder, just like you learned in grade school. Only now, once we have seen how many 26s go into the number, we are going to throw all the 26s away and just keep the remainder. For example, to encipher the last letter of the preceding example, I multiplied 25 by 3 to get 75. Then I divided 75 by 26:

$$
\begin{array}{r}
2 \\
26\overline{)75} \\
-52 \\
\hline
23
\end{array}
$$

The quotient is 2, which I can throw away, and the remainder is 23, which is the number I need for my ciphertext. Another way of thinking about it is that the division with remainder shows that $75 = 2 \times 26 + 23$; that is, 75 is twice 26 with 23 left over. But 26 is the same as 0 modulo 26, so 75 is the same as $2 \times 0 + 23 = 23$ modulo 26.

How many keys does the multiplicative cipher have? At first glance, you might expect 26 again, including one stupid key. But hold on a moment—multiplying by 26 modulo 26 is the same as multiplying by 0. And multiplying by 0 is *bad*. Not just stupid, but bad. A multiplicative cipher with a key of 0 looks like this:

plaintext	number	times 0	ciphertext
a	1	0	Z
b	2	0	Z
⋮	⋮	⋮	⋮
y	25	0	Z
z	26	0	Z

So if we encrypt a message with this cipher, it comes out as

plaintext:	a	r	e	a	l	l	y	b	a	d	k	e	y
numbers:	1	18	5	1	12	12	25	2	1	4	11	5	25
times 0:	0	0	0	0	0	0	0	0	0	0	0	0	0
ciphertext:	Z	Z	Z	Z	Z	Z	Z	Z	Z	Z	Z	Z	Z

There's no way on earth to decrypt that! So we can't use that key.

Are there any other keys we can't use? Think about multiplying by 2—we know that any number multiplied by 2 is even. A multiplicative cipher with a key of 2 looks like this:

plaintext	number	times 2	ciphertext
a	1	2	B
b	2	4	D
⋮	⋮	⋮	⋮
l	12	24	X
m	13	26	Z
n	14	2	B
o	15	4	D
⋮	⋮	⋮	⋮
y	25	24	X
z	26	26	Z

That's better than multiplying by 0, but it still presents a problem when deciphering: a ciphertext B could be plaintext a or plaintext n; similarly, there are two plaintext letters for every other ciphertext letter. The same thing will happen with every other even key, so that makes 13 bad keys so far, and 13 left. There's one more bad key—take a moment to try and find it. So there are actually only 12 good keys for a multiplicative cipher, including multiplication by 1, the stupid key.

We've talked about enciphering a message with a multiplicative cipher but not really about deciphering it. Remember that to decipher a message, you need to do the opposite from enciphering it. To decrypt a Caesar cipher, you shift 3 letters left instead of shifting right. To decrypt a shift cipher, you shift k letters left. What about a multiplicative cipher? Well, you could just write out the whole table and use it backward, and in practice you probably would most of the time. But for very short messages, you might not want to write out the whole table. How can you reverse a multiplication?

The everyday answer is to divide. The opposite of multiplying by 3 is dividing by 3. That works fine for some of the letters in our multiplicative cipher with key 3. Ciphertext C becomes 3, which divided by 3 becomes 1, which is plaintext a. Ciphertext F becomes 6, which divided by 3 is 2, which is b. But what about A? It becomes 1, which divided by 3 is $\frac{1}{3}$, which isn't a letter. The solution is in the wraparound. The number 1 is the same as 27 modulo 26, so we could also say A becomes 27, which divided by 3 is 9, which is i. Likewise B could be not just 2 but also 28 and 54, and 54 divided by 3 is 18, so B corresponds to r.

ciphertext	number	divided by 3	plaintext
B	2	$\frac{2}{3}$	(not a letter)
B	28	$9\frac{1}{3}$	(not a letter)
B	54	18	r

This sort of trial and error works but is not much more efficient than writing out the table. For example, suppose your key is 15 instead of 3 for a moment. What plaintext letter does ciphertext B correspond to? Modulo 26, B could be any of the numbers 2, 28, 54, 80, 106, 132, 158, 184, 210, ...

ciphertext	number	divided by 15	plaintext
B	2	$\frac{2}{15}$	(not a letter)
B	28	$1\frac{13}{15}$	(not a letter)
B	54	$3\frac{9}{15}$	(not a letter)
B	80	$5\frac{5}{15}$	(not a letter)
B	106	$7\frac{1}{15}$	(not a letter)
B	132	$8\frac{12}{15}$	(not a letter)
B	158	$10\frac{8}{15}$	(not a letter)
B	184	$12\frac{4}{15}$	(not a letter)
B	210	14	n

It takes 9 tries before you find a value that's divisible by 15, and there's nothing to assure you that it won't be even worse for other letters. What would be really useful is a whole number that works modulo 26 like $\frac{1}{3}$ does for ordinary numbers. We could call this number $\overline{3}$. Then multiplying by $\overline{3}$ modulo 26 would be the same as multiplying by $\frac{1}{3}$ modulo 26, which is the same as dividing by 3 modulo 26.

Why might we think that $\overline{3}$ exists? If we look back at our example multiplicative cipher with key 3 from earlier, its deciphering table would look like this:

ciphertext	number	divided by 3 modulo 26	plaintext
A	1	9	i
B	2	18	r
C	3	1	a
D	4	10	j
⋮	⋮	⋮	⋮
Y	25	17	q
Z	26	26	z

It appears that perhaps dividing by 3 modulo 26 is the same as multiplying by 9 modulo 26. If this is true, then to decipher another letter, say E, we could calculate as follows:

ciphertext	number	times $\overline{3}$ = times 9	plaintext
E	5	19	s

Once I know what $\bar{3}$ is, then I can calculate this without using trial and error or searching through the encryption table.

If k is the key to a multiplicative cipher, can we be sure $\bar{k}$ exists? If so, how do we find it? Answering these questions will take us on a little detour, which, strangely enough, starts back at our "bad keys" for our multiplicative cipher.

We discovered that these bad keys are 2, 4, 6, 8, 10, 12, 14, 16, 18, 20, 22, 24, 26, and one more, which I will now reveal is 13. (You should check that this is, in fact, bad.) What these numbers have in common is that they are all multiples of 2, 13, or both. And $2 \times 13 = 26$, which is not a coincidence. If we were working with Julius Caesar's 21-letter alphabet (i.e., modulo 21), then the bad keys would be multiples of 3 or 7 (or both), since $21 = 3 \times 7$. Romanian has 28 letters and $28 = 2 \times 2 \times 7$, so the bad keys would be multiples of 2 or 7 (or both). In Danish, Norwegian, and Swedish, which have 29 letters, 29 would be the only bad key.

What we have done with these letters (26, 21, 28, 29) is to break them up into their smallest irreducible components, the **prime numbers**. This process, which is called **factoring**, can always be done in one and only one way. This was known at least as long ago as the fourth century BCE, when **Euclid** put it in his *Elements*. What we want to know is whether our key and our modulus have a **common divisor**, that is, a number that divides them both.   The number 1 always divides both numbers, but that's considered trivial and doesn't count for this purpose. Euclid's *Elements* also tells us how to find a common divisor very efficiently by finding the **greatest common divisor**, or GCD, which is just what it sounds like. The method for calculating the GCD is known as the **Euclidean algorithm**, although we don't really know whether Euclid invented it or borrowed it from someone else. An **algorithm** is a well-defined method for doing something which always produces a specific correct answer for each input, such as a computer program.

Here's an example of the Euclidean algorithm in action, calculating the GCD of 756 and 210.

$$756 = 3 \times 210 + 126$$

$$210 = 1 \times 126 + 84$$

$$126 = 1 \times 84 + 42$$

$$84 = 2 \times 42 + 0$$

Each step is a division with a whole-number quotient and a remainder, just like we did earlier. The end result is that the greatest common divisor of 756 and 210 is 42, the last nonzero remainder.

We can use this algorithm to tell whether 6 is a bad key modulo 26 by calculating the GCD of 26 and 6.

$$26 = 6 \times 4 + 2$$

$$4 = 2 \times 2 + 0$$

We see that 2 is a bad key, since 2 divides 6 and 2 divides 26. What if we have a good key instead, like 3?

$$26 = 3 \times 8 + 2$$

$$3 = 2 \times 1 + 1$$

$$2 = 1 \times 2 + 0$$

The greatest whole number that divides both 3 and 26 is 1, which doesn't count, so 3 is a good key.

You might be wondering why we are bothering with Euclid's algorithm instead of just factoring the numbers and looking for prime factors in common. There are two answers to that question: First, we will eventually see that this algorithm is faster than factoring for large numbers. Second, once we have done the Euclidean algorithm, we can do a neat little trick to get $\overline{3}$.

Our next goal is to write 1 with a "3 times something" part and a "26 times something" part. We will write the equations of the Euclidean algorithm with 3 and 26 moved to the right-hand side, and every time we see part of the right-hand side without a 3 or a 26 in it, we will use a previous line to replace it with 3s and 26s.

$26 = 3 \times 8 + 2$:

$2 = \boxed{26} - \left(\boxed{3} \times 8\right)$ A 26 part and a 3 part, so OK.

$ = \left(\boxed{26} \times 1\right) - \left(\boxed{3} \times 8\right)$ Make both parts look the same.

$3 = 2 \times 1 + 1$:

$1 = \boxed{3} - (2 \times 1)$ Last part has no 26, so not OK.

$$ = \boxed{3} - \left(\overbrace{\left(\boxed{26} - \left(\boxed{3} \times 8\right)\right)}^{= \,2 \text{ by a previous line}} \times 1\right)$$

$ = \left(\boxed{3} \times 1\right) + \left(\boxed{3} \times 8\right) - \left(\boxed{26} \times 1\right)$ 3 parts and 26 parts.

$ = \left(\boxed{3} \times 9\right) - \left(\boxed{26} \times 1\right)$ Collect 3s and 26s.

We have now written 1 with a 3 part and a 26 part. Why do we want to do this? Well, we want to work modulo 26, and 26 is the same as 0 modulo 26, so

$$1 = (3 \times 9) - (26 \times 1)$$

means that

$$1 \equiv (3 \times 9) - (0 \times 1) \quad \text{modulo 26,}$$

or

$$1 \equiv 3 \times 9 \quad \text{modulo 26,}$$

or

$$\frac{1}{3} \equiv 9 \quad \text{modulo 26.}$$

So now we have confirmed that 9 is the number $\overline{3}$, which acts like $\frac{1}{3}$ modulo 26. Again, it might seem that we could have figured this out faster by trial and error. But for large numbers, this way really is much faster.

ciphertext	number	times 9	plaintext
A	1	9	i
⋮	⋮	⋮	⋮
E	5	19	s
⋮	⋮	⋮	⋮

Incidentally, the technical term for $\overline{3}$ is the **multiplicative inverse** of 3 modulo 26. The general idea of **inverses** is terribly important in many branches of mathematics. We've now seen **additive inverses—** that is, negatives—and multiplicative inverses, and we will see other examples in the future. A good thing to notice about inverses in modular arithmetic is that, unlike in ordinary arithmetic, there isn't usually any qualitative difference between a number and it's inverse. That is, in ordinary arithmetic, 2 is a positive number and -2 is a negative number, but modulo 26, $-2 \equiv 24$. So 2 and 24 are arithmetic inverses, but neither is particularly "negative." Likewise, in ordinary arithmetic, 3 is a whole number and $\frac{1}{3}$ is a fraction, but modulo 26, 3 and 9 are multiplicative inverses, despite neither being "fractional." This is characteristic of situations where there are only finitely many numbers that are considered distinct. Another way of looking at it is that there is no real distinction between forward and backward in these situations. Likewise, there is no mathematical difference between an arbitrary encryption and an arbitrary decryption for ciphers that use these operations—once you have figured out the inverse, you can "go forward to go backward." This will be sufficiently important in later sections that you might want to think about it a bit before going on.

1.4 AFFINE CIPHERS

Now we have a shift cipher with 26 good keys, 1 of which is stupid, and a multiplicative cipher with 12 good keys, 1 of which is stupid. Both of these are pretty easy for Eve to attack with a **brute-force attack**, meaning that she just tries every possible key until she gets the right one. Even if Alice and Bob can choose either type of cipher, that still leaves Eve only 38 choices to try. But what if Alice and Bob could use more than one cipher at the same time?

This has the potential to get complicated enough so that we'll introduce a little more mathematical notation. We'll use P to stand for any number between 1 and 26 that represents a plaintext letter and C to stand for a number that represents a ciphertext letter. We'll still use k to stand for a key. Encrypting using a shift cipher with a key of k can be written as

$$C \equiv P + k \quad \text{modulo 26,}$$

and using a multiplicative cipher with a key of k can be written as

$$C \equiv kP \quad \text{modulo 26.}$$

Similarly, decrypting in the shift cipher case looks like

$$P \equiv C - k \quad \text{modulo 26,}$$

and, in the multiplicative cipher case, looks like

$$P \equiv \bar{k}C \quad \text{modulo 26.}$$

What if Alice tries to encrypt using two different shift cipher keys, say k and m?* Is that twice as secure? It would look like

$$C \equiv P + k + m \quad \text{modulo 26.}$$

Unfortunately for Alice and Bob, from Eve's point of view this looks exactly the same as encrypting once using the key $k + m$, so Eve will break the cipher just as easily if she tries a brute-force attack. The same thing will happen if Alice uses two different multiplicative cipher keys. But what if she uses one of each? Suppose Alice first multiplies the plaintext by k and then adds m to get the ciphertext:

$$C \equiv kP + m \quad \text{modulo 26.}$$

Bob will decrypt by first subtracting m and then multiplying by $\bar{k}$:

$$P \equiv \bar{k}(C - m) \quad \text{modulo 26.}$$

Notice that Bob has to not only reverse the operations, but also reverse their order! If this seems unintuitive, think about getting dressed and undressed. To get dressed, you have to put on your socks first, and then

---

*Cryptographers sometimes use m to stand for a second cipher key because it comes after k and the letter l looks too much like the number 1.

your shoes. To get undressed, you have to remove them both, but in the opposite order. Otherwise bad things happen.

This combination gives us a new kind of cipher, which is technically called an **affine cipher**, although I sometimes prefer to just call it a $kP + m$ cipher. There are 12 choices for k and 26 choices for m, so there are $12 \times 26 = 312$ different keys for this cipher. This is getting to be enough to make Eve's brute-force attack a little difficult, although it is still not very hard if she has access to a computer.

The idea of combining two ciphers to make a **product cipher** is a fairly obvious one and goes back quite a long time in history. The idea that one can combine any decimation method (i.e., multiplicative cipher; see Section 1.3) with any shift cipher (i.e., additive cipher, see Section 1.2) goes back at least as far as the 1930s. It's worth mentioning one much older cipher that is a particular form of a $kP + m$ cipher. This is called the **atbash** cipher, and it's at least as old as the Biblical Book of Jeremiah. Like the decimation method, it starts by writing out the plaintext alphabet. Below it, the ciphertext alphabet is the same alphabet written backward. We'll use the modern English alphabet instead of the Hebrew alphabet:

plaintext:	a	b	c	d	e	f	g	h	i	j	k	l	m
ciphertext:	Z	Y	X	W	V	U	T	S	R	Q	P	O	N

plaintext:	n	o	p	q	r	s	t	u	v	w	x	y	z
ciphertext:	M	L	K	J	I	H	G	F	E	D	C	B	A

So why is this a form of a $kP + m$ cipher? When we translate the numbers into letters, we get

plaintext:	a	b	c	d	e	f	g	h	i	j	$\cdots$	y	z
numbers:	1	2	3	4	5	6	7	8	9	10	$\cdots$	25	26
some operation?:	26	25	24	23	22	21	20	19	18	17	$\cdots$	2	1
ciphertext:	Z	Y	X	W	V	U	T	S	R	Q	$\cdots$	B	A

We see that the ciphertext obeys the rule

$$C \equiv 27 - P \quad \text{modulo } 26.$$

Of course we can also write that as

$$C \equiv (-1)P + 27 \quad \text{modulo } 26,$$

and modulo 26 that's the same as

$$C \equiv 25P + 1 \quad \text{modulo } 26.$$

So this is a $kP + m$ cipher with the key $k = 25$, $m = 1$.

1.5 ATTACK AT DAWN: CRYPTANALYSIS
OF SIMPLE SUBSTITUTION CIPHERS

If we continue along this path of making our operations modulo 26 more and more complicated, we could eventually figure out a way to specify where every single plaintext letter goes individually. So a can go to any of the 26 ciphertext letters. Then we could send b to any ciphertext letter different from the ciphertext for a, so there are 25 choices. There are 24 ciphertext letters still unused for c, then 23 for d, and so on, until we have only one letter left for z. A cipher of this kind is called a **monoalphabetic monographic substitution cipher**, **monographic** meaning that it makes substitutions one letter at a time and **monoalphabetic** meaning that the substitution rule is the same for every letter in the message. That's a pretty unwieldy name and it's a pretty common cipher, so to save time I'm just going to call it a **simple substitution cipher**. All told there are $26 \times 25 \times 24 \times \cdots \times 3 \times 2 \times 1 = 403{,}291{,}461{,}126{,}605{,}635{,}584{,}000{,}000$ ways to make this kind of cipher, which includes all three of the ciphers we have discussed as well as the cryptogram puzzles that one finds in many daily papers. That's way too many keys to attack by brute force. Unfortunately for Alice and Bob, Eve has a much better attack available to her.

A very effective way of breaking simple substitution ciphers is called **letter frequency analysis**. This technique goes back at least as far as the ninth-century Arab scholar **Abu Yusuf Yaqub ibn Ishaq al-Sabbah al-Kindi**. The idea is simply that some letters in English, Arabic, or any other human language are used more often than others. For example, in a typical English text, the letter e will occur about 13% of the time, far more than any other. If Eve has a piece of ciphertext where a letter, say R, occurs about 13% of the time and more often than any other letter, then there's a good chance that R ($C = 18$) represents e ($P = 5$). If the cipher is an additive cipher, then Eve knows that

$$5 + k \equiv 18 \quad \text{modulo 26,}$$

so there is a very good chance that the key is $k = 13$.

If Eve has another type of cipher, such as an affine cipher, this might not be enough information. In this case, she might need to guess another letter, such as t, which occurs about 8% of the time, or a, which occurs about 7% of the time. For example, if Eve guesses that R represents e and F represents a, then she knows that

$$5k + m \equiv 18 \quad \text{modulo 26,}$$

$$1k + m \equiv 6 \quad \text{modulo 26.}$$

Now Eve has two equations in two unknowns. Subtracting them gives

$$4k \equiv 12 \quad \text{modulo 26.}$$

If the number 4 had an inverse modulo 26, then Eve could multiply each side by that inverse to cancel out the 4 and find k. Unfortunately, the GCD of 4 and 26 is 2, so 4 doesn't have an inverse. This means that our equation has either no solutions or more than one solution. If there are no solutions, it means in this case that Eve probably made a bad guess from the letter frequencies and she should try again. But in this case it turns out that there are two solutions, $k = 3$ and $k = 16$, and in either case m must be $6 - 1k$ modulo 26. So the possibilities are $k = 3$ and $m = 3$ or $k = 16$ and $m = 16$. Eve can then try to decrypt using each combination and see if she gets readable text. Since a, t, and several other letters have similar frequencies, it's possible that neither one is correct, in which case Eve has to go back to the beginning and try to guess e and a again. It might take a few guesses, but in the end Eve should be able to determine the correct key a lot faster than using brute force.

The one big caveat to this technique is that you need to have enough ciphertext to work with. The frequencies I have mentioned are only averages, and short messages may very well have radically different letter frequencies. Just imagine trying to decrypt the message "Zola is taking zebras to the zoo," for instance. We will see how this problem can compound itself when cryptanalyzing more complicated substitution ciphers in the future.

1.6 JUST TO GET UP THAT HILL: POLYGRAPHIC SUBSTITUTION CIPHERS

There are a couple of obvious ways to make ciphers on which letter frequency analysis doesn't work—you could change the substitution rule so that it's different at different places in the message (**polyalphabetic**) or you could make the substitutions work on more than one letter at a time (**polygraphic**). Both have their places in modern cryptography, but we are going to turn now to polygraphic ciphers.

The first thing you need to decide on in a polygraphic cipher is a **block size**. Ciphers with block size 2 are digraphic, those of block size 3 are trigraphic, and so on. Digraphic ciphers were proposed as early as the sixteenth century, although the first practical ones date from the nineteenth century. In 1929, **Lester S. Hill** invented the **Hill cipher**, which can be used with any block size. We will illustrate with a block size of 2. Divide up the plaintext into 2-letter blocks. If there are unfilled spaces in the last block, fill them with any random letters—these are called **nulls**, or **padding**.

ja   ck   ya   nd   ji   ll   ya   nd   ev   ex

Let the first letter in each plaintext block be P_1 and the second letter be P_2. Then calculate two ciphertext letters using the formulas

$$C_1 \equiv k_1 P_1 + k_2 P_2 \quad \text{modulo } 26,$$

$$C_2 \equiv k_3 P_1 + k_4 P_2 \quad \text{modulo } 26,$$

where k_1, k_2, k_3, and k_4 are numbers between 1 and 26, which together make up the key. For example, if the key is 3, 5, 6, 1, then the formulas are

$$C_1 \equiv 3P_1 + 5P_2 \quad \text{modulo } 26,$$

$$C_2 \equiv 6P_1 + 1P_2 \quad \text{modulo } 26.$$

If the plaintext is

plaintext:	ja	ck	ya	nd	ji	ll	ya	nd	ev	ex
numbers:	10, 1	3, 11	25, 1	14, 4	10, 9	12, 12	25, 1	14, 4	5, 22	5, 24

then the numbers for the first two letters of the ciphertext are

$$C_1 \equiv 3 \times 10 + 5 \times 1 \equiv 9 \quad \text{modulo } 26,$$

$$C_2 \equiv 6 \times 10 + 1 \times 1 \equiv 9 \quad \text{modulo } 26.$$

The x at the end of the plaintext is a null.

For the rest of the message we have

plaintext:	ja	ck	ya	nd	ji	ll	ya	nd	ev	ex
numbers:	10, 1	3, 11	25, 1	14, 4	10, 9	12, 12	25, 1	14, 4	5, 22	5, 24
Hill formulas:	9, 9	12, 3	2, 21	10, 10	23, 17	18, 6	2, 21	10, 10	21, 0	5, 2
ciphertext:	II	LC	BU	JJ	WQ	RF	BU	JJ	UZ	EB

Notice that the j of jacky gets mapped to an I, but the j of jilly gets mapped to a W. Likewise the two l's of jill get mapped to different letters, but the j and the a of jacky both end up as I's. This, of course, is because the letters are not encrypted individually, but as pairs. Also notice that yand gets mapped to BUJJ both times.

In order to decipher the message, Bob needs to solve a system of two equations in two unknowns:

$$C_1 \equiv k_1 P_1 + k_2 P_2 \quad \text{modulo } 26,$$

$$C_2 \equiv k_3 P_1 + k_4 P_2 \quad \text{modulo } 26.$$

There are lots of ways to do this; one way is to multiply the top equation by k_4 and the bottom equation by k_2 and then subtract. For instance, to decrypt the last block of our example, Bob observes that

$$5 \equiv 3P_1 + 5P_2 \quad \text{modulo } 26,$$

$$2 \equiv 6P_1 + 1P_2 \quad \text{modulo } 26,$$

which he can make into

$$1 \times 5 \equiv (1 \times 3)P_1 + (1 \times 5)P_2 \quad \text{modulo } 26,$$

$$5 \times 2 \equiv (5 \times 6)P_1 + (5 \times 1)P_2 \quad \text{modulo } 26$$

and subtract to get

$$1 \times 5 - 5 \times 2 \equiv (1 \times 3 - 5 \times 6)P_1 \quad \text{modulo } 26.$$

Similarly, Bob can multiply the top equation by k_3 and the bottom equation by k_1, which gives him

$$6 \times 5 \equiv (6 \times 3)P_1 + (6 \times 5)P_2 \quad \text{modulo 26,}$$

$$3 \times 2 \equiv (3 \times 6)P_1 + (3 \times 1)P_2 \quad \text{modulo 26.}$$

This time he takes the bottom minus the top to get

$$3 \times 2 - 6 \times 5 \equiv (3 \times 1 - 6 \times 5)P_2 \quad \text{modulo 26.}$$

Notice that in both cases there is a -27 on the right-hand side, which is $k_1k_4 - k_2k_3$. This number is called the **determinant** of the system. If the greatest common divisor of the determinant and 26 is 1, then the determinant has a multiplicative inverse, and Bob can multiply each side of his equations by that inverse to find P_1 and P_2. This is very similar to the case of ordinary arithmetic, where two equations in two unknowns can always be solved as long as the determinant of the system is not equal to zero.

In our example, the determinant is -27, as we said, which is the same as 25 modulo 26. If Bob runs through the Euclidean algorithm, he will find that

$$\overline{25} \equiv 25 \quad \text{modulo 26,}$$

so he gets

$$P_1 \equiv ((1 \times 5) - (5 \times 2)) \times 25 \quad \text{modulo 26,}$$

$$P_2 \equiv ((3 \times 2) - (6 \times 5)) \times 25 \quad \text{modulo 26,}$$

which finally reduces to

$$P_1 \equiv 5 \quad \text{modulo 26,} \qquad P_2 \equiv 24 \quad \text{modulo 26,}$$

or ex.

In general, if $k_1k_4 - k_2k_3$ has an inverse, then the solution to

$$C_1 \equiv k_1P_1 + k_2P_2 \quad \text{modulo 26,}$$

$$C_2 \equiv k_3P_1 + k_4P_2 \quad \text{modulo 26}$$

is

$$P_1 \equiv \overline{(k_1k_4 - k_2k_3)}(k_4C_1 - k_2C_2) \quad \text{modulo 26,}$$

$$P_2 \equiv \overline{(k_1k_4 - k_2k_3)}(-k_3C_1 + k_1C_2) \quad \text{modulo 26.}$$

The general form of this method for solving a system of several equations in the same number of unknowns is usually known as **Cramer's rule**, named for **Gabriel Cramer**. Cramer was an eighteenth-century Swiss mathematician who did much work studying systems of equations and the curves they describe. The same rule seems to have been published slightly earlier by **Colin Maclaurin** in Scotland. Cramer's rule is not the fastest way of solving large systems of equations, but it's certainly good enough for the block sizes one is likely to use in a Hill cipher.

Notice that if we give new names to the numbers

$$m_1 = \overline{(k_1k_4 - k_2k_3)}(k_4),$$

$$m_2 = \overline{(k_1k_4 - k_2k_3)}(-k_2),$$

$$m_3 = \overline{(k_1k_4 - k_2k_3)}(-k_3), \quad \text{and}$$

$$m_4 = \overline{(k_1k_4 - k_2k_3)}(k_1),$$

then we can write

$$P_1 \equiv m_1C_1 + m_2C_2 \quad \text{modulo 26,}$$

$$P_2 \equiv m_3C_1 + m_4C_2 \quad \text{modulo 26.}$$

We can think of this system of equations as an inverse of the original system, and we can think of m_1, m_2, m_3, m_4 as a sort of "inverse key" for the original encryption key k_1, k_2, k_3, k_4. In our example this key would be 25×1, 25×-5, 25×-6, 25×3, or 25, 5, 6, 23 modulo 26. Once Bob has worked this out, the process of decryption works exactly the same as encryption. This is another example of the idea of going forward to go backward that we talked about in Section 1.3.

It's a little involved to work out exactly how many good keys (i.e., keys where the determinant has an inverse) there are for a Hill cipher, but it's about 45,000 for a block size of 2 and about 52,000,000,000 for a block size of 3, so a brute-force attack is getting to be rather difficult. Also note that Bob needs to be aware that there may be nulls at the end of his message. This ought to be clear when he reads it.

In 1931, Hill followed up his original cipher with several extensions. The most important one is now generally known as the **affine Hill cipher**, because it combines the original Hill cipher with an addition step, just like we combined the multiplicative and additive ciphers to get the affine cipher. If we let the block size be 2 again, the new formulas are

$$C_1 \equiv k_1 P_1 + k_2 P_2 + m_1 \quad \text{modulo } 26,$$

$$C_2 \equiv k_3 P_1 + k_4 P_2 + m_2 \quad \text{modulo } 26,$$

where the key now consists of six numbers, k_1, k_2, k_3, k_4, m_1, and m_2, all between 1 and 26. Once again, this is a good key as long as the greatest common divisor of the determinant $k_1 k_4 - k_2 k_3$ and 26 is 1. (The new key numbers m_1 and m_2 can be anything.) To decrypt, Bob just needs to subtract m_1 from C_1 and m_2 from C_2 and then solve the system as before.

A letter frequency analysis no longer works on a polygraphic cipher, because, as you can see from the example, the same letter in the plaintext doesn't always go to the same letter in the ciphertext. Therefore, the whole idea of guessing which letter is e fails. On the other hand, we also saw that the same plaintext block always goes to the same ciphertext block, and in the case of block size 2 or 3, it is possible to exploit this. For example, the most common digraph, or 2-letter block, is "th," which occurs, according to one study, approximately 2.5% of the time. The most common trigraph (3-letter block) is "the," which occurs, by the same study, just under 1% of the time. Eve could use facts like these to do a digraph or trigraph frequency analysis and perhaps break a digraphic or trigraphic substitution cipher. However, for larger block sizes this quickly gets very difficult, as there are a lot of possible blocks and not a lot of difference between the frequencies of the various blocks. Even in 1929, Hill managed to construct a machine that used a set of gears to mechanically encipher texts using block size 6 and was thus essentially unbreakable using frequency analysis. Unfortunately for Hill, his machine never caught on.

The Hill ciphers were never used much—they were too unwieldy to use by hand, and cryptography via mechanical devices went in the direction of polyalphabetic substitution ciphers instead. Hill's idea of using systems of equations has regained substantial importance with the advent of digital computers in cryptography, but from a modern point of view, these ciphers used by themselves have the problem that they are badly vulnerable to a type of attack that is rather different from the ones we have talked about so far.

1.7 KNOWN-PLAINTEXT ATTACKS

So far, all of the cryptanalytic attacks we have discussed are **ciphertext-only attacks**, where all that Eve knows is the ciphertext message she has intercepted passing between Alice and Bob. But suppose that somehow Eve has gotten hold of both the plaintext and ciphertext of some message (or part of a message) that Alice has sent. Then she can try a **known-plaintext attack**, where she knows both the plaintext and the ciphertext and the goal is to get the key. Once she has the key, she can find out the content of not just the message she has, but other messages or parts of messages sent with the same key.

In the case of block size 2 and the original Hill cipher, suppose Eve recovers four letters of plaintext, P_1, P_2, P_3, and P_4, and the matching letters of ciphertext, C_1, C_2, C_3, and C_4. Then she knows

$$C_1 \equiv k_1 P_1 + k_2 P_2 \quad \text{modulo 26},$$

$$C_2 \equiv k_3 P_1 + k_4 P_2 \quad \text{modulo 26},$$

$$C_3 \equiv k_1 P_3 + k_2 P_4 \quad \text{modulo 26},$$

$$C_4 \equiv k_3 P_3 + k_4 P_4 \quad \text{modulo 26}.$$

From Eve's point of view, only the key numbers are unknowns, so she has four equations in four unknowns, and she can solve the system to recover the key.

In the earlier example, if Eve managed to recover the last two blocks of plaintext, she will know

$$21 \equiv k_1 5 + k_2 22 \quad \text{modulo 26},$$

$$0 \equiv k_3 5 + k_4 22 \quad \text{modulo 26},$$

$$5 \equiv k_1 5 + k_2 24 \quad \text{modulo 26},$$

$$2 \equiv k_3 5 + k_4 24 \quad \text{modulo 26}.$$

This is really two sets of equations,

$$21 \equiv k_1 5 + k_2 22 \quad \text{modulo 26,}$$

$$5 \equiv k_1 5 + k_2 24 \quad \text{modulo 26}$$

and

$$0 \equiv k_3 5 + k_4 22 \quad \text{modulo 26,}$$

$$2 \equiv k_3 5 + k_4 24 \quad \text{modulo 26.}$$

Eve could solve each set with Cramer's rule in the same way that Bob solved his equations in the previous section. For the first set, the rule gives

$$k_1 \equiv \overline{(5 \times 24 - 22 \times 5)}(24 \times 21 - 22 \times 5) \quad \text{modulo 26,}$$

$$k_2 \equiv \overline{(5 \times 24 - 22 \times 5)}(-5 \times 21 + 5 \times 5) \quad \text{modulo 26.}$$

If you finish the calculations, you will see

$$k_1 \equiv 3 \quad \text{modulo 26,} \qquad k_2 \equiv 5 \quad \text{modulo 26.}$$

Similarly, the second set gives Eve

$$k_3 \equiv \overline{(5 \times 24 - 22 \times 5)}(24 \times 0 - 22 \times 2) \quad \text{modulo 26,}$$

$$k_4 \equiv \overline{(5 \times 24 - 22 \times 5)}(-5 \times 0 + 5 \times 2) \quad \text{modulo 26.}$$

which gives her the last two key numbers:

$$k_3 \equiv 6 \quad \text{modulo 26,} \qquad k_4 \equiv 1 \quad \text{modulo 26.}$$

In general, Eve will need to recover only as many blocks of plaintext as there are letters in a block. So it's almost as easy to break the Hill cipher using a known-plaintext attack as it is to decipher a message. This is considered unacceptable, so the Hill cipher is never used in its original form. The idea of using a system of equations for polygraphic encryption, however, forms a piece of many modern ciphers.

1.8 LOOKING FORWARD

I warned you in the preface to this book that some of the ciphers I discuss in this book are considered obsolete in today's world, and that includes all the ciphers in this chapter and the next two, more or less. For one thing, they all work on the letters of the alphabet, and in the modern world we want to encrypt numbers, pictures, sounds, and all sorts of other things. That's not really a serious problem, since we know how to represent all these types of information using numbers, and we can easily adjust our ciphers to use numbers instead of letters. Additive and multiplicative ciphers are vulnerable to brute-force attacks because they don't have enough keys, and with computers available to help break ciphers, affine ciphers don't really have enough keys either. Perhaps more importantly, all monoalphabetic monographic substitution ciphers are vulnerable to letter frequency attacks. Monoalphabetic substitution ciphers are significant in modern times because they are the basis for, among other things, the polyalphabetic substitution ciphers we discuss in Chapter 2. To understand that chapter, you need to understand this one first. Polyalphabetic substitution ciphers are no longer considered state of the art in security either, but we'll get to that at the end of Chapter 2.

Polygraphic substitution ciphers are resistant to frequency analysis if the block size is large enough. In fact, one of the two main types of modern cipher, the block cipher (which we define in Chapter 5), is sometimes thought of as a type of polygraphic substitution cipher acting on an alphabet of just 0 and 1. The examples we have seen so far, the Hill cipher and affine Hill cipher, are vulnerable to known-plaintext attacks, as I have shown. So these particular polygraphic ciphers are not considered secure. As I mentioned, however, these two ciphers are used as building blocks in modern block ciphers, including the current US government standard in block ciphers, which I describe in Chapter 4. So you can't understand modern block ciphers without the affine Hill cipher, and you can't properly understand that without the additive, multiplicative, and affine ciphers.

I should point out that the cryptanalysis techniques discussed in this chapter, while not state of the art, are still very important in understanding modern cryptanalytic techniques. Letter frequency is not relevant with regard to a modern block cipher, but frequency attacks certainly are. The differential attacks discussed in Chapter 4, for instance, rely heavily on the same sorts of statistical frequency calculations as letter frequency attacks, but applied to the differences between ciphertexts rather than the ciphertexts themselves. Similarly, the linear attacks that

I mention in Chapter 4 are more sophisticated versions of the known-plaintext attack I showed you against the Hill cipher. Modern ciphers do not consist solely of the types of equations that the Hill and affine Hill ciphers do, but they can sometimes be approximated by such equations. Linear cryptanalysis takes advantage of this fact.

Finally, you might be wondering whether the concepts and notation of modular arithmetic were really necessary, or whether there were easier ways to describe the ciphers in this chapter. Additive, multiplicative, and affine ciphers were in fact used and analyzed perfectly well before anyone thought to describe them with modular arithmetic. The Hill and affine Hill ciphers, on the other hand, were invented with modular arithmetic in mind and are harder to deal with without those concepts. Even more importantly, modular arithmetic is critical for understanding the exponential ciphers and public-key ciphers of Chapters 6, 7, and 8.

■ ■ ■ ■ ■ ■ **2** ■ ■ ■ ■ ■ ■

Polyalphabetic Substitution Ciphers

2.1 HOMOPHONIC CIPHERS

Polygraphic ciphers, which work on more than one letter at a time, are one way to make ciphers that resist a straightforward letter frequency analysis. As we have seen, they can be difficult or impossible to do by hand, even with 3-letter blocks, and somewhat cumbersome even with machines. A polyalphabetic cipher, on the other hand, still works on 1 letter at a time like a monoalphabetic cipher, but it changes the substitution rule from letter to letter. This can be as simple as giving Alice, the encipherer, more than one ciphertext option for some or all plaintext letters, which she can choose from at whim. This is called a **homophonic** cipher—in linguistics, homophones are 2 letters or groups of letters that are spelled differently but pronounced the same. In cryptography, homophones are letters or groups of letters that are written differently in the ciphertext but deciphered the same.

As with many other aspects of cryptography, the ideas behind homophonic ciphers seem to have been first explored by the Arabs. The first known cipher that explicitly uses homophones as a central technique, however, appeared in Italy, having been prepared in 1401 by a cipher secretary of the Duchy of Mantua. This cipher appears to simply be a version of the atbash cipher, with the addition of 12 extra symbols: 3 each for the letters a, e, o, and u, which were high-frequency letters in fifteenth-century Italian. A representation of this idea with modern English letters and typographical symbols might look like this:

plaintext:	a	b	c	d	e	f	g	h	i	j	k	l	m
ciphertext:	Z	Y	X	W	V	U	T	S	R	Q	P	O	N
	!				@								
	%				&								
	)				-								

plaintext:	n	o	p	q	r	s	t	u	v	w	x	y	z
ciphertext:	M	L	K	J	I	H	G	F	E	D	C	B	A
		#						$					
		*						(					
		=						+					

One suspects that this didn't improve the security of such a simple cipher by very much, but the idea is sound: if the ciphertext letters corresponding to high-frequency plaintext letters are randomly divided up between multiple options, a straightforward letter frequency analysis becomes rather difficult. When used properly, the cipher shown here will produce a ciphertext in which no letter comes even near to the 13% frequency that one expects for the ciphertext letter corresponding to e. Instead, there will be four different symbols (V, @, &, and -), which each occur with just over 4% frequency. Lots of other letters also occur with 4% frequency, so this doesn't help the cryptanalyst much. This works only if Alice really picks one of the four symbols at random. A common pitfall is for a sloppy encipherer to primarily use only one of the options (say V, which might be more convenient on a keyboard) and only occasionally use the others—this will pretty much destroy the usefulness of the homophones.

It is not clear how much was known in Europe at this point in time about letter frequency analysis. The fact that the Mantuan cipher gives homophones only for vowels, which are high frequency, leads one to suspect that they knew something about the subject. We don't know for sure because unlike in the Arab world, where cryptography was mostly an academic pursuit, in Renaissance Europe it was a deadly serious part of diplomacy and its secrets were kept well guarded. It would not be until 1466 or 1467 that a description of frequency analysis would appear in print in Europe, by Leon Battista Alberti, whom we shall meet shortly. And due, perhaps, to the stereotypical conservatism of diplomats, the first ciphers with homophones for consonants as well as vowels did not appear until the mid-1500s.

2.2 COINCIDENCE OR CONSPIRACY?

So far we have been assuming Kerckhoffs' principle without too much reflection when we take the role of Eve. Often, however, Eve doesn't even need to steal the system in order to make some good guesses about how it works. For instance, how might Eve guess that a homophonic system is being used? True, it will generally have more than 26 characters. But perhaps the message is in a language other than English, or perhaps not all the possible ciphertext letters actually appear in the message. Can we tell what is going on?

Making a table of the frequency of each letter in the ciphertext is a good first step. Suppose Eve has intercepted this ciphertext:

QBVDL	WXTEQ	GXOKT	NGZJQ	GKXST	RQLYR
XJYGJ	NALRX	OTQLS	LRKJQ	FJYGJ	NGXLK
QLYUZ	GJSXQ	GXSLQ	XNQXL	VXKOJ	DVJNN
BTKJZ	BKPXU	LYUNZ	XLQXU	JYQGX	NTYQG
XKXQJ	KXULK	QJNQN	LQBYL	OLKKX	SJYQG
XNGLU	XRSBN	XOFUL	YDSXU	GJNSX	DNVTY
RGXUG	JNLEE	SXLYU	ESLYY	XUQGX	NSLTD
GQXKB	AVBKX	JYYBR	XYQNQ	GKXZ	LNYBS
LRPBA	VLQXK	JLSOB	FNGLE	EXYXU	LSBYD
XWXKF	SJQQS	XZGJS	XQGXF	RLVXQ	BMXXK
OTQKX	VLJYX	UQBZG	JQXZL	NG	

Alice has removed the spaces from her plaintext and divided it up into 5-letter groups in order to make things harder for Eve by obscuring any short, common words. Eve starts by counting how often each letter appears and what percentage each letter takes up of the 322 letters total (see Table 2.1).

There are only 23 distinct characters in the ciphertext, which could mean that Eve is dealing with a language with less than 26 letters, or that Alice used some sort of polygraphic system which doesn't need all of the characters, or just that some letters in the plaintext don't appear.

TABLE 2.1.
Letter frequencies observed in our ciphertext

Letter	Number of Occurrences	Percent Frequency
A	3	.9
B	14	4.3
D	6	1.9
E	6	1.9
F	5	1.6
G	23	7.1
J	22	6.8
K	19	5.9
L	30	9.3
M	1	.3
N	20	6.2
O	7	2.2
P	2	.6
Q	30	9.3
R	9	2.8
S	17	5.3
T	9	2.8
U	13	4.0
V	8	2.5
W	2	.6
X	47	14.6
Y	21	6.5
Z	8	2.5

How does Eve's table compare with the expected frequencies in English text? See Table 2.2.

It seems reasonably plausible that what we have is a simple substitution cipher that just doesn't happen to have some of the lowest-frequency letters in its plaintext. If homophones were being used, we would expect to see more low-frequency letters and fewer (if any) high-frequency ones. It would be nice if we could make this observation more quantitative, though.

The tool for that is called the **index of coincidence**, and it was invented by **William Friedman**, easily one of the most important figures in early twentieth-century cryptology. Friedman never set out to be a cryptologist. He studied genetics in college and graduate school and

Table 2.2.
Letter frequencies in English text compared with our ciphertext

Letter	Percent Frequency in English Text	Letter	Percent Frequency in Our Ciphertext
e	12.7	X	14.6
t	9.1	L	9.3
a	8.2	Q	9.3
o	7.5	G	7.1
i	7.0	J	6.8
n	6.7	Y	6.5
s	6.3	N	6.2
h	6.1	K	5.9
r	6.0	S	5.3
d	4.3	B	4.3
l	4.0	U	4.0
c	2.8	R	2.8
u	2.8	T	2.8
m	2.4	V	2.5
w	2.4	Z	2.5
f	2.2	O	2.2
g	2.0	D	1.9
y	2.0	E	1.9
p	1.9	F	1.6
b	1.5	A	.9
v	1.0	P	.6
k	.8	W	.6
j	.2	M	.3
x	.2		
q	.1		
z	.1		

was invited to join the Department of Genetics at the Riverbank Laboratories, an organization founded and run by an eccentric Illinois millionaire. Friedman got involved in cryptology when he was asked to help with photography for a group attempting to find hidden ciphers in the works of Shakespeare. Although he eventually concluded that no such ciphers were present, he found both his future wife and his future career in the Riverbank cryptology group. Friedman left Riverbank to join the US Army during World War I and eventually moved to the National Security Agency when it was formed after World War II. His wife, Elizebeth, had her own distinguished career in the meantime, solving ciphers for the US Coast Guard, the Treasury Department, and several other government agencies.

When he invented the index of coincidence, Friedman was considering the chance that if you pick two letters at random, they will be the same. First, suppose you are picking from a large number of English letters distributed at random, so that each letter appears equally often. The chance that the first letter you pick will be an a is 1/26, and the chance that you also pick an a the second time is also 1/26. In probability, if you want to know the chance of two independent things both happening, you multiply the probabilities, so the chance that you will pick two letters that are a is $(1/26) \times (1/26) = 1/26^2$. Likewise, the chance that you will pick two that are b is $1/26^2$, the chance that you will pick two that are c is $1/26^2$, and so on. What is the chance that you will pick two of the same letter, regardless of which letter it is? If you want to know the probability of either of two mutually exclusive things happening, you add the probabilities, so the chance that you will pick two of any letter is

$$\underbrace{\frac{1}{26^2}}_{\text{two are "a"}} + \underbrace{\frac{1}{26^2}}_{\text{two are "b"}} + \underbrace{\frac{1}{26^2}}_{\text{two are "c"}} + \cdots + \underbrace{\frac{1}{26^2}}_{\text{two are "z"}} = 26 \times \frac{1}{26^2} = \frac{1}{26} \approx .038.$$

The chance of picking two identical letters out of a selection of text is called the index of coincidence of that text, so the index of coincidence of random text (with English letters) is about .038, or 3.8%.

Now, suppose you are picking from a large amount of actual English text. We know that the chance that you will pick the letter a is about 8.2%, or .082. So the chance that you will pick two letters that are a is $(.082)^2$. The chance that you will pick two that are b is about $(.015)^2$, the chance that you will pick two that are c is about $(.028)^2$, and so on. The total probability that you will pick two of the same letter is

$$\underbrace{(.082)^2}_{\text{two are "a"}} + \underbrace{(.015)^2}_{\text{two are "b"}} + \underbrace{(.028)^2}_{\text{two are "c"}} + \cdots + \underbrace{(.001)^2}_{\text{two are "z"}} \approx .066.$$

In other words, the index of coincidence of actual English text is about .066, or 6.6%.

The first thing Friedman realized is that this number won't change if you apply a simple substitution cipher to the text—the order in which the numbers are added will change, but the total won't. So if our ciphertext was encrypted with a simple substitution cipher, we would expect the index of coincidence to be about .066, and if the cipher had homophones, we would expect it to be substantially different. In fact, because there would be less variation in the frequencies, we would expect the index to be between .038 and .066, since .038 is the index if all 26 letters are the same, and this turns out to be the minimum possible index for an alphabet of 26 letters.

Let's compute the index of coincidence for our ciphertext. The chance that we will pick an A, according to our table, is 3/322, since there are 3 letters that are A out of 322 letters total. For our second pick, we will assume that we don't pick exactly the same A again, although we can pick one of the other two letters that are A. Then the chance of picking an A the second time is 2/321, since there are 2 letters that are A left out of 321 letters left. The chance of picking two letters that are A is $(3/322) \times (2/321)$. Similarly, the chance of picking two that are B is $(14/322) \times (13/321)$, and so on. The index of coincidence for our ciphertext is

$$\frac{3}{322} \times \frac{2}{321} + \frac{14}{322} \times \frac{13}{321} + \cdots + \frac{8}{322} \times \frac{7}{321} \approx .068.$$

This is definitely not closer to .038 than .066, so it's a very good bet that we have a simple substitution cipher. Friedman called this test the **phi test** to distinguish it from other tests using the index of coincidence, which we will see later. You may want to amuse yourself by trying to solve the cipher using the techniques of Section 1.5.

On the other hand, the following ciphertext can be calculated to have an index of coincidence of approximately .046—not as low as random text, but much lower than a simple substitution cipher, even despite having more than 26 characters, which tends to raise the index.

IW*CI	W@G*L	&H&L(	ASN*A	E)U&V	$CNPC
SIW*E	DDSA@	LTCIH	!(A#C	V%EIW	*!#HA
*IW@N	TAEHR	$CI(C	JTS!C	SHDS#	SIW@S
DVW@R	G$HH*	SIW*W	)JH@(	CUGDC	IDUIW
*&AIP	GWTUA	TLS$L	CIW*D	IWTG!	#HATW
TRG$H	H*SQT	U$G*I	W@S)D	GHWTR	APBDG
*S%EI	W@WDB	@HIG@	IRWWX	H&CV+	XHWVG
*LLXI	WW#HE	G)VG@	HHI#A	AEGTH	@CIAN
W*L!H	Q%I!L	)DAAN	R)BTI	B)K#C	VXC#I
HDGQX	ILXIW	IW@VA	*&B!C	SIWTH	E**S$
UA(VW	I				

Again, feel free to try and cryptanalyze this—I'll even give you a hint. The cipher is an additive cipher plus homophones added for the vowels, very similar to the Mantuan cipher. Thus you should probably look for ciphertext letters that correspond to high-frequency plaintext consonants.

2.3 ALBERTI CIPHERS

A cipher with homophones is polyalphabetic in the sense that some or all letters have more than one possible substitution rule. However, the name polyalphabetic seems to imply that more than one entire ciphertext "alphabet" should be in use. To make this work without a very large number of symbols, Alice needs to do something more systematic than pick a ciphertext letter at random from a list. **Leon Battista Alberti**, an Italian Renaissance author, artist, architect, athlete, philosopher, and all-around Renaissance man, wrote the first known description of a method by which she can do that.

Alberti's *De Componendis Cifris*, or *A Treatise on Ciphers*, was written in 1466 or early 1467, and its 25 handwritten pages make up Europe's earliest known scholarly work on cryptography and cryptanalysis. This book explains for the first time in Europe how to do letter frequency analysis, it discusses the use of nulls and homophones, and it introduces Alberti's **cipher disk**, which could be considered the first true polyalphabetic system as well as the first cipher machine for substitution ciphers.

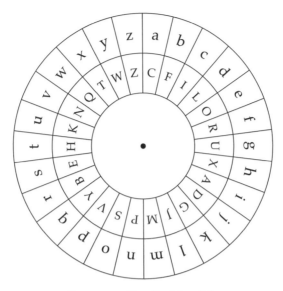

Figure 2.1. Alberti's cipher disk.

The cipher disk consists of two circular plates (made of copper, in Alberti's case), a larger stationary plate and a smaller movable one, held together with a pin through the center, as shown in Figure 2.1. A ring around the outside of each plate is divided into as many cells as there are letters of the alphabet. I will use the English alphabet, so each ring has 26 cells, and the disks are made so that all 52 cells can be seen at once. The plaintext is written in the outer ring, in the usual order, and a ciphertext alphabet is written in the inner ring, "not in regular order like the stationary characters, but scattered at random." If the inner ring did not move, we would have a classic monoalphabetic substitution cipher.

Alberti explains how we can coordinate the movements of the rings in order to bring new alphabets into play. Alice and Bob agree on either a plaintext or a ciphertext letter as the "index." Alice starts the message by writing a letter of the other alphabet—this indicates that the disk should be rotated so that the index is next to this letter.

Time for an example: suppose the ciphertext alphabet is arranged in this order:

ciphertext: C F I L O R U X A D G J M
             P S V Y B E H K N Q T W Z

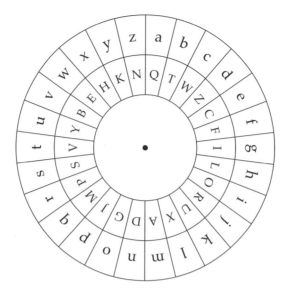

Figure 2.2. Alberti's cipher disk rotated to a different position.

If C is the index letter, then starting the message with the key letter a would indicate that the disks are placed as in Figure 2.1.

Alice can now encrypt the plaintext Leon Battista Alberti as

aJOSPFCHHAEHCCJFOBHA

Alberti says after 3 or 4 words, one should rotate the disks, indicating this by a new key letter in the ciphertext. For instance, Alice can choose e as the next key letter, meaning that she (and Bob) rotate the disks to the position shown in Figure 2.2.

So the full plaintext, Leon Battista Alberti, Father of Western Cryptography, might come out as

aJOSPFCHHAEHCCJFOBHAeFQVLCPGFECSVCPDWPKJVGIPQJLK

Bob, who must have an identical cipher disk, decrypts the message using first the position where C is next to a and later the position where C is next to e.

Eve has a problem if she wants to cryptanalyze the cipher. Even if she knows that Alberti's system is being used and even if she knows that the lowercase letters are key letters, if she doesn't know the order of the ciphertext alphabet, she can't decrypt in the way that Bob can. If Alice has changed disk positions often enough, and the message isn't so long that she has to reuse disk positions too often, then Eve won't have enough text from any one disk position to do a frequency attack either. On the other hand, if Eve knows that the key letters are an indication of how much to rotate the cipher disk, she may be able to compensate for the rotation first and then break the cipher using a frequency attack. Furthermore, if Alice changes the rotation only every 3 or 4 words, as Alberti suggests, a lot of information contained in the patterns of repeated letters within words will be preserved, and Eve may be able to use that. A better cipher would both change alphabets more frequently and make it less obvious where and how the changes occurred.

Before we go on, I can't resist pointing out yet another application of modular arithmetic in our example, even though it's an anachronism. The rotation of the cipher disk is equivalent to an enciphering step using an additive cipher, so this can be seen as a combination of an additive cipher and whatever kind of cipher produced the alphabet on the inner ring. In our case this was a multiplicative cipher, so we have a combination of an additive cipher and a multiplicative cipher, like the $kP + m$ ciphers we saw in Section 1.4. Unlike those ciphers, this time the addition is done before the multiplication.

plaintext:	a	b	c	d	e	f	g	h	i	j	$\cdots$	y	z
numbers:	1	2	3	4	5	6	7	8	9	10	$\cdots$	25	26
plus 22:	23	24	25	26	1	2	3	4	5	6	$\cdots$	21	22
rotated plaintext:	w	x	y	z	a	b	c	d	e	f	$\cdots$	u	v
rotated numbers times 3:	17	20	23	26	3	6	9	12	15	18	$\cdots$	11	14
ciphertext:	Q	T	W	Z	C	F	I	L	O	R	$\cdots$	K	N

2.4 IT'S HIP TO BE SQUARE: *TABULA RECTA*, OR VIGENÈRE SQUARE CIPHERS

Although Alberti gets credit for the first European book on cryptography and the first printed book on architecture, someone else gets credit for the first printed book on cryptography. That was **Johannes Trithemius**, or Johannes of Trittenheim. Trithemius was abbot of

the Benedictine abbey of Sponheim, now in the state of Rhineland-
Palatinate, Germany. Trithemius was an important author in the late
fifteenth and early sixteenth centuries, and a founder of both cryptography and library science in Europe. He was also intensely interested in
alchemy, astrology, demons, spirits, and other occult matters, to a degree that was controversial at the time and may seem positively bizarre
to us now. In many cases it is difficult to tell whether Trithemius is writing about cryptography, magic, or both, and many of his writings were
dismissed for centuries as divorced from reality. Recently, evidence has
surfaced suggesting that many, if not all, of Trithemius' stranger writings were, in fact, covers for more examples of cryptography and other
concealed writing.

Be that as it may, Trithemius is best known among cryptography
buffs today for his *tabula recta*, or "proper table," sometimes also called
a square table, letter square, or tableau. Imagine, for starters, an Alberti cipher disk with the ciphertext alphabet in the same order as the
plaintext. We start with the disk rotated one position to get an additive
cipher:

plaintext:	a	b	c	d	e	f	g	⋯	t	u	v	w	x	y	z
ciphertext:	B	C	D	E	F	G	H	⋯	U	V	W	X	Y	Z	A

Then rotate two positions, then three, then four, … until eventually we
get back to the starting position.

plaintext:	a	b	c	d	e	f	g	⋯	t	u	v	w	x	y	z
ciphertext:	B	C	D	E	F	G	H	⋯	U	V	W	X	Y	Z	A
ciphertext:	C	D	E	F	G	H	I	⋯	V	W	X	Y	Z	A	B
ciphertext:	D	E	F	G	H	I	J	⋯	W	X	Y	Z	A	B	C
⋮								⋮							
ciphertext:	Y	Z	A	B	C	D	E	⋯	R	S	T	U	V	W	X
ciphertext:	Z	A	B	C	D	E	F	⋯	S	T	U	V	W	X	Y
ciphertext:	A	B	C	D	E	F	G	⋯	T	U	V	W	X	Y	Z

This table has all the same information as the cipher disk but in
a form that shows it all at once. More importantly, Trithemius used his

table in a rather different way from Alberti. Instead of Alice shifting to a new disk position of her choosing at a time of her choosing, Trithemius suggests that she change ciphertext alphabets every single letter and do it by an orderly procession down the table, starting over when she gets to the bottom. This is called a **progressive** system. Progressive systems have some advantages; they obliterate patterns of repeated letters in words like attack or meeting, and they don't leave telltale key letters in the ciphertext. On the other hand, this system doesn't have a key at all. In modern terms that's a no-no, as we saw in Section 1.2. Trithemius recognized that the alphabets could go in various different orders; in addition to the tabula recta, he also presented a *tabula aversa* (reversed table), where the alphabets ran the other direction, and various other tables in various orders. None of these systems seem to be keyed, however.

We need to go back to Italy to see how a key can be added to this system. This seems to have been first suggested by **Giovan Battista Bellaso**, about whom not a lot else is known. He was apparently a secretary for one or more cardinals of the Catholic Church, which would certainly have given him occasion to study ciphers and secret writing. He published three short books on cryptography in 1553, 1555, and 1564, each of which contains various versions of a polyalphabetic cipher. Rather than using alphabets in the standard order, Bellaso used **reciprocal** alphabets, meaning that the role of the encryption and decryption alphabets can be swapped without changing the cipher. Thus decryption follows the same process as encryption, which is convenient in practice. The atbash cipher from Section 1.4 is one example of this, and so are Trithemius' tabula aversa alphabets.

For simplicity, we will stick with Trithemius' tabula recta. Bellaso's innovation was essentially to add a column of key letters down the side of the table:

	a	b	c	d	e	f	g	⋯	t	u	v	w	x	y	z
A	B	C	D	E	F	G	H	⋯	U	V	W	X	Y	Z	A
B	C	D	E	F	G	H	I	⋯	V	W	X	Y	Z	A	B
C	D	E	F	G	H	I	J	⋯	W	X	Y	Z	A	B	C
⋮								⋮							
X	Y	Z	A	B	C	D	E	⋯	R	S	T	U	V	W	X
Y	Z	A	B	C	D	E	F	⋯	S	T	U	V	W	X	Y
Z	A	B	C	D	E	F	G	⋯	T	U	V	W	X	Y	Z

Alice and Bob agree on a **keyword**, or **keyphrase**, which Alice writes above the plaintext, repeating as necessary:

keyphrase:	T	R	E	T	E	S	T	E	D	I	L	E	O	N	E	T	R	E
plaintext:	s	p	o	r	t	i	n	g	h	i	s	c	l	o	t	h	e	s

Then Alice encrypts each letter of the plaintext using the cipher alphabet corresponding to the appropriate key letter:

keyphrase:	T	R	E	T	E	S	T	E	D	I	L	E	O	N	E	T	R	E
plaintext:	s	p	o	r	t	i	n	g	h	i	s	c	l	o	t	h	e	s
ciphertext:	M	H	T	L	Y	B	H	L	L	R	E	H	A	C	Y	B	W	X

Note that as with the polygraphic ciphers of Section 1.6, the same letter in the plaintext will become a different letter in the ciphertext, depending on position. For example, the three plaintext letters that are s become M, E, and X, respectively. We call this form of polyalphabetic cipher a **repeating-key cipher** for obvious reasons. The key doesn't have to be a word or phrase, but that's the most common form.

As I mentioned earlier, Bellaso used a more complicated system with ciphertext and key alphabets in various orders. From a mathematical point of view, using the tabula recta as I have set it up has the advantage that it is easily represented using modular arithmetic:

keyphrase:	T	R	E	T	E	S	T	E	D	I	L	E	O	N	E	T	R	E
numbers:	20	18	5	20	5	19	20	5	4	9	12	5	15	14	5	20	18	5
plaintext:	s	p	o	r	t	i	n	g	h	i	s	c	l	o	t	h	e	s
numbers:	19	16	15	18	20	9	14	7	8	9	19	3	12	15	20	8	5	19
ciphertext:	M	H	T	L	Y	B	H	L	L	R	E	H	A	C	Y	B	W	X
numbers:	13	8	20	12	25	2	8	12	12	18	5	8	1	3	25	2	23	24

As you can see, the ciphertext numbers are merely the key numbers plus the plaintext numbers modulo 26. This idea presages modern digital stream ciphers, which we shall see in Chapter 5.

Poor Bellaso: his invention quickly became well known, but he himself never got recognition for it. Already in 1564, Bellaso himself was writing that somebody was "sporting his clothes and divesting him of his labors and honors." That somebody seems to have been Giovanni Battista Della Porta, who in 1563 published essentially the same system that Bellaso did in 1553 without giving Bellaso any credit. Until very recently, scholars of cryptography seem to have overlooked Bellaso's 1553 book or confused it with his 1564 book and thus gave Della Porta credit for the repeating-key cipher. Even worse, sometime in the nineteenth century, the repeating-key tabula recta cipher got credited to Blaise de Vigenère, whom we shall meet in Section 5.3. In 1586, Vigenère described the tabula recta, the repeating-key cipher, and their combination but never claimed to have invented any of them. Despite this, this simplified version of Bellaso's cipher is commonly known even today as the Vigenère cipher, and the tabula recta is often called the Vigenère square.

2.5 HOW MANY IS MANY? DETERMINING THE NUMBER OF ALPHABETS

What the majority of the polyalphabetic ciphers we've discussed have in common is a repeating key: there are only so many alphabets used by the system, and eventually they repeat after a shorter or longer number of letters. This number of letters is called the **period** of the cipher; in Bellaso's cipher, for example, this would just be the length of the keyphrase. Before we see how to break a repeating-key system, we might ask how to tell if we are dealing with a repeating-key cipher. Luckily, we can use the same tool as we did for homophone ciphers, namely, the index of coincidence. Consider our repeating-key tabula recta cipher from before but with a rather shorter key:

keyword:	L	E	O	N	L	E	O	N	L	E	O	N	L	E	O
plaintext:	t	h	e	c	a	t	o	n	t	h	e	m	a	t	b
ciphertext:	F	M	T	Q	M	Y	D	B	F	M	T	A	M	Y	Q

keyword:	N	L	E	O	N	L	E	O	N	L
plaintext:	a	t	t	e	d	a	g	n	a	t
ciphertext:	O	F	Y	T	R	M	L	C	O	F

What would we expect the index of coincidence to be? We will pretend for now that the key letters are chosen at random; when they make up an English word or phrase, that will somewhat affect the calculations.

Suppose the period of the cipher is ℓ. Then we can arrange the ciphertext letters into ℓ columns, one for each letter of the key, and if we have n letters total, then there are approximately n/ℓ letters in each column. For instance, in the preceding example, we would have

column:	I	II	III	IV
key letter:	L	E	O	N
ciphertext:	F	M	T	Q
	M	Y	D	B
	F	M	T	A
	M	Y	Q	O
	F	Y	T	R
	M	L	C	O
	F			

with $n = 25$ and $\ell = 4$. If we pick two letters from the same column, they are enciphered the same way, so the probability that they are the same should be approximately .038. On the other hand, if we pick two letters from different columns, encrypted with two different randomly chosen ciphers, then the chance that they are the same should be .066. There are n ways to pick the first letter. If the second letter is in the same column, there are $(n/\ell - 1)$ ways to pick it and a chance of about .038 that it is the same. If the second letter is in a different column, then there are $(n - n/\ell)$ ways to pick it and a chance of about .066 that it is the same. There are $n \times (n - 1)$ total ways to pick two letters, so the chance that they are the same—or the index of coincidence—should be about

$$\frac{n \times (n/\ell - 1) \times .038 + n \times (n - n/\ell) \times .066}{n \times (n - 1)}$$

$$= \frac{n/\ell - 1}{n - 1} \times .038 + \frac{n - n/\ell}{n - 1} \times .066,$$

which some experimentation should convince you is, in fact, between .038 and .066. If $\ell = 1$, then the cipher is monoalphabetic and the index is .038, while if $\ell = n$ then the index is .066; every plaintext letter has been enciphered with a potentially different randomly chosen alphabet, and the ciphertext is effectively random. In our example $n = 25$ and $\ell = 4$, so we would expect the index to be about $(5.25/24) \times .038 + (18.75/24) \times .066$, or about .060, although with such a short ciphertext, it probably won't be very close.

Once we know that we are dealing with a polyalphabetic cipher with a repeating key, the first step in breaking it is to find the period. As is often the case, especially in a subject so fraught with secrecy, the most commonly useful technique for finding the period was invented independently by two different people at roughly the same time—in this case, the mid-nineteenth century. One was **Charles Babbage**, who dabbled in many aspects of science, mathematics, and engineering but is best known today for coming up with the idea of the programmable computer. Unfortunately, while Babbage intended to publish his work on polyalphabetic ciphers, he never got around to it. The man who did publish this method was **Friedrich Kasiski**. Unlike Babbage, Kasiski does not seem to have made much of an impact outside of this one very important contribution. He was a major in the Prussian army but does not seem to have particularly engaged in cryptography during his service. After retiring from active duty, he wrote a short book that generally focuses on this particular technique.

So what is the technique, now generally known as the **Kasiski test**? The central idea is that if the repetition of the key and a repetition in the plaintext happen to line up, they will cause a repetition in the ciphertext. Let's consider again our previous example:

keyword:	L	E	O	N	L	E	O	N	L	E	O	N	L	E	O
plaintext:	t	h	e	c	a	t	o	n	t	h	e	m	a	t	b
ciphertext:	F	M	T	Q	M	Y	D	B	F	M	T	A	M	Y	Q

keyword:	N	L	E	O	N	L	E	O	N	L
plaintext:	a	t	t	e	d	a	g	n	a	t
ciphertext:	O	F	Y	T	R	M	L	C	O	F

The plaintext letters at repeat 4 times. The first 2 times happen to line up in the same point of the key, but the third and fourth times don't line up with the first two. Thus the first 2 occurrences of at are both encrypted to MY, although the third and fourth are encrypted to OF instead.

Now suppose Eve has only the ciphertext. The Kasiski examination starts by looking for groups of letters that are repeated—in this case she sees FMT, MY, and OF. Next she finds the number of letters between the start of the first group of a repetition and the start of the second. In this case, all 3 groups repeat 8 letters apart. From that Eve can conclude that the period is a factor of 8. (Which it is—it's 4.)

In longer ciphers the test is both more complicated and more effective. Let's look at an example ciphertext:

HXJVX	DMTUX	NUOGB	USUHZ	LFWXK	FFJKX
KAGLB	AFJGZ	IKIXK	ZUTMX	YAOMA	LNBGD
HZEHY	OMWBG	NZPMA	PZHMH	KAPGV	LASMP
POFLA	LTBWI	LQQXW	PZUHM	OQCHH	RTFKL
PEUXK	DMTKX	HPJGZ	IGUBM	OMEGH	WUDMN
YQTHK	JAOOX	YEBMB	VZTBG	PFBGW	DTBMB
ZFIXN	ZQPYT	IAPDM	OAVZA	AMMBV	LIJMA
VGUIB	JFVKX	ZASVH	UHFKL	HFJHG	

Assuming this was enciphered with one of the systems we have seen so far, Eve would probably first ask, Is this monoalphabetic or poly-alphabetic? The ciphertext has an index of coincidence of 0.044, which is squarely between 0.038 and 0.066. In addition, no ciphertext letter has a frequency above 8%, and there are exactly 26 ciphertext letters used, so this is either a rather unusual homophone cipher or a repeating-key cipher. The underlined letters in the ciphertext above are the repetitions

Eve finds in a Kasiski examination. There seem to be an awful lot of repetitions of 2-letter groups, so Eve is going to ignore them for the moment.

Eve makes a table of the repeated groups, their positions, and the interval between the repetitions.

repetition	first position	second position	interval
DMT	6	126	120
JGZI	38	133	95
FKL	118	228	110
BMB	163	178	15
MBV	164	203	39

Other than 1, there isn't any possible period that is a factor of all these intervals. However, all except the last one have a factor of 5. In fact, 5 is the greatest common divisor of 120, 95, 110, and 15, so it is extremely likely that the period of the cipher is 5. The last repetition, MBV, apparently happened by pure chance rather than through the process we described earlier.

If Eve isn't satisfied with the results of her Kasiski examination, there are a couple of other things she could try. One is to try to match her observed index of coincidence with the formula we saw earlier:

$$.044 = \frac{235/\ell - 1}{234} \times .038 + \frac{235 - 235/\ell}{234} \times .066.$$

Solving this for ℓ gives

$$\ell = \frac{235 \times .028}{234 \times .044 - 0.038 \times 235 + .066} \approx 4.6.$$

This certainly confirms the Kasiski value of 5. By itself, it could indicate that the period is 4 or 5 or, if you had bad luck, maybe 3 or 6—I wouldn't rely on it by itself unless you had *lots* of ciphertext. On the other hand, it can be very useful if you are not sure whether to include some of the Kasiski repetitions—in this case it definitely indicates that you should throw out the interval of 39, which would require the period to be 1. Alternatively, you might suspect that the real key length is a factor of the result of the Kasiski examination, as in the example on page 44. These two tests, in fact, work together very nicely—the Kasiski examination might be off by a whole number factor, and the index of coincidence gives you only the approximate size, but using both will usually pin things down.

The last thing that Eve could try is the **kappa test**, which was Friedman's original index of coincidence test. The kappa test checks to see if two ciphertexts were encrypted using the same polyalphabetic cipher, regardless of whether the key repeats. Consider two samples of plaintext:

sample 1:	h	e	r	e	i	s	e	d	w	a	r	d	b	e	a	r	c
sample 2:	t	h	e	p	i	g	l	e	t	l	i	v	e	d	i	n	a

sample 1:	o	m	i	n	g	d	o	w	n	s	t	a	i	r	s	n	o
sample 2:	v	e	r	y	g	r	a	n	d	h	o	u	s	e	i	n	t

sample 1:	w	b	u	m	p	b	u	m	p	b	u	m	p	o	n	t
sample 2:	h	e	m	i	d	d	l	e	o	f	a	b	e	e	c	h

If you pick a position at random, what would you expect the chance to be that the letter at that position in the first plaintext is the same as the letter in that position in the second plaintext? Once, again, it's the chance that they are both "a" plus the chance that they are both "b," and so on, so if the plaintexts are made up of ordinary English text, you would expect the chance to be about .066, as usual. So if there are 50 letters in each sample, you would expect about $.066 \times 50 = 3.3$ coincidences. (In fact, there are 3, as shown by the underlined letters.)

Now suppose we have two samples of randomly generated letters:

sample 1:	u	c	z	j	t	t	c	t	k	e	t	x	q	y	h	m	x
sample 2:	q	h	e	a	w	y	a	o	r	l	q	e	q	e	k	w	z

sample 1:	v	s	t	v	s	n	e	p	k	n	u	y	q	u	o	n	a
sample 2:	i	e	i	e	o	j	s	u	n	v	b	q	z	q	z	w	i

sample 1:	i	n	p	z	o	k	t	g	p	n	o	x	b	f	m	u
sample 2:	h	o	t	e	d	q	f	g	e	b	e	k	a	t	i	k

Now we expect the chance of a coincidence to be .038, so there should be about $.038 \times 50 = 1.9$ coincidences; in fact, there are 2.

Now suppose we encrypt each plaintext from the first pair with the *tabula recta*, using the same repeating key.

keyword:	C	H	R	I	S	T	O	P	H	E	R	C	H	R	I	S	T
sample 1:	h	e	r	e	i	s	e	d	w	a	r	d	b	e	a	r	c
ciphertext 1:	K	M	J	N	B	M	T	T	E	F	J	G	J	W	J	K	W
sample 2:	t	h	e	p	i	g	l	e	t	l	i	v	e	d	i	n	a
ciphertext 2:	W	P	W	Y	B	A	A	U	B	Q	A	Y	M	V	R	G	U

keyword:	O	P	H	E	R	C	H	R	I	S	T	O	P	H	E	R	C
sample 1:	o	m	i	n	g	d	o	w	n	s	t	a	i	r	s	n	o
ciphertext 1:	D	C	Q	S	Y	G	W	O	W	L	N	P	Y	Z	X	F	R
sample 2:	v	e	r	y	g	r	a	n	d	h	o	u	s	e	i	n	t
ciphertext 2:	K	U	Z	D	Y	U	I	F	M	A	I	J	I	M	N	F	W

keyword:	H	R	I	S	T	O	P	H	E	R	C	H	R	I	S	T
sample 1:	w	b	u	m	p	b	u	m	p	b	u	m	p	o	n	t
ciphertext 1:	E	T	D	F	J	Q	K	U	U	T	X	U	H	X	G	N
sample 2:	h	e	m	i	d	d	l	e	o	f	a	b	e	e	c	h
ciphertext 2:	P	W	V	B	X	S	B	M	T	X	D	J	W	N	V	B

The same coincidences are still there. So if two English ciphertexts are encrypted with the same key, we still expect the percentage of coincidences to be about 6.6%.

On the other hand, if we encrypt the ciphertexts with different keys, then there's no particular reason the coincidences should be anything other than random:

keyword 1:	C	H	R	I	S	T	O	P	H	E	R	C	H	R	I	S	T
sample 1:	h	e	r	e	i	s	e	d	w	a	r	d	b	e	a	r	c
ciphertext 1:	K	M	J	N	B	M	T	T	E	F	J	G	J	W	J	K	W
keyword 2:	E	E	Y	O	R	E	E	E	Y	O	R	E	E	E	Y	O	R
sample 2:	t	h	e	p	i	g	l	e	t	l	i	v	e	d	i	n	a
ciphertext 2:	Y	M	D	E	A	L	Q	J	S	A	A	A	J	I	H	C	S

keyword 1:	O	P	H	E	R	C	H	R	I	S	T	O	P	H	E	R	C
sample 1:	o	m	i	n	g	d	o	w	n	s	t	a	i	r	s	n	o
ciphertext 1:	D	C	Q	S	Y	G	W	O	W	L	N	P	Y	Z	X	F	R
keyword 2:	E	E	E	Y	O	R	E	E	E	Y	O	R	E	E	E	Y	O
sample 2:	v	e	r	y	g	r	a	n	d	h	o	u	s	e	i	n	t
ciphertext 2:	A	J	W	X	V	J	F	S	I	G	D	M	X	J	N	M	I

keyword 1:	H	R	I	S	T	O	P	H	E	R	C	H	R	I	S	T
sample 1:	w	b	u	m	p	b	u	m	p	b	u	m	p	o	n	t
ciphertext 1:	E	T	D	F	J	Q	K	U	U	T	X	U	H	X	G	N
keyword 2:	R	E	E	E	Y	O	R	E	E	E	Y	O	R	E	E	Y
sample 2:	h	e	m	i	d	d	l	e	o	f	a	b	e	e	c	h
ciphertext 2:	Z	J	R	N	C	S	D	J	T	K	Z	Q	W	J	H	M

And indeed, the percentage of coincidences is about 3.8%, as it would be for two random samples of letters.

But how can Eve use this to determine the length of the key? Let's look at the example from page 44 yet another time, but this time also slide the plaintext 4 steps to the right:

keyword 1:	L	E	O	N	L	E	O	N	L	E	O	N	L	E	O
plaintext 1:	t	h	e	c	a	t	o	n	t	h	e	m	a	t	b
ciphertext 1:	F	M	T	Q	M	Y	D	B	F	M	T	A	M	Y	Q
keyword 2:					L	E	O	N	L	E	O	N	L	E	O
plaintext 2:					t	h	e	c	a	t	o	n	t	h	e
ciphertext 2:					F	M	T	Q	M	Y	D	B	F	M	T

keyword 1:	N	L	E	O	N	L	E	O	N	L	E	O	N	L
plaintext 1:	a	t	t	e	d	a	g	n	a	t	t	h	e	c
ciphertext 1:	O	F	Y	T	R	M	L	C	O	F	Y	W	S	O
keyword 2:	N	L	E	O	N	L	E	O	N	L	E	O	N	L
plaintext 2:	m	a	t	b	a	t	t	e	d	a	g	n	a	t
ciphertext 2:	A	M	Y	Q	O	F	Y	T	R	M	L	C	O	F

We usually say the text is **displaced** rather than *slid* and refer to the **displacement** of the lower text, since *slide* and *shift* have other common meanings in cryptography. I've listed the two positions of the key as two keywords on separate lines, but they are really the same. So the two

TABLE 2.3.
Results of the kappa test for our ciphertext

Displacement	Coincidences	Index
1	7	.030
2	10	.043
3	9	.038
4	11	.047
5	14	.060
6	15	.064
7	15	.064
8	9	.038
9	11	.047
10	14	.060
11	10	.043
12	3	.013
13	14	.060
14	12	.051
15	17	.072

different positions of plaintext are effectively enciphered with the same key and thus should obey the rule of having approximately 6.6% coincidences. If I had displaced them by 3 steps instead—or 5—they would have behaved like plaintexts encrypted with different keys and should have had approximately 3.8% coincidences. On the other hand, if the displacement was 8, or 12, the keys would have lined up again and the index of coincidence should rise again.

So going back to our mystery ciphertext from page earlier, Eve can try the kappa test with the ciphertexts displaced different amounts and see what the percentage of coincidences are, as shown in Table 2.3.

Displacements of 6 and 7 both look promising, but they can't both be right, and 5 is not far behind. If 6 were the key length, then 12 should have a large number of coincidences, so that's definitely out. If 7 were the key length, then 14 should have a large number of coincidences, and it's not bad but not great. On the other hand, if 5 were the key length, then 10 and 15 should both have a large number of coincidences, and 15 is quite large. Like the Kasiski test, the kappa test has the possibility of being off by a whole-number factor, so it makes sense to combine it with the estimate of 4.6 we got from the index of coincidence formula.

Once again, the two tests together strongly indicate that the period is 5. The kappa test is a good choice if the Kasiski test doesn't seem to be working—maybe Alice has been careful to avoid repeated words in her plaintext. She can't avoid the index of coincidence, though!

2.6 SUPERMAN IS STAYING FOR DINNER:
SUPERIMPOSITION AND REDUCTION

To continue our example, now that Eve knows that the key repeats every 5 letters, what's next? This means that she can separate the ciphertext letters from page 46 into five different columns, each one using a different key letter and, therefore, enciphered with a different alphabet, as in Table 2.4.

TABLE 2.4.
Superimposition of the ciphertext   (*continued*)

I	II	III	IV	V		I	II	III	IV	V
H	X	J	V	X		P	E	U	X	K
D	M	T	U	X		D	M	T	K	X
N	U	O	G	B		H	P	J	G	Z
U	S	U	H	Z		I	G	U	B	M
L	F	W	X	K		O	M	E	G	H
F	F	J	K	X		W	U	D	M	N
K	A	G	L	B		Y	Q	T	H	K
A	F	J	G	Z		J	A	O	O	X
I	K	I	X	K		Y	E	B	M	B
Z	U	T	M	X		V	Z	T	B	G
Y	A	O	M	A		P	F	B	G	W
L	N	B	G	D		D	T	B	M	B
H	Z	E	H	Y		Z	F	I	X	N
O	M	W	B	G		Z	Q	P	Y	T
N	Z	P	M	A		I	A	P	D	M
P	Z	H	M	H		O	A	V	Z	A
K	A	P	G	V		A	M	M	B	V
L	A	S	M	P		L	I	J	M	A
P	O	F	L	A		V	G	U	I	B
L	T	B	W	I		J	F	V	K	X
L	Q	Q	X	W		Z	A	S	V	H
P	Z	U	H	M		U	H	F	K	L
O	Q	C	H	H		H	F	J	H	G
R	T	F	K	L						

Arranging the ciphertext like this is called **superimposition** of the different rows. Each of these columns should have been monoalphabetically enciphered using the same cipher alphabet, which we could confirm with the phi test; in fact, the corresponding indices are 0.054, 0.077, 0.057, 0.093, and 0.061, which is pretty good for the amount of ciphertext we have.

Eve has now **reduced the ciphertext to monoalphabetic terms**. If she has enough ciphertext, she can attack each column separately. Suppose Eve knows that Alice and Bob are using our particular version of the repeating-key cipher, where each key letter indicates a particular additive cipher. Then all she has to do is identify the ciphertext letter corresponding to e in each column. The most frequent letter in each column is L for column I, A for column II, J for III, M for IV, and X for V. Calculating the shifts on this basis gives key letters GVEHS, and decrypting using this key gives

abene	wqome	gyjyi	nwpzg	ejrpr	yjece
debdi	tjeyg	bodpr	syoee	rejeh	erwyk
adzzf	hqrtn	gdkeh	idceo	dekyc	eenew
isadh	exwop	eulpd	idpzt	huxzo	kxacs
iippr	wqoce	ateyg	bkptt	hqzyo	pyyeu
ruozr	cejge	riwei	odotn	ijwyd	wxwei
sjdpu	sukqa	bekvt	heqrh	tqhtc	emeeh
okpai	cjqce	senno	nlacs	ajezn	

Obviously that's not the correct plaintext. Eve could start systematically switching some of the columns to the second-most-frequent letter until it starts to look right, but there are some more clever things she could try. It should be clear that when each column is correctly deciphered, it is more likely to have high-frequency plaintext letters in it than low-frequency—after all, that's what high frequency means. Friedman pointed out that one way to measure frequency would be to add up the frequencies of the letters in each column. The columns with the highest sums are most likely to be right. The numbers we arrive at for each column are approximately 2.9, 1.9, 2.1, 2.4, and 3.1. So, the first and fifth columns are most likely to be correct. Added evidence comes from

TABLE 2.5.
Sums of the letter frequencies for each
possible key applied to our ciphertext

Key Letter	Frequency Sum
A	2.2
B	1.7
C	1.2
D	1.5
E	1.9
F	1.8
G	1.9
H	2.2
I	1.6
J	1.0
K	1.6
L	3.3
M	2.0
N	1.6
O	1.4
P	1.6
Q	1.5
R	2.1
S	2.1
T	1.6
U	1.7
V	2.0
W	2.0
X	1.8
Y	1.8
Z	2.0

the fact that among the low-frequency letters, all the instances of q, x, and z that we see are in the middle three columns.

Eve could now try some other options for these columns, but since the number of letters in each column is a little small, it could take her three or four tries by trying to match high-frequency letters in each column. If she suspected that the columns were encrypted with affine ciphers and required two pairs of matching letters in each column to solve, she would probably want to continue in that direction. However, since she knows that the columns are encrypted with additive ciphers, it's not that hard to just do a brute-force search with each possible key and see which ones give us the highest-frequency plaintexts. Even before computers, this was considered quite feasible, and with a modern computer, it's a snap. Table 2.5 shows the sums of the frequencies for each possible key.

Eve sees that key letter L gives far and away the highest sum, so that's probably the second key letter. Proceeding in this way gives all five key letters as GLASS, which makes her feel a lot better since she was wondering just what a GVEHS was, anyway. Of course, the proof of the pudding is in the decrypting—try it and see if your answer makes sense!

2.7 PRODUCTS OF POLYALPHABETIC CIPHERS

Can we improve the security of a repeating-key polyalphabetic cipher by reencrypting using a second key? Based on Section 1.4, you are probably guessing not. Suppose after Alice encrypts her message with the keyword GLASS, she decides to encrypt again with the keyword QUEEN.

keyword:	G	L	A	S	S	G	L	A	S	S	G	L	A	S	S	G	L	A	S
plaintext:	a	l	i	c	e	w	a	s	b	e	g	i	n	n	i	n	g	t	o
first ciphertext:	H	X	J	V	X	D	M	T	U	X	N	U	O	G	B	U	S	U	H

keyword:	Q	U	E	E	N	Q	U	E	E	N	Q	U	E	E	N	Q	U	E	E
first ciphertext:	h	x	j	v	x	d	m	t	u	x	n	u	o	g	b	u	s	u	h
second ciphertext:	Y	S	O	A	L	U	H	Y	Z	L	E	P	T	L	P	L	N	Z	M

This is still an encryption using a repeating-key polyalphabetic cipher with a length of 5, and it can still be attacked by converting it to 5 monoalphabetic ciphers using the techniques of Sections 2.5 and 2.6. Then it's just a question of what type of monoalphabetic ciphers we have been using. If they are both additive, like in the repeating-key tabula recta cipher we used in the example, then the result will be additive. In our example, encrypting once with the keyword GLASS and again with the keyword QUEEN is the same as encrypting once total with the keyword obtained like this:

keyword 1:	G	L	A	S	S
numbers:	7	12	1	19	19
keyword 2:	Q	U	E	E	N
numbers:	17	21	5	5	14
sum (modulo 26):	24	7	6	24	7
final keyword:	X	G	F	X	G

If the monoalphabetic ciphers are both multiplicative, or affine, the result will be multiplicative, or affine. So the only extra security in the product of 2 repeating-key ciphers with the same key length is the small amount that comes from having a harder-to-guess keyword (like XGFXG). That probably isn't worth the extra trouble.

What if you take the product of 2 repeating-key ciphers with different key lengths? Maybe this time, Alice first encrypts with the keyword RABBIT and then reencrypts with the keyword CURIOUSER:

keyword 1:	R	A	B	B	I	T	R	A	B	B	I	T	R	A	B	B	I	T	R
plaintext:	a	l	i	c	e	w	a	s	b	e	g	i	n	n	i	n	g	t	o
first ciphertext:	S	M	K	E	N	Q	S	T	D	G	P	C	F	O	K	P	P	N	G

keyword 2:	C	U	R	I	O	U	S	E	R	C	U	R	I	O	U	S	E	R	C
first ciphertext:	s	m	k	e	n	q	s	t	d	g	p	c	f	o	k	p	p	n	g
second ciphertext:	V	H	C	N	C	L	L	Y	V	J	K	U	O	D	F	I	U	F	J

This is still a repeating-key cipher, but how often does it repeat? It repeats only when the two keywords both end in the same place, and you can see by looking at the example that that will happen every 18 letters. The reason for this is that 18 is the **least common multiple**, or LCM, of 6 and 9. The LCM and the GCD are related by a very nice formula:

$$\text{LCM}(a, b) = \frac{a \times b}{\text{GCD}(a, b)}.$$

In our example,

$$\text{LCM}(6, 9) = \frac{6 \times 9}{\text{GCD}(6, 9)} = \frac{54}{3} = 18.$$

So if you know the GCD of two numbers, say, from using the Euclidean algorithm, it's very easy to figure out their LCM.

And since the ciphers are additive, once again we can figure out what the equivalent 18-letter keyword would be.

keyword 1:	R	A	B	B	I	T	R	A	B
numbers:	18	1	2	2	9	20	18	1	2
keyword 2:	C	U	R	I	O	U	S	E	R
numbers:	3	21	18	9	15	21	19	5	18
sum (modulo 26):	21	22	20	11	24	15	11	6	20
final keyword:	U	V	T	K	X	O	K	F	T

keyword 1:	B	I	T	R	A	B	B	I	T
numbers:	2	9	20	18	1	2	2	9	20
keyword 2:	C	U	R	I	O	U	S	E	R
numbers:	3	21	18	9	15	21	19	5	18
sum (modulo 26):	5	4	12	1	16	23	21	14	12
final keyword:	E	D	L	A	P	W	U	N	L

So, we have made some progress in the security of our cipher. As long as Eve doesn't guess what Alice did, Alice has achieved the security of an 18-letter keyword using only 15 letters from a 6-letter word and a 9-letter word. In fact, we could have done even a little better—we could achieve an 18-letter repeat using only a 2-letter word and a 9-letter word, since 18 is also the LCM of 2 and 9:

$$\text{LCM}(2, 9) = \frac{2 \times 9}{\text{GCD}(2, 9)} = \frac{18}{1} = 18.$$

Repeating-key encryption was rediscovered several times between the sixteenth and nineteenth centuries, and product ciphers with 2 keywords of 2 different lengths probably were too. In particular, in 1854 a nineteenth-century hopeful named **John Hall Brock Thwaites** publicly challenged Charles Babbage to break a cipher that turned out to be a repeating-key tabular recta encryption using the keywords TWO and COMBINED. Babbage succeeded in breaking the cipher with the help of his youngest son. Although he did not publish a full account of his methods, apparently he used the principles of modular arithmetic, which would make him the first person to do so.

2.8 PINWHEEL MACHINES AND ROTOR MACHINES

The history of machines used to perform or aid in encryption is a long one, stretching back perhaps to the ancient Greek scytale, which we will meet in Section 3.1. It continues through Leon Alberti and Lester Hill up to the present day. Along the way, luminaries appear, such as **Thomas Jefferson**, third president of the United States, and **Sir Charles Wheatstone**, the British scientist, engineer, and inventor best known now for his contribution to the Wheatstone bridge used for measuring electrical resistance. The heyday of cipher machines was the mid–twentieth century from around the end of World War I to the development of modern computers. Hill dates from this period, but as we saw, his ideas did not get much practical use. Much more important were two other types of machines, which, like Hill's, used gears to drive their encryption.

The type that is most similar to the ciphers we have looked at so far is the one that was invented later. These are the **pinwheel** machines, which use a large number of gears turning more or less independently. Each gear has irregularly spaced pins on it (hence the name pinwheel), which produce through mechanical or electrical means the equivalent of a repeating-key polyalphabetic substitution. The period of each pinwheel is different and the set is designed to give a very large combined period.

The first pinwheel device seems to have been invented by **Boris Caesar Wilhelm Hagelin**, a Swedish engineer and employee of Emanuel Nobel, nephew of the Nobel prize originator. In 1922 Hagelin was assigned to look after the Nobel interests in the Swedish firm Aktiebolaget Cryptograph, and in 1925 he invented the first of a very successful line of pinwheel-based cipher machines, the B-21. Other well-known models in this line were the C-36, used by the French Army before and during World War II, the M-209, used by the US Armed Forces for tactical purposes during World War II on a huge scale and continuing through the Korean War, and the C-52/CX-52, used by more than 60 countries during the Cold War.

Figure 2.3. The C-36.

Figure 2.4. Left: Pin and guide arm in inactive position. Right: Pin and guide arm in active position.

The C-36 (Figure 2.3) is a good example of the series. There are five pinwheels, with 25, 23, 21, 19, and 17 pins, respectively. Note that the greatest common divisor of each pair of these numbers is 1, so the combined period is $25 \times 23 \times 21 \times 19 \times 17 = 3{,}900{,}225$. Each pin can stick out to the right of the wheel, in which case it is "active," or to the left ("inactive"). See Figure 2.4. One position on each wheel (the "basic pin") controls a flat rod, or "guide arm," which is also pushed into an active or inactive position. There is also a "cage" containing 25 bars arranged in a rotating horizontal cylinder. Each bar has a lug in one of seven places; the positions were fixed in the original C-36 but movable in the revised

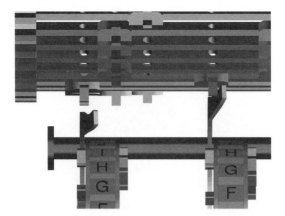

Figure 2.5. Left: Inactive guide arm. Right: Active guide arm engaging lug.

C-362. Five of the places correspond to the pinwheels, while the other two are inactive.

To encrypt a letter, the plaintext is set on an indicating disk and a handle is pushed, causing the cage to be rotated. Active lugs on the bars engage active guide arms, causing the corresponding bar to stick out to the left. See Figure 2.5. Each bar thus activated causes the final ciphertext wheel to turn one place. This results in a final ciphertext letter of

$$C \equiv 1 + (ax_1 + bx_2 + cx_3 + dx_4 + ex_5) - P \quad \text{modulo } 26,$$

where x_i is the number of bars whose lugs are in the position of the ith pinwheel and a, b, c, d, and e are 0 or 1, depending on whether the basic pin for that letter is active or not.

After the ciphertext letter is printed, each pinwheel rotates one pin forward and the guide arms and bars reset for the next letter. Taking the rotation of the pinwheels into account, the nth letter is encrypted according to the substitution

$$C_n \equiv 1 + (a_n x_1 + b_n x_2 + c_n x_3 + d_n x_4 + e_n x_5) - P_n \quad \text{modulo } 26,$$

where x_i is as before. Now a_n is 0 or 1, depending on whether the pin corresponding to n modulo 17 is set to the active position on the first pinwheel, b_n depends on whether the pin corresponding to n modulo 19 is active on the second pinwheel, and so on. You can see that this is equivalent to a repeating-key substitution with a period of 17 and the "keyword"

TABLE 2.6.
Sample lug and pin settings

Bar	Lug Position	Pin Number	Wheel: 1	2	3	4	5
1	1	1	0	1	0	0	1
2	2	2	0	1	0	1	1
3	2	3	0	0	1	0	0
4	3	4	1	0	0	1	1
5	3	5	0	0	1	1	0
6	3	6	1	0	1	0	0
7	4	7	1	1	0	0	0
8	4	8	1	1	1	1	1
9	4	9	0	0	1	1	1
10	4	10	1	1	0	0	1
11	4	11	1	0	0	0	0
12	4	12	1	1	0	0	0
13	4	13	0	1	1	1	1
14	5	14	0	1	1	1	0
15	5	15	1	0	0	0	1
16	5	16	0	0	1	1	0
17	5	17	1	0	0	0	1
18	5	18	0	1	0	0	
19	5	19	1	1	1	1	
20	5	20	0	1	1		
21	5	21	1	1	1		
22	5	22	1	0			
23	5	23	1	1			
24	5	24	1				
25	5	25	0				

$$a_1 x_1, a_2 x_1, a_3 x_1, \ldots, a_{17} x_1,$$

followed by one with a period of 19 and the keyword:

$$b_1 x_2, b_2 x_2, b_3 x_2, \ldots, b_{17} x_2, b_{18} x_2, b_{19} x_2,$$

and so on.

For example, consider the lug and pin settings shown in Table 2.6. These produce the keywords (read vertically) and final ciphertext numbers shown in Table 2.7.

TABLE 2.7.
Keywords and final ciphertext produced by the lug and pin settings

Position	ax_1	bx_2	cx_3	dx_4	ex_5	Total Active Bars	Ciphertext (modulo 26)
1	0	2	0	0	12	14	$15 - P_1$
2	0	2	0	7	12	21	$22 - P_2$
3	0	0	3	0	0	3	$4 - P_3$
4	1	0	0	7	12	20	$21 - P_4$
5	0	0	3	7	0	10	$11 - P_5$
6	1	0	3	0	0	4	$5 - P_6$
7	1	2	0	0	0	3	$4 - P_7$
8	1	2	3	7	12	25	$26 - P_8$
9	0	0	3	7	12	22	$23 - P_9$
10	1	2	0	0	12	15	$16 - P_{10}$
11	1	0	0	0	0	1	$2 - P_{11}$
12	1	2	0	0	0	3	$4 - P_{12}$
13	0	2	3	7	12	24	$25 - P_{13}$
14	0	2	3	7	0	12	$13 - P_{14}$
15	1	0	0	0	12	13	$14 - P_{15}$
16	0	0	3	7	0	10	$11 - P_{16}$
17	1	0	0	0	12	13	$14 - P_{17}$
18	0	2	0	0	12	14	$15 - P_{18}$
19	1	2	3	7	12	25	$26 - P_{19}$
20	0	2	3	0	0	5	$6 - P_{20}$
21	1	2	3	7	12	25	$26 - P_{21}$
22	1	0	0	0	0	1	$2 - P_{22}$
23	1	2	0	7	0	10	$11 - P_{23}$
24	1	2	3	7	0	13	$14 - P_{24}$
25	0	2	0	0	12	14	$15 - P_{25}$

So a sample encryption using these settings might look like the following:

key numbers:	15	22	4	21	11	5	4	26	23	16	2	4
plaintext:	b	o	r	k	b	o	r	k	b	o	r	k
plaintext numbers:	2	15	18	11	2	15	18	11	2	15	18	11
key minus plaintext:	13	7	12	10	9	16	12	15	21	1	10	19
ciphertext:	M	G	L	J	I	P	L	O	U	A	J	S

The key settings on the C-36 include the selection of active pins on the pinwheels, the lug positions (for models with movable lugs), and the starting position of the wheels at the beginning of the encipherment.

The more widely used M-209 had several improvements, including 6 pinwheels instead of 5, for a total period of $26 \times 25 \times 23 \times 21 \times 19 \times 17 =$

101,405,850, and 27 bars instead of 25. In addition, each bar had 2 lugs instead of 1, which could be set to the positions of any 0, 1, or 2 pinwheels. If both lugs on the same bar engaged active pins, however, the action was the same as if only one was engaged. This makes the enciphering equation somewhat more complicated. Most well-known pinwheel machines outside the Hagelin series were teletypewriter machines related to those we will encounter in Section 4.6. These included most of the German World War II ciphers that the British called Fish, such as the Lorenz SZ 40 and SZ 42 and the Siemens and Halske T52 *Geheimschreiber.*

Since the Hagelin cipher machines basically perform multiple polyalphabetic repeating-key encryption, the cryptanalysis methods from Section 2.7 are also relevant here. Other helpful techniques result from the extremely long period and the fact that each keyword has only two different letters. For each ciphertext position, each of the 5 basic pins is either active or inactive, so there are $2^5 = 32$ different positions. If we focus on one of the wheels, say, wheel 1, then the positions where that pin is active will be encrypted with one of $2^4 = 16$ alphabets, possibly not all different. The positions where that pin is inactive will be encrypted with one of a different set of $2^4 = 16$ alphabets. This pattern will repeat every 25 letters as wheel 1 turns. So, if we superimpose rows of 25 letters of ciphertext, the columns will fall into two different groups, which we can often distinguish statistically. Furthermore, once the two groups are distinguished, the two corresponding letter frequency patterns should be shifted by exactly x_1, the number of bars whose lugs are in the position of the first wheel. Proceeding in this way, we can determine the pin and lug settings for each wheel. A known-plaintext attack on the Hagelin machines is also worth considering, since it might be fairly common to find several messages using the same pin and lug settings but different wheel starting positions. Given the plaintext and ciphertext at position n of the message we can easily recover the corresponding key number using the equation

$$C_n \equiv k_n - P_n \quad \text{modulo 26.}$$

Due to the extremely long period, we will probably need to recover the pin and lug settings in order to decipher messages with other wheel starting positions. In this situation we can superimpose the key numbers rather than the actual ciphertext. Now the columns with active basic pins will have larger key numbers compared to the columns with inactive basic pins, and we can proceed more or less as in the ciphertext-only case.

Another type of geared cipher machine developed in the twentieth century uses a set of disks called **rotors**. Rotor machines seem to have been invented independently at least three and possibly as many as five times during the early twentieth century—it is still not entirely clear which inventors truly worked independently and which borrowed the idea from others—or perhaps stole it outright. Recent research suggests that the first priority should go to two first sea lieutenants in the Dutch Navy, **Theo A. van Hengel** and **R.P.C. Spengler**, who were posted in the Dutch East Indies during World War I. Unfortunately for the two lieutenants, the Dutch Navy seems to have held up their patent application for reasons that are now unclear. Before the matter was settled, van Hengel and Spengler were scooped by four others: **Edward Hugh Hebern**, who started work on a rotor machine in the United States in 1917 and filed a patent in 1921, **Arthur Scherbius**, who filed a patent in Germany in 1918, **Hugo Alexander Koch**, who filed a patent in the Netherlands in 1919, and **Arvid Gerhard Damm**, who filed in Sweden, also in 1919. There is some evidence that Koch had access to an early draft of van Hengel and Spengler's patent application, and he may have shared it with Scherbius, with whom he later had a close business relationship. Hebern, Damm, and perhaps Scherbius appear to have come up with their inventions independently of the Dutch inventors.

The idea of a rotor is that it performs a monoalphabetic substitution electrically, by means of wires. Each side of the disk has one contact for each letter of the alphabet, and a complicated set of wires connects each contact on the left side with one contact on the right, as shown in Figure 2.6. So far this is just an electrical version of Alberti's cipher disk. The difference is in the behavior when the rotor is rotated.

Suppose, for instance, that the rotor is wired to perform a multiplicative cipher with a key of 3. Then we have a table:

plaintext:	a	b	c	d	e	f	g	h	i	j	⋯	y	z
numbers:	1	2	3	4	5	6	7	8	9	10	⋯	25	26
times 3:	3	6	9	12	15	18	21	24	1	4	⋯	23	26
ciphertext:	C	F	I	L	O	R	U	X	A	D	⋯	W	Z

Figure 2.6. A rotor taken apart to show its wiring.

We also have a formula,

$$C \equiv 3P \quad \text{modulo } 26,$$

and a schematic diagram, shown in Figure 2.7. Now suppose we rotate the rotor one place, as in Figure 2.8. Note that the plaintext and ciphertext letters don't move—only the wires do. We can think of this as doing a shift, then a multiplication, and then a shift *back*. The shift back is what makes this different from Alberti's disk.

plaintext:	a	b	c	d	e	f	g	h	i	$\cdots$	x	y	z
numbers:	1	2	3	4	5	6	7	8	9	$\cdots$	24	25	26
shifted plaintext:	b	c	d	e	f	g	h	i	j	$\cdots$	y	z	a
numbers plus 1:	2	3	4	5	6	7	8	9	10	$\cdots$	25	26	1
times 3:	6	9	12	15	18	21	24	1	4	$\cdots$	23	26	3
shifted ciphertext:	F	I	L	O	R	U	X	A	D	$\cdots$	W	Z	C
minus 1:	5	8	11	15	17	21	23	26	3	$\cdots$	22	25	2
final ciphertext:	E	H	K	N	Q	T	W	Z	C	$\cdots$	V	Y	B

Following this through gives the formula

$$C \equiv 3(P + 1) - 1 \quad \text{modulo } 26.$$

In general, when the rotor is rotated k places, the formula will be

$$C \equiv 3(P + k) - k \quad \text{modulo } 26.$$

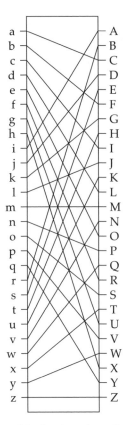

Figure 2.7. A rotor, schematically.

That's kind of interesting but hardly earth shattering. Even when we hook the rotor up to a mechanism so that it rotates automatically one place for each plaintext letter, we really just get a version of Trithemius' progressive cipher. It's a little better than Trithemius' system because the rotor wiring gives us a key, but the fact that the rotor comes back to the beginning every 26 letters (i.e., the period is 26) makes it pretty easy to attack.

It gets really interesting when we add another rotor that turns at a *different* rate. There are several ways to arrange this, but the most common is probably to make the second rotor turn one position whenever the first rotor finishes a complete rotation. In our example, the second rotor will move once every 26 letters.

Figure 2.8. The same rotor, rotated one place.

The second rotor does an additional substitution, as shown in Figure 2.9. If, for instance, it's wired to perform a multiplicative cipher with a key of 5, then for the first 26 letters the final substitution will be

$$C \equiv 5(3(P + k) - k) \quad \text{modulo 26,}$$

as shown, for example, in Figures 2.9 and 2.10. For the second 26 letters, the final substitution will be

$$C \equiv 5((3(P + k) - k) + 1) - 1 \quad \text{modulo 26,}$$

as shown, for example, in Figures 2.11 and 2.12. Mathematicians use the symbol $\lfloor x \rfloor$ to mean round x down to the nearest whole number. In this

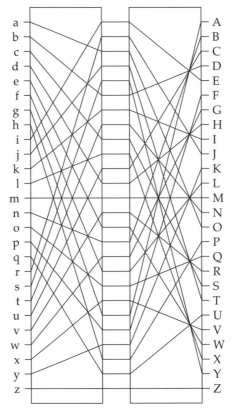

Figure 2.9. Two rotors.

notation, at the kth letter the second rotor has turned $\lfloor k/26 \rfloor$ places, and the substitution will be

$$C \equiv 5((3(P+k) - k) + \lfloor k/26 \rfloor) - \lfloor k/26 \rfloor \quad \text{modulo 26.}$$

With two rotors, it will be $26^2 = 676$ letters before both rotors come back to the beginning, so the period will be 676. This is a much more secure cipher than one rotor. A third rotor could be added, which turns one position every time the second rotor makes a full rotation. Then at the kth letter the first rotor will have turned k places, the second rotor will have turned $\lfloor k/26 \rfloor$ places, and the third rotor will have turned $\lfloor k/26^2 \rfloor$ places. As we add more rotors, the period gets longer and the substitution equations get more complicated—with s rotors the period is 26^s and the equations are nested s levels deep.

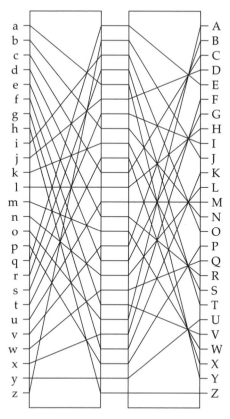

Figure 2.10. The same two rotors with the first one rotated, ready to encrypt the second letter.

Nevertheless it is possible to break a rotor system. The earliest successful techniques for doing so were worked out by Allied cryptanalysts before and during World War II, notably at the Polish Cipher Bureau before the invasion of Poland and later at Bletchley Park, the location of the Government Code and Cypher School in Great Britain. The Poles discovered that the German military had adopted a cipher machine known as the **Enigma**. This was a modified version of the rotor system invented by Scherbius and sold commercially by his firm. The basic military version had three rotors, which could be inserted in any order. There was also a reflector rotor at the far end of the three, which made yet another substitution and then sent the electrical current back through the first three rotors, giving a total of 7 substitutions. The

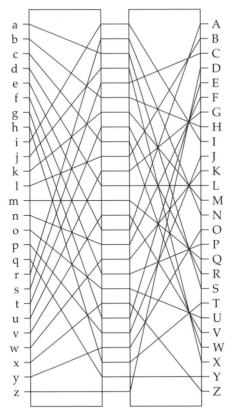

Figure 2.11. The same two rotors with the second one rotated, ready to encrypt the twenty-sixth letter.

reflector also made the cipher reciprocal. Finally, there was a plugboard between the keyboard and the rotors, which added yet another substitution. (See Figure 2.13 for the complete system.) The key settings for the Enigma included the order of the rotors, the initial position of each rotor, the position on each rotor that would cause the next rotor to turn, and the settings of the plugboard.

The first step in breaking a rotor system is to figure out how the rotors are wired. Early efforts took advantage of peculiarities of the enciphered key indicators used by the Germans to specify the starting rotor positions, mistakes made by the Enigma operators, and information secretly bought from a German source. Later, captured machines, rotors, and instructions also played a part in determining rotor wirings. Once

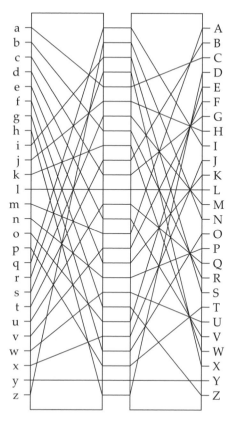

Figure 2.12. The same two rotors with both rotated, ready to encrypt the twenty-seventh letter.

the rotor wirings were figured out, determining the key settings was basically a matter of using enciphered indicators and probable words to rule out some settings as being impossible and mounting a brute-force attack on the rest. A more modern type of attack on the wiring and key settings of a rotor system uses a known-plaintext attack and the fact that we can select sets of letters whose encryption settings are all the same, with the exception of the position of a given rotor.

Of the six people involved in the development of the rotor machine that we mentioned previously, none really managed to profit from sales of the invention. Van Hengel and Spengler challenged Koch's patent until 1923, when their final appeal was denied. Rather suspiciously, the chair of the committee of appeal was the same man who was

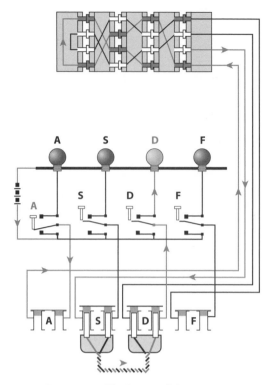

Figure 2.13. The Wiring of the Enigma.

minister of the navy when their initial request to apply for a patent was delayed. Hebern formed a company to market his machines and in the late 1920s and early 1930s succeeded in selling a small number to the US Navy. Government cryptographers who had seen Hebern's machine then proceeded to develop a more secure version of their own, the widely used SIGABA. Hebern was not compensated for his contribution. His lawsuits were settled by his estate in 1958 for a small fraction of what he probably deserved. Koch never built a machine based on his design; he eventually sold his patent rights to Scherbius and died before seeing the Enigma become a success. Scherbius himself set up a company that sold a few Enigmas commercially and a few to the German armed forces, but he also died before Hitler's vast expansion of the German military created the enormous demand for the Enigma that would eventually occur.

And Arvid Damm also formed a company and also died before it became a success. This company was in fact the same Aktiebolaget Cryptograph which was later taken over by Boris Hagelin. Hagelin gave up on rotor machines and switched to the pinwheel machines we saw above. He and his firm then made millions of dollars from commercial machines sold both before and after World War II, machines sold to the French military before the German invasion—and, of course, the US military's M-209.

2.9 LOOKING FORWARD

I said at the end of Chapter 1 that in Chapter 5 we will divide modern ciphers into two types, block ciphers and stream ciphers, and that a block cipher can be thought of as a type of polygraphic cipher. Similarly, it's not too much of a stretch to say that a stream cipher is a kind of polyalphabetic cipher. In fact, the autokey ciphers, which were the earliest ciphers that could be called stream ciphers, were developed at the same time as the polygraphic ciphers discussed in this chapter and by the same people. The ciphers in this chapter mostly have keys that repeat after a short or long period. We will see that the goal of modern stream ciphers is to have an extremely long period or, preferably, no repetition at all. And like modern block ciphers, modern stream ciphers act on an "alphabet" of 0 and 1 instead of the letters humans use for writing.

As for the particular ciphers in this chapter, homophonic ciphers are interesting because they are an early form of probabilistic encryption, where the same plaintext and key might result in different ciphertexts, depending on some random factor. We will see another example of this in Chapter 8. I've introduced Alberti ciphers largely as a link between homophonic ciphers and tabula recta ciphers. Changing the alphabet every letter is just more secure than changing it at random intervals.

Progressive systems are also mainly important as a precursor of repeating-key tabula recta ciphers. Pinwheel machines and rotor machines are simply repeating-key ciphers with extremely long periods and were generally considered state of the art in cipher security until the development of modern electronics in the middle of the twentieth century. Even then, the earliest electronic cipher devices were still basically attempts to produce very long repeating keys and combine them with an alphabet of 0s and 1s.

As far as cryptanalysis is concerned, the most important idea in this section is almost certainly the index of coincidence. The phi test and the kappa test act on letters of the alphabet and their frequencies and therefore can't be directly applied to modern ciphers. However, the index of coincidence is vitally important as one of the earliest uses in cryptanalysis of the idea of correlation. To use correlation, the cryptanalyst performs a statistical comparison of two different sets of frequencies or of one set of frequencies to an altered version of itself. The goal is to try to find patterns that give information about the encryption process. In Chapter 5, I will mention an attack where the frequency distribution of an intermediate value in the cipher is compared to the ciphertext values in order to get information about the key. This is what is specifically known to cryptanalysts as a correlation attack. The idea of correlation is used in other areas as well. For instance, the keystream produced by a stream cipher should pass certain randomness tests in order to be difficult to cryptanalyze. One of the tests is that the autocorrelation, or correlation between the keystream and shifted version of itself, should be as small as possible. Other versions of the same idea involve comparing plaintext frequencies to ciphertext frequencies or different ciphertext frequencies to each other. Finding patterns in these frequency comparisons is one of the important tools used to attack modern ciphers.

The Kasiski test is less commonly used against modern stream ciphers, since, as I said, the goal is to have little or no repetition. Sometimes the period turns out not to be as long as it should be, though. Or maybe a large fraction of the key repeats even though the whole key doesn't. In that case the Kasiski test works just as well against modern ciphers as against tabula recta ciphers. For similar reasons, the version of superimposition with reduction to monoalphabets used in this chapter is going to be used pretty rarely against modern ciphers. We will see other versions of superimposition in Chapter 5, however, and those will be powerful tools against stream ciphers, especially when the ciphers are not used properly. Sadly, this still happens on a regular basis.

3

Transposition Ciphers

3.1 THIS IS SPARTA! THE SCYTALE

All the ciphers we have looked at so far are substitution ciphers; one letter or group of letters is substituted for another. Now we are going to look at a different idea—instead of changing letters, we are "just" going to move them around.

Like the substitution cipher, this idea goes back at least to classical times. The first recorded instance may be the **scytale**. (Scytale rhymes with Italy, although it's Greek, not Italian. The *c* is usually not pronounced in English, although "skytalē" would be a more accurate transliteration of the ancient Greek word.) This cipher device was supposedly used by the ancient Spartans at least as far back as the Spartan general **Lysander**, although there is some question whether the whole idea was made up at a later date.

Scytale literally means *staff*, or *stick*, and the first description of how such a thing might have been used as a cryptographic device comes from the Roman historian Plutarch, several centuries after Lysander:

> The dispatch-scroll is of the following character. When the ephors [the city council of Sparta, more or less] send out an admiral or a general, they make two round pieces of wood exactly alike in length and thickness, so that each corresponds to the other in its dimensions, and keep one themselves, while they give the other to their envoy. These pieces of wood they call "scytalae." Whenever, then, they wish to send some secret and important message, they make a scroll of parchment long and narrow, like a leathern strap, and wind it round their "scytale," leaving no vacant space thereon, but covering its surface all round with the parchment. After doing this, they write what they wish on the parchment,

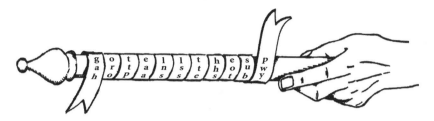

Figure 3.1. A scytale.

just as it lies wrapped about the "scytale"; and when they have written their message, they take the parchment off, and send it, without the
piece of wood, to the commander. He, when he has received it, cannot
other[wise] get any meaning of it,—since the letters have no connection,
but are disarranged,—unless he takes his own "scytale" and winds the
strip of parchment about it, so that, when its spiral course is restored perfectly, and that which follows is joined to that which precedes, he reads
around the staff, and so discovers the continuity of the message. And the
parchment, like the staff, is called "scytale," as the thing measured bears
the name of the measure.

This is really best described by a picture, such as Figure 3.1.

Alice and Bob can accomplish more or less the same thing without
using a stick of wood, of course. Suppose the stick is just large enough
around to fit 3 letters and just long enough to fit 11 turns of the strip of
paper. Then Alice essentially has a 3×11 grid in which she writes the
plaintext letters. If the message doesn't fill the grid, Alice can use nulls
for the remaining spaces.

$$
\begin{array}{c c c c c c c c c c c}
\rightarrow & g & o & t & e & l & l & t & h & e & s & p & \rightarrow \\
\rightarrow & a & r & t & a & n & s & t & h & o & u & w & \rightarrow \\
\rightarrow & h & o & p & a & s & s & e & s & t & b & y & \rightarrow
\end{array}
$$

Note that if Alice were writing on an actual scytale, each column
would be a different turn of the paper. Then, reading down the columns
instead of across, or unwinding the paper, she gets the ciphertext

$$\downarrow \quad \downarrow \quad \downarrow \quad \downarrow \quad \downarrow \quad \downarrow \quad \downarrow \quad \downarrow \quad \downarrow \quad \downarrow \quad \downarrow$$

g	o	t	e	l	l	t	h	e	s	p
a	r	t	a	n	s	t	h	o	u	w
h	o	p	a	s	s	e	s	t	b	y

$$\downarrow \quad \downarrow \quad \downarrow \quad \downarrow \quad \downarrow \quad \downarrow \quad \downarrow \quad \downarrow \quad \downarrow \quad \downarrow \quad \downarrow$$

or

GAHOR OTTPE AALNS LSSTT EHHSE OTSUB PWYAZ

When the scytale cipher is done using a rectangle like this, it's usually called **columnar transposition**. It's traditional to arbitrarily divide the ciphertext into five-letter blocks, like we did in Section 2.2. The last block is padded with nulls in order to disguise the true length of the message. As we will see in a few paragraphs, it should not be obvious to Eve which letters are nulls.

Does this cipher have a key? According to Plutarch, we need "two round pieces of wood exactly alike in length and thickness," one for Alice and one for Bob. As far as we can tell, it's not so much the length as the thickness that really needs to match. If Eve tries to decrypt the ciphertext using a stick that is, say, four letters around instead of three, then when she winds up the paper she gets

$$\downarrow \quad \downarrow \quad \downarrow \quad \downarrow \quad \downarrow \quad \downarrow \quad \downarrow \quad \downarrow \quad \downarrow$$

G	R	P	L	S	E	E	U	Y
A	O	E	N	S	H	O	B	A
H	T	A	S	T	H	T	P	Z
O	T	A	L	T	S	S	W	

$$\downarrow \quad \downarrow \quad \downarrow \quad \downarrow \quad \downarrow \quad \downarrow \quad \downarrow \quad \downarrow \quad \downarrow$$

In other words, the ciphertext is written down the columns and read across the rows. You can see this doesn't make any sense when Eve reads it across.

On the other hand, when Bob decrypts the ciphertext using the correct stick or the correct grid, he writes the ciphertext down the columns and gets

```
  ↓   ↓   ↓   ↓   ↓   ↓   ↓   ↓   ↓   ↓   ↓   ↓
  G   O   T   E   L   L   T   H   E   S   P   A
  A   R   T   A   N   S   T   H   O   U   W   Z
  H   O   P   A   S   S   E   S   T   B   Y
  ↓   ↓   ↓   ↓   ↓   ↓   ↓   ↓   ↓   ↓   ↓   ↓
```

Since the last column is only a partial column, he knows it must be nulls. He throws it away and then reads the plaintext across the rows without difficulty.

So, the key to the scytale is the circumference of the stick, or equivalently, the number of rows in the grid—in this case, 3. Note that if Alice had not padded the message with nulls, the key would have been fairly easy for Eve to guess. She would know that the 33 letters in the ciphertext completely filled a rectangular grid, so there are only four possibilities: 1×33, 3×11, 11×3, and 33×1. The first and last of these are trivial, so that makes it pretty easy for Eve.

3.2 RAILS AND ROUTES: GEOMETRIC TRANSPOSITION CIPHERS

Of course, once you have thought of the idea of writing your message in the rows of a rectangle, there are lots of other things you can do besides just reading off the columns. Colonel **Parker Hitt**, the author of a US Army manual on cryptography during World War I, listed the following methods of reading the message out of the rectangle, noting that each one can be started at any of the four corners: simple horizontal (this includes the trivial cipher if you start at the upper left), simple vertical (this includes the scytale if you start at the upper left), alternate horizontal (alternating left to right and right to left), alternate vertical, simple diagonal, alternate diagonal, spiral clockwise, and spiral counterclockwise. These methods are shown in Figure 3.2, which is taken from Hitt's manual.

In addition to transpositions based on rectangles, Friedman's 1941 manual adds ciphers based on trapezoids, triangles, crosses, and zigzags. You might have come across some of these yourself, including the **rail fence cipher**, in which the message is written on two (or sometimes more) lines in a zigzag and read off by rows.

(a) Simple horizontal:

```
ABCDEF FEDCBA STUVWX XWVUTS
GHIJKL LKJIHG MNOPQR RQPONM
MNOPQR RQPONM GHIJKL LKJIHG
STUVWX XWVUTS ABCDEF FEDCBA
```

(b) Simple vertical:

```
AEIMQU DHLPTX UQMIEA XTPLHD
BFJNRV CGKOSW VRNJFB WSOKGC
CGKOSW BFJNRV WSOKGC VRNJFB
DHLPTX AEIMQU XTPLHD UQMIEA
```

(c) Alternate horizontal:

```
ABCDEF FEDCBA XWVUTS STUVWX
LKJIHG GHIJKL MNOPQR RQPONM
MNOPQR RQPONM LKJIHG GHIJKL
XWVUTS STUVWX ABCDEF FEDCBA
```

(d) Alternate vertical:

```
AHIPQX DELMTU XQPIHA UTMLED
BGJORW CFKNSV WROJGB VSNKFC
CFKNSV BGJORW VSNKFC WROJGB
DELMTU AHIPQX UTMLED XQPIHA
```

(e) Simple diagonal:

```
ABDGKO GKOSVX OKGDBA XVSOKG
CEHLPS DHLPTW SPLHEC WTPLHD
FIMQTV BEIMQU VTQMIF UQMIEB
JNRUWX ACFJNR XWURNJ RNJFCA
```

```
ACFJNR JNRUWX RNJFCA XWURNJ
BEIMQU FIMQTV UQMIEB VTQMIF
DHLPTW CEHLPS WTPLHD SPLHEC
GKOSVX ABDGKO XVSOKG OKGDBA
```

(f) Alternate diagonal:

```
ABFGNO GNOUVX ONGFBA XVUONG
CEHMPU FHMPTW UPMHEC WTPMHF
DILQTV BEILQS VTQLID SQLIEB
JKRSWX ACDJKR XWSRKJ RKJDCA
```

```
ACDJKR JKRSWX RKJDCA XWSRKJ
BEILQS DILQTV SQLIEB VTQLID
FHMPTW CEHMPU WTPMHF UPMHEC
GNOUVX ABFGNO XVUONG ONGFBA
```

(g) Spiral, clockwise:

```
ABCDEF LMNOPA IJKLMN DEFGHI
PQRSTG KVWXQB HUVWXO CRSTUJ
OXWVUH JUTSRC GTSRQP BQXWVK
NMLKJI IHGFED FEDCBA APONML
```

(h) Spiral, counterclockwise:

```
APONML NMLKJI IHGFED FEDCBA
BQXWVK OXWVUH JUTSRC GTSRQP
CRSTUJ PQRSTG KVWXQB HUVWXO
DEFGHI ABCDEF LMNOPA IJKLMN
```

Figure 3.2. Methods of transposition using a rectangle.

```
plaintext:   t e  a l  p i  t r  o p  e i  e t
             h r  i s  l t  e f  r r  s d  n
ciphertext:     TEALP  ITROP  EIETH  RISLT  EFRRS  DN
```

Hitt notes that the rail fence cipher "permits of no variation [i.e., doesn't have a key] and is therefore read almost as easily as straight text when the method is known." In fact, he points out that none of the purely geometric ciphers are very secure because "they do not depend on a key which can be readily and frequently changed."

A slightly fancier—and slightly more secure—variation on the rectangular idea is the **route cipher**, where some sort of key tells you how to read your message out of the rectangle. Historically, this was often used as a sort of code-cipher hybrid, where entire words were written in the spaces of the rectangular grid. This was supposedly used by the Earl of Argyll in his 1685 uprising against King James II, but it is best known to Americans from its use by the Union Army for telegraph transmission in the Civil War. The following example was sent by Abraham Lincoln on June 1, 1863:

GUARD ADAM THEM THEY AT WAYLAND BROWN FOR KISSING VENUS CORRESPONDENTS AT NEPTUNE ARE OFF NELLY TURNING UP CAN GET WHY DETAINED TRIBUNE AND TIMES RICHARDSON THE ARE ASCERTAIN AND YOU FILLS BELLY THIS IF DETAINED PLEASE ODOR OF LUDLOW COMMISSIONER

According to the cipher key that was then in use by the War Department, the keyword GUARD meant that the words should be arranged in a grid with seven rows of five words each and that they should be written in this route: up the first column, down the second, up the fifth, down the fourth, up the third, with a null word at the end of each column. This gives us

	~~kissing~~	~~Commissioner~~		~~Times~~
For	*Venus*	Ludlow	Richardson	and
Brown	correspondents	of	the	Tribune
Wayland	at	*odor*	are	detained
at	*Neptune*	please	ascertain	why
they	are	detained	and	get
them	off	if	you	can
Adam	*Nelly*	~~this~~	~~fills~~	~~up~~
			~~belly~~	~~turning~~

As you can see, the nulls were often chosen to appear to make sense—perhaps humorously!—with words in the previous or following column. This cipher also specified that Venus was a codeword for Colonel, Wayland meant captured, odor meant Vicksburg, Neptune was Richmond, Adam was President of the United States, and Nelly meant that the message was sent at 4:30 p.m. Once the grid is filled, it should be clear that the last three words of the last full row are also nulls, making the following plaintext:

> For Colonel Ludlow. Richardson and Brown, correspondents of the Tribune, captured at Vicksburg, are detained at Richmond. Please ascertain why they are detained and get them off if you can. The President, 4:30 p.m.

3.3 PERMUTATIONS AND PERMUTATION CIPHERS

There are other types of transpositions that are not primarily based on geometric figures or objects, either 2- or 3-dimensional. Scribes throughout the centuries have probably amused themselves by jumbling up the letters in words, but the first systematic description of transposition without geometry seems to be by the same al-Kindi mentioned in Section 1.5, who describes various means of transpositions within words and within lines.

These methods were expanded on by **Taj ad-Din Ali ibn ad-Duraihim ben Muhammad ath-Tha'alibi al-Mausili**, who described 24 variations of transposition ciphers, including writing each word backward and reversing alternate letters of the message. In an English example of the latter method, Alice would write the plaintext "Drink to the rose from a rosy red wine" as follows:

plaintext:	dr	in	kt	ot	he	ro	se	fr	om	ar	os	yr	ed	wi	ne
ciphertext:	RD	NI	TK	TO	EH	OR	ES	RF	MO	RA	SO	RY	DE	IW	EN

What's the big deal here? We are seeing the first explicit example of a **permutation cipher**. A **permutation** in mathematics is any specified way of rearranging the order of the elements of some set. For example, consider the cipher which takes

ruby wine

to

UYBR IENW

In each group of four letters, the first ciphertext position holds the second plaintext letter, the second position holds the fourth letter, the third letter stays in the same position, and the fourth ciphertext position holds the first plaintext letter. There are several ways used by mathematicians to notate this permutation, but one common one is

$$\begin{pmatrix} 1 & 2 & 3 & 4 \\ 2 & 4 & 3 & 1 \end{pmatrix}.$$

Likewise, ibn ad-Duraihim's permutation could be written as

$$\begin{pmatrix} 1 & 2 \\ 2 & 1 \end{pmatrix}.$$

The key to a permutation cipher is just the permutation used. One common way of choosing and remembering a permutation is by a keyword. Alice writes the letters of the keyword above the plaintext, much like in the tabula recta cipher of Section 2.4:

keyword:	TALE	TALE	TALE	TALE	TALE	TALE	TALE	TALE	TALE	TALE
plaintext:	theb	attl	eand	thes	word	thep	aper	andt	hepe	nllu

She then assigns numbers to the letters of the keyword in alphabetical order:

	4132	4132	4132	4132	4132	4132	4132	4132	4132	4132
keyword:	TALE	TALE	TALE	TALE	TALE	TALE	TALE	TALE	TALE	TALE
plaintext:	theb	attl	eand	thes	word	thep	aper	andt	hepe	nllu

Note that the numbers 4132 give us another way of representing our permutation.

The length of the keyword determines the length of each group (in this case, 4 letters) and in each group, the letters of the ciphertext are read off in order of the numbers corresponding to the key letters.

	4132	4132	4132	4132	4132	4132	4132	4132	4132	4132
keyword:	TALE	TALE	TALE	TALE	TALE	TALE	TALE	TALE	TALE	TALE
plaintext:	theb	attl	eand	thes	word	thep	aper	andt	hepe	nllu
ciphertext:	HBET	TLTA	ADNE	HSET	ODRW	HPET	PREA	NTDA	EEPH	LULN

Before Alice sends the message to Bob, she removes or changes the groupings of the letters to make it harder for Eve to guess the length of the permutation. The final cipher text is then

HBETT LTAAD NEHSE TODRW HPETP REANT DAEEP HLULN

What about deciphering? For that, you have to "unpermute" the ciphertext letters. This should remind you of the idea of inverses that we mentioned at the end of Section 1.3; in fact every permutation has an **inverse permutation** that reverses its effects. Here's one way to find it: if Bob has the enciphering permutation

$$\begin{pmatrix} 1 & 2 & 3 & 4 \\ 2 & 4 & 3 & 1 \end{pmatrix},$$

he starts by exchanging the rows,

$$\begin{pmatrix} 2 & 4 & 3 & 1 \\ 1 & 2 & 3 & 4 \end{pmatrix},$$

and then he sorts the columns by the top row:

$$\begin{pmatrix} 1 & 2 & 3 & 4 \\ 4 & 1 & 3 & 2 \end{pmatrix}.$$

So, the inverse of the permutation

$$\begin{pmatrix} 1 & 2 & 3 & 4 \\ 2 & 4 & 3 & 1 \end{pmatrix}$$

is the permutation in which the first position holds the original fourth letter, the second position holds the first, the third letter stays in the same position, and the fourth position holds the original second letter.

You might practice by deciphering the ciphertext

HDETS REEKO NTSEM WELLW

which was encrypted using the same key (corresponding to the keyword TALE) as before.

It's worth asking whether there are any bad keys for permutation ciphers. Consider the expression

$$\begin{pmatrix} 1 & 2 & 3 & 4 \\ 4 & 1 & 1 & 3 \end{pmatrix}.$$

This appears to be telling us that the first ciphertext position gets the fourth plaintext letter, both the second and third positions get the first letter, the fourth position gets the third letter, and the second letter apparently gets thrown away.

For example,

garb agei ngar bage outx

would become

BGGR IAAE RNNA EBBG XOOT

This isn't technically a permutation but a more general thing, which mathematicians call a **function** from positions to letters. It doesn't have an inverse, because once we've thrown away the second letter, we can't generally get it back. Luckily, it's pretty easy to tell the difference between a function that is a permutation and one that isn't—just make sure each letter is used exactly once.

So if all permutations are good keys, how many are there? If we use 4-letter groups, then the first letter can go in the first, second, third, or fourth position. The second letter can go in any of the 3 positions that are left, the third letter has 2 positions it can go in, and the last letter has only 1 choice. So there are $4 \times 3 \times 2 \times 1 = 24$ permutations on groups of 4 letters. In general, if you have n-letter groups, then there are

$$n \times (n-1) \times (n-2) \times \cdots \times 3 \times 2 \times 1$$

permutations, including the **trivial permutation**, which produces the trivial cipher, as usual. Mathematicians use the symbol n to represent this number and call it the **factorial** of n, or just n factorial. Factorials get pretty big pretty fast; for example $12 = 479,001,600$, so there are 479,001,600 different permutation ciphers using 12-letter groups. As with the other ciphers we have discussed, there are better methods of breaking permutation ciphers than brute force, and we shall see some of them in Sections 3.6 and 3.7.

Now, having implied that functions that are not permutations are not useful for encryption and decryption, I should say that that's not quite true. However, we have to do something about the fact that some letters get thrown away. The solution is to encrypt using an **expansion function**, which leaves us with more letters than we started with. Then it's okay if some of them get thrown away when we decrypt. For example, consider the cipher which takes

westw ardho

to

SEWTEW DROHRA

In our previous notation, this would correspond to the function

$$\begin{pmatrix} 1 & 2 & 3 & 4 & 5 & 6 \\ 3 & 2 & 5 & 4 & 2 & 1 \end{pmatrix}.$$

Note that there has to be a number in the top row for each letter in the *ciphertext*, which is more than there are in the plaintext. One or more of the numbers in the top row will not appear in the bottom row, but there has to be a number in the bottom row for every number in the *plaintext*. Mathematicians generally call a function like this an **onto function**, but expansion function is a good description for cryptography. This sort of encryption is very useful when Alice needs a certain number of letters for some reason, like a particular step in a product cipher, or if she just wants to confuse Eve but wants to use something less random than nulls.

How does Bob decrypt a cipher like this? In the case of decryption, Bob really does want to throw away some of the letters since they are duplicates. For example, he could decrypt the preceding ciphertext using the function

$$\begin{pmatrix} 1 & 2 & 3 & 4 & 5 \\ 6 & 2 & 1 & 4 & 3 \end{pmatrix}$$

This time there is a number in the top row for each number in the *plaintext*, and the bottom row can skip some numbers from the ciphertext. It is important that no number in the bottom row be repeated, though; otherwise we would have some letter in the ciphertext used twice. Mathematicians call a function like this a **one-to-one function**, and cryptographers call it a **compression function**. Notice that since the ciphertext letters in the second position and the fifth position are always going to be the same, Bob could equally well decrypt using the function

$$\begin{pmatrix} 1 & 2 & 3 & 4 & 5 \\ 6 & 5 & 1 & 4 & 3 \end{pmatrix}.$$

This is related to the fact that expansion functions, since they are not permutations, don't really have true inverses. We will talk about that a little more after we talk about products of permutations in the next section.

3.4 PERMUTATION PRODUCTS

I hope by now you can begin to guess, more or less, what will happen if we encrypt twice using two different permutation ciphers. Let's see what happens if after Alice encrypts her message with the keyword TALE, which is equivalent to the permutation

$$\begin{pmatrix} 1 & 2 & 3 & 4 \\ 2 & 4 & 3 & 1 \end{pmatrix},$$

she encrypts again with the keyword POEM, which is equivalent to the permutation

$$\begin{pmatrix} 1 & 2 & 3 & 4 \\ 3 & 4 & 2 & 1 \end{pmatrix}.$$

	4132	4132	4132	4132	4132
keyword:	TALE	TALE	TALE	TALE	TALE
plaintext:	theb	attl	eand	thes	word
first ciphertext:	HBET	TLTA	ADNE	HSET	ODRW

	4312	4312	4312	4312	4312
keyword:	POEM	POEM	POEM	POEM	POEM
first ciphertext:	hbet	tlta	adne	hset	odrw
second ciphertext:	ETBH	TALT	NEDA	ETSH	RWDO

	4132	4132	4132	4132	4132
keyword:	TALE	TALE	TALE	TALE	TALE
plaintext:	thep	aper	andt	hepe	nllu
first ciphertext:	HPET	PREA	NTDA	EEPH	LULN

	4312	4312	4312	4312	4312
keyword:	POEM	POEM	POEM	POEM	POEM
first ciphertext:	hpet	prea	ntda	eeph	luln
second ciphertext:	ETPH	EARP	DATN	PHEE	LNUL

If Eve could look at both the ciphertext and the plaintext, she would figure out quickly that this is the same as if Alice had encrypted with the key

$$\begin{pmatrix} 1 & 2 & 3 & 4 \\ 3 & 1 & 4 & 2 \end{pmatrix}.$$

Mathematicians often express this using a product notation:

$$\begin{pmatrix} 1 & 2 & 3 & 4 \\ 2 & 4 & 3 & 1 \end{pmatrix} \times \begin{pmatrix} 1 & 2 & 3 & 4 \\ 3 & 4 & 2 & 1 \end{pmatrix} = \begin{pmatrix} 1 & 2 & 3 & 4 \\ 3 & 1 & 4 & 2 \end{pmatrix}.$$

While we are on the subject, it's worth noting that

$$\begin{pmatrix} 1 & 2 & 3 & 4 \\ 2 & 4 & 3 & 1 \end{pmatrix} \times \begin{pmatrix} 1 & 2 & 3 & 4 \\ 3 & 4 & 2 & 1 \end{pmatrix}$$

is *not* the same as

$$\begin{pmatrix} 1 & 2 & 3 & 4 \\ 3 & 4 & 2 & 1 \end{pmatrix} \times \begin{pmatrix} 1 & 2 & 3 & 4 \\ 2 & 4 & 3 & 1 \end{pmatrix}.$$

In other words, permutation products are not necessarily **commutative**. If you don't believe me, try encrypting our plaintext using the keyword POEM first and then the keyword TALE. You should get a different answer. This makes combining permutation ciphers a little different from some of the other ciphers we have looked at.

We can also think about the product of a permutation with its inverse. For example,

$$\begin{pmatrix} 1 & 2 & 3 & 4 \\ 2 & 4 & 3 & 1 \end{pmatrix} \times \begin{pmatrix} 1 & 2 & 3 & 4 \\ 4 & 1 & 3 & 2 \end{pmatrix} = \begin{pmatrix} 1 & 2 & 3 & 4 \\ 1 & 2 & 3 & 4 \end{pmatrix}.$$

In general, the product of a permutation and its inverse is the trivial permutation. This makes sense, since encrypting followed by decrypting should return the message to the original state. Likewise,

$$\begin{pmatrix} 1 & 2 & 3 & 4 \\ 4 & 1 & 3 & 2 \end{pmatrix} \times \begin{pmatrix} 1 & 2 & 3 & 4 \\ 2 & 4 & 3 & 1 \end{pmatrix} = \begin{pmatrix} 1 & 2 & 3 & 4 \\ 1 & 2 & 3 & 4 \end{pmatrix},$$

which makes sense since we would expect the inverse of the inverse to be the original permutation. This is one case where permutation products are commutative.

Expansion and compression functions don't behave so nicely. Doing an encryption followed by a decryption does give us the trivial permutation

$$\begin{pmatrix} 1 & 2 & 3 & 4 & 5 & 6 \\ 3 & 2 & 5 & 4 & 2 & 1 \end{pmatrix} \times \begin{pmatrix} 1 & 2 & 3 & 4 & 5 \\ 6 & 2 & 1 & 4 & 3 \end{pmatrix} = \begin{pmatrix} 1 & 2 & 3 & 4 & 5 \\ 1 & 2 & 3 & 4 & 5 \end{pmatrix}.$$

But this time reversing the order gives us something else:

$$\begin{pmatrix} 1 & 2 & 3 & 4 & 5 \\ 6 & 2 & 1 & 4 & 3 \end{pmatrix} \times \begin{pmatrix} 1 & 2 & 3 & 4 & 5 & 6 \\ 3 & 2 & 5 & 4 & 2 & 1 \end{pmatrix} = \begin{pmatrix} 1 & 2 & 3 & 4 & 5 & 6 \\ 1 & 2 & 3 & 4 & 2 & 6 \end{pmatrix}.$$

Once again, you should try this on a message. The technical distinction is that expansion and compression functions only have **one-sided inverses** instead of true, two-sided inverses. This is related to why there can be two decryption functions for the same encryption function, or vice versa, as we saw in Section 3.3. The practical effect is that you should encrypt only with expansion functions and decrypt only with compression functions, not the other way around.

But to answer our original question of this section, the upshot is that—just like combining two repeating-key polyalphabetic ciphers—combining two permutation ciphers with the same group length gives you another permutation cipher with the same group length. What if you combine permutation ciphers with different group lengths? For example, after Alice encrypts her message with the keyword TALE, she could encrypt again with the keyword POETRY.

	4132	4132	4132	4132	4132	4132
keyword:	TALE	TALE	TALE	TALE	TALE	TALE
plaintext:	theb	attl	eand	thes	word	thep
first ciphertext:	HBET	TLTA	ADNE	HSET	ODRW	HPET

	321546	321546	321546	321546		
keyword:	POETRY	POETRY	POETRY	POETRY		
first ciphertext:	hbettl	taadne	hsetod	rwhpet		
second ciphertext:	EBHTTL	AATNDE	ESHOTD	HWREPT		

	4132	4132	4132	4132	4132	4132
keyword:	TALE	TALE	TALE	TALE	TALE	TALE
plaintext:	aper	andt	hepe	nllu	xgar	bage
first ciphertext:	PREA	NTDA	EEPH	LULN	GRAX	AEGB

	321546	321546	321546	321546		
keyword:	POETRY	POETRY	POETRY	POETRY		
first ciphertext:	preant	daeeph	lulngr	axaegb		
second ciphertext:	ERPNAT	EADPEH	LULGNR	AXAGEB		

In this example Alice needed to add some extra nulls to make things come out even.

Is this the same as encrypting with one permutation? If you look closely, you'll see that it can't be a permutation on 4-letter groups, because some letters "leak" across from one group to another. The same is true for the 6-letter groups. However, since the two keywords line up every 12 letters, this is actually the same as a permutation on 12-letter groups. In fact, we can write both our keyword permutation ciphers as permutations on 12 letters: the permutation cipher corresponding to TALE, which we wrote with a key of

$$\begin{pmatrix} 1 & 2 & 3 & 4 \\ 2 & 4 & 3 & 1 \end{pmatrix},$$

could also be written with a key of

$$\begin{pmatrix} 1 & 2 & 3 & 4 & 5 & 6 & 7 & 8 & 9 & 10 & 11 & 12 \\ 2 & 4 & 3 & 1 & 6 & 8 & 7 & 5 & 10 & 12 & 11 & 9 \end{pmatrix}$$

and the cipher corresponding to POETRY, which we would ordinarily write with a key of

$$\begin{pmatrix} 1 & 2 & 3 & 4 & 5 & 6 \\ 3 & 2 & 1 & 5 & 4 & 6 \end{pmatrix},$$

could also be written with a key of

$$\begin{pmatrix} 1 & 2 & 3 & 4 & 5 & 6 & 7 & 8 & 9 & 10 & 11 & 12 \\ 3 & 2 & 1 & 5 & 4 & 6 & 9 & 8 & 7 & 11 & 10 & 12 \end{pmatrix}$$

Then the key for the product cipher would be the product of the two permutations, namely,

$$\begin{pmatrix} 1 & 2 & 3 & 4 & 5 & 6 & 7 & 8 & 9 & 10 & 11 & 12 \\ 2 & 4 & 3 & 1 & 6 & 8 & 7 & 5 & 10 & 12 & 11 & 9 \end{pmatrix}$$
$$\times \begin{pmatrix} 1 & 2 & 3 & 4 & 5 & 6 & 7 & 8 & 9 & 10 & 11 & 12 \\ 3 & 2 & 1 & 5 & 4 & 6 & 9 & 8 & 7 & 11 & 10 & 12 \end{pmatrix}$$
$$= \begin{pmatrix} 1 & 2 & 3 & 4 & 5 & 6 & 7 & 8 & 9 & 10 & 11 & 12 \\ 3 & 4 & 2 & 6 & 1 & 8 & 10 & 5 & 7 & 11 & 12 & 9 \end{pmatrix}$$

Permutation ciphers with different group lengths still behave very much like repeating-key ciphers, in that the length of the product key is the least common multiple of the lengths of the original keys. As in Section 2.7, as long as Eve doesn't guess what Alice did, Alice has achieved the security of a 12-letter keyword using only 10 letters. Eve might realize that the letters don't get mixed up as thoroughly with these product ciphers as they would with a true 12-letter keyword permutation cipher. It's possible to keep alternating 4-letter keywords and 6-letter keywords until the letters are mixed up as much as you want, but in that case Alice and Bob are probably better off just using a 12-letter keyword.

3.5 KEYED COLUMNAR TRANSPOSITION CIPHERS

Now that we've spent some time examining permutation and keyword permutation ciphers, I have to tell you that there doesn't seem to be much documentation for anyone using them in practice. The reason is probably that whenever someone started to work out a keyword permutation cipher, that person immediately realized that it would be at least as secure for no more effort to make a product cipher by combining a permutation with a columnar transposition.

Let's look at one of our previous permutation encryption examples again but this time arrange the text slightly differently:

plaintext	ciphertext
4132	
TALE	
theb	HBET
attl	TLTA
eand	ADNE
thes	HSET
word	ODRW
thep	HPET
aper	PREA
andt	NTDA
hepe	EEPH
nllu	LULN

This seems like a convenient way for Alice to keep track of where she is in the plaintext. She has to write the keyword out only once, and if she reads across the rows on the right-hand side of this table, she gets the same ciphertext as before. However, since the ciphertext letters are now in a rectangle, it seems logical to apply columnar transposition and read down the columns instead. This gives her the following ciphertext.

HTAHO HPNEL BLDSD PRTEU ETNER EEDPL TAETW TAAHN

Maybe you noticed that Alice didn't actually need the columns on the right-hand side of the preceding display: all she really has to do is read down the columns on the left-hand side but in the order according to the key. First she reads the column numbered 1, then 2, 3, and 4. This product cipher is called **keyed columnar transposition** and appears, apparently for the first time, in a work on cryptography by **John Falconer**. Falconer was a seventeenth-century English cryptographer at the Court of James II about whom not much is known. His work was published posthumously in 1685. After that, ciphers based at least in part on keyed columnar transposition were in serious use somewhere in the world more or less continuously until at least the 1950s.

In order for Bob to decrypt the message quickly, he can start by writing the keyword and column numbers across the top of a blank table. He can figure out the correct number of rows by dividing the total number of letters by the length of the key, and then he writes the ciphertext down the columns *in the order specified by the key*. Finally, he reads off the plaintext across the rows.

In terms of security, keyed columnar transpositions are not actually that much more secure than permutation ciphers. The key for a columnar transposition is the number of rows—or the number of columns, since if you know the approximate length of the message, knowing one is the same as knowing the other. In keyed columnar transposition the number of columns depends only on the length of the key for the permutation cipher. So, there are exactly the same number of keyed columnar transposition keys as permutation keys. And, as we shall see in Sections 3.6 and 3.7, there are other attacks on permutation ciphers that apply about equally well to keyed columnar transposition ciphers.

There is one big advantage to keyed columnar transpositions over permutations. Remember that the product of two additive ciphers is another additive cipher, two multiplicative ciphers make a multiplicative cipher, two affine ciphers give you another affine cipher, and two permutation ciphers give you another permutation cipher, although it might be of a different key length. But the product of two keyed columnar transpositions is not a keyed columnar transposition and is considerably harder to break in general than a single such transposition.

To see why, let's consider a very short message with only 9 letters and a very short key of only 3 positions.

3	1	2
a	g	r
e	a	t
w	a	r

First we apply the following permutation.

key:	312	312	312
plaintext:	agr	eat	war
first ciphertext:	GRA	ATE	ARW

We could think of this as the permutation

$$\begin{pmatrix} 1 & 2 & 3 & 4 & 5 & 6 & 7 & 8 & 9 \\ 2 & 3 & 1 & 5 & 6 & 4 & 8 & 9 & 7 \end{pmatrix}$$

on nine letters. Then we apply columnar transposition.

first ciphertext	second ciphertext
GRA	GAA
ATE	RTR
ARW	AEW

We could also think of this as a permutation on nine letters:

$$\begin{pmatrix} 1 & 2 & 3 & 4 & 5 & 6 & 7 & 8 & 9 \\ 1 & 4 & 7 & 2 & 5 & 8 & 3 & 6 & 9 \end{pmatrix}$$

Notice that since we happen to have a square, applying the transposition twice would cancel itself out. This permutation happens to be its own inverse.

Now, for the sake of example, let's apply another columnar transposition with the inverse key. This time I'll condense the steps, since you probably have the idea.

2	3	1
G	A	A
R	T	R
A	E	W

Keep in mind what we are doing: we've applied two permutations that are inverses of each other, alternated with two columnar transpositions that are also inverses of each other. You might expect everything to cancel out. But it doesn't: the final ciphertext is

ARW GRA ATE

which is not the same as the original plaintext.

How did this happen? Remember that the order in which we combine permutations makes a difference, unlike for addition or multiplication. So the fact that we alternate keyed permutations with columnar transpositions means that nothing cancels out and we end up with a slightly more complicated transposition cipher, which can get very much more complicated if the two rectangles are not the same size. (See Sidebar 3.1 for more details.)

■ ■ ■ SIDEBAR 3.1. FUNCTIONAL NIHILISM ■ ■ ■

If you are paying close attention, you might notice that while our example of double keyed columnar transposition on a 3×3 square didn't produce plaintext, it actually does produce something you can read without a columnar transposition. The easiest way to see why uses the function notation I will explain in Section 4.3, so feel free to postpone reading this sidebar until after you have read that section.

First of all, we will use Greek letters to stand for permutations, as mathematicians often do. In particular, since π is the Greek letter corresponding to p, it's often used for permutations. It has nothing to do here with the number 3.1459 ..., which is the ratio of the perimeter of a circle to its diameter. We will also use σ to stand for the permutation corresponding to a scytale cipher since σ is the first letter in scytale.

Let's let π_n stand for a permutation cipher with a key length of n—the exact key doesn't matter. The one we used on our 3×3 square, for example, would be π_3. And let's let σ_{mn} be the scytale cipher that writes the plaintext in m rows and reads off the

ciphertext in n columns. The inverse of a permutation cipher with
key length n is some other permutation cipher with key length
n, which we'll denote π_n^{-1}. According to the "shoes-and-socks"
principle, the inverse of writing the plaintext in m rows and
reading the ciphertext in n columns is first writing the ciphertext
in n columns and then reading the plaintext in m rows. But this is
the same as writing in n rows and reading in m columns. So, the
inverse of σ_{mn} is σ_{nm}, and the inverse of σ_{33}, or any scytale cipher
with a square rectangle, is itself. We observed this earlier without
using the notation.

Now let's look at our example. First, we wrote the plaintext
in rows and applied the permutation π_3. Then we read it off in
columns, which is σ_{33}. Then we wrote it in rows again and applied
π_3^{-1}. Lastly, we applied σ_{33} again. So the final ciphertext was

$$C_1 C_2 \cdots C_9 = \sigma_{33} \pi_3^{-1} \sigma_{33} \pi_3 (P_1 P_2 \cdots P_9)$$

Does that tell us anything? We've pretty firmly established
that $\pi_3^{-1} \sigma_{33}$ is not the same as $\sigma_{33} \pi_3^{-1}$, which is why the two
scytale ciphers and the two permutations don't all cancel. But let's
think about σ_{33} some more. Remember that we can think of it as
writing in three columns and reading in three rows or writing in
three rows and reading in three columns. So what we are really
doing is just swapping rows and columns. Thinking of it that way,
we see that

$$\sigma_{33} \pi_3^{-1} \sigma_{33}$$

means swap the columns and rows, permute the columns, and
swap the columns and rows again. If you try this, it should
become clear that the final result is to swap the *rows* of the text
with which we are working. So

$$C_1 C_2 \cdots C_9 = \sigma_{33} \pi_3^{-1} \sigma_{33} \pi_3 (P_1 P_2 \cdots P_9)$$

means permute the columns using π_3 and then permute the rows
using π_3^{-1}. In fact, there isn't really a columnar transposition
going on at all. You should test this with our example:

plaintext			ciphertext		
a	g	r	A	R	W
e	a	t	G	R	A
w	a	r	A	T	E

Incidentally, a transposition that permutes the rows of a rectangle as well as the columns is often called a **Nihilist transposition cipher**. According to Kerckhoffs, a transposition using a square and permuting both rows and columns by the same key, without exchanging rows and columns, was one of the methods used by Russian Nihilists to send secret messages in the 1870s and 1880s. We will call the more general version, with any rectangle, two keys, and the exchange of rows and columns, a **Nihilist columnar transposition**. An analysis like the one we just did will show you that a product of two Nihilist columnar transpositions using the same completely filled square grid will be another Nihilist columnar transposition. So will two Nihilist columnar transpositions on different completely filled nonsquare rectangles, but only if the number of columns in the first rectangle is the same as the number of rows in the second. As far as security is concerned, it turns out that the techniques of Sections 3.6 and 3.7 will still break everything except for the order of the rows, which is pretty easy to sort out once you have the correct plaintext on each row. So this cipher isn't generally considered worth the trouble.

If the number of columns in the first rectangle is *not* the same as the number of rows in the second, then you have a true double columnar transposition, which can be very hard to break.

This idea of a **double keyed columnar transposition** cipher (often called **double transposition** for short) seems to have come into common use somewhat before World War I. Although not impossible to break, as we shall see in Section 3.7, it is generally considered the most secure transposition cipher that can reliably be done purely by hand and continued to be used into World War II, especially by Allied secret agents in the field and resistance fighters in occupied Europe.

3.6 DETERMINING THE WIDTH OF THE RECTANGLE

The steps in cryptanalyzing the transposition ciphers we have discussed are very much parallel to the steps in cryptanalyzing a repeating-key cipher: first Eve needs to make sure she knows what kind of cipher she really has, then she finds the length of the key, and finally she uses superimposition to find the key itself. Luckily, the first step is pretty easy. Since transposition ciphers move letters around without changing them, the frequencies of the letters will be the same in the ciphertext as they are in the plaintext. This is usually pretty obvious; if there is any doubt one could use the various index of coincidence tests that we saw in Section 2.2 and will see in Section 5.1.

The key for the scytale cipher is the number of rows or, alternatively, the number of columns since it's easy to get one from the other. As we said in Section 3.1, that's pretty easy to find out. If Eve knows the total number of squares in the grid, she just needs to find all the possible numbers of rows and columns that give her the right rectangle and try writing the ciphertext down the columns until she gets readable plaintext across the rows. If Alice has sensibly padded the message with nulls, then Eve might get stuck. If so, she would next throw away the last letter of the message and try again, and so on.

If Eve suspects she has a permutation cipher or a keyed columnar transposition cipher, she starts the same way. She guesses the number of rows and columns and writes the ciphertext either across (for a permutation cipher) or down (for a columnar transposition). The number of columns is the length of the permutation or keyword being used for the key. In this case, it's not quite as easy for her to tell whether she has the right-size grid. One test that she can use is to see if the proportions of vowels and consonants on each row is approximately correct.

Suppose we pick some letters from English text at random. According to our letter-frequency tables, approximately 38.1% of them will be vowels. So if you pick 10 letters at random, the average number of vowels you will get is 3.81, and the most likely outcome is that you will get 4. This won't always happen: sometimes you will get a few more, sometimes a few less. In fact, it's more likely that we will *not* get exactly 4. How likely are 4 vowels? Well, first, we can list the number of patterns of vowels and consonants we could have:

$$\text{VVVVCCCCCC}$$
$$\text{VVVCVCCCCC}$$
$$\text{VVVCCVCCCC}$$
$$\text{VVVCCCVCCC}$$
$$\vdots$$

That's going to take a while, but if you finish it, you will get 210 possible patterns.

Now consider the first pattern, which is 4 vowels followed by 6 consonants. The chance of picking the first vowel is .381, the same for the second, the third, and so on. The chance of picking the first consonant is .619, the same for the second, and so on. So the total chance of that pattern is $.381 \times .381 \times .381 \times .381 \times .619 \times .619 \times .619 \times .619 \times .619 \times .619 \approx .00119$.

If you think about it, the chance of the second pattern is the same, and so on for every other pattern, so the total chance of picking exactly 4 vowels is about $210 \times .00119 \approx .249$. In other words, we will get *exactly* 4 vowels only about a quarter of the time. But we will very often get *approximately* 4 vowels. How do we quantify that?

Statisticians have long had a way to measure how close to the average we expect to get in situations like this; it's now known as **variance**. The idea is that we will pick our 10 letters at random several times, say, 100 times, and each time we will calculate the difference between what we actually get and what we would get on the average. Some of the differences will be positive and some negative, and we don't want these to cancel out. Cryptographers originally took the absolute value of the difference, but it turns out to be easier to predict what will happen mathematically if we square the difference instead. Then we take the average of these squared differences. Ordinarily that would mean dividing by 100, but again in this special situation it turns out to be easier to predict what will happen if we divide by 1 less, or 99. And that's the variance.

What do we expect the variance for the number of vowels to be? Statisticians have shown that it should be about the average chance of a vowel times the average chance of a consonant times the number of letters we picked each time, or $.381 \times (1 - .381) \times 10 \approx 2.358$ in this case.

That holds only if the letters are picked at random. If instead we actually pick 100 ten-letter English words, then the variance will be different. For starters, if we pick 10 letters at random, there's a small chance (about .8%) that we won't pick any vowels, whereas if we pick a 10-letter word, there's virtually no chance that we will pick one with no a's, e's, i's, o's, or u's. In general, the variance for actual English text is going to be much smaller than that for randomly chosen letters from English text.

How does this help Eve cryptanalyze a transposition cipher? Suppose she has the following ciphertext.

OHIVR	SVAHT	BLRHL	HLBIT	MBETM	NOEIO
ITETK	ROWTN	ATHIG	NSDEN	UPBLN	TSEMA
TADAA	ERARI	AOWSA	YIAPT	NAEOW	BCDRE
WAHMT	GEDER	HFDDT	EAEHA	TEHME	IELBO
HIUSI	EKIUE	UHESL	MTKSE	CREP	

She suspects it is from a keyed columnar transposition. There are 144 letters here, which means a lot of divisors she could try: 1, 2, 3, 4, 6, 8, 9, 12, 16, 18, 24, 36, 48, 72, and 144. However, it's pretty uncommon to have less than 4 or more than 20 columns in a cipher like this, especially if the key comes from a keyword. So, Eve can probably narrow this down some. Six seems like a good number of letters for a keyword, so let's start there. Then there are 24 rows, so Eve writes the ciphertext down the columns of her table. She counts the number of vowels on each row. Since there are 6 columns, the average number of vowels in 6 letters is $6 \times .381 \approx 2.286$, and she records the square of the difference between these two numbers on each row (Table 3.1).

The grand total turns out to be about 40.787; dividing that by 17 (the number of rows minus 1) gives about 5.098 for the variance. What does that mean? If Eve guessed right about the number of columns, then reading across the rows should give her—well, not plaintext yet because the columns aren't in the right order. However, each row would be the right letters for plaintext—just in the wrong order. On the other hand, if she guessed wrong, then everything is hopelessly jumbled. If Eve guessed wrong, the variance will be more like the variance for random sets of 6 letters, which is $.381 \times (1 - .381) \times 6 \approx 1.415$, and if Eve guessed right,

TABLE 3.1.
Calculating the variance for our ciphertext

						Vowels	Expected	Square of Difference
O	M	E	W	E	H	3	2.286	$(.714)^2 \approx .510$
H	N	N	S	D	I	1	2.286	$(-1.286)^2 \approx 1.654$
I	O	U	A	E	U	6	2.286	$(3.714)^2 \approx 13.794$
V	E	P	Y	R	S	1	2.286	$(-1.286)^2 \approx 1.654$
R	I	B	I	H	I	3	2.286	$(.714)^2 \approx .510$
S	O	L	A	F	E	3	2.286	$(.714)^2 \approx .510$
V	I	N	P	D	K	1	2.286	$(-1.286)^2 \approx 1.654$
A	T	T	T	D	I	2	2.286	$(-.286)^2 \approx .0818$
H	E	S	N	T	U	2	2.286	$(-.286)^2 \approx .0818$
T	T	E	A	E	E	4	2.286	$(1.714)^2 \approx 2.938$
B	K	M	E	A	U	3	2.286	$(.714)^2 \approx .510$
L	R	A	O	E	H	3	2.286	$(.714)^2 \approx .510$
R	O	T	W	H	E	2	2.286	$(-.286)^2 \approx .0818$
H	W	A	B	A	S	2	2.286	$(-.286)^2 \approx .0818$
L	T	D	C	T	L	0	2.286	$(-2.286)^2 \approx 5.226$
H	N	A	D	E	M	2	2.286	$(-.286)^2 \approx .0818$
L	A	A	R	H	T	2	2.286	$(-.286)^2 \approx .0818$
B	T	E	E	M	K	2	2.286	$(-.286)^2 \approx .0818$
I	H	R	W	E	S	2	2.286	$(-.286)^2 \approx .0818$
T	I	A	A	I	E	5	2.286	$(2.714)^2 \approx 7.366$
M	G	R	H	E	C	1	2.286	$(-1.286)^2 \approx 1.654$
B	N	I	M	L	R	1	2.286	$(-1.286)^2 \approx 1.654$
E	S	A	T	B	E	3	2.286	$(.714)^2 \approx .510$
T	D	O	G	O	P	2	2.286	$(-.286)^2 \approx .0818$

it will be more like the variance for English text, which is much smaller. Since the variance Eve got is even larger than for random letters, she concludes that she guessed wrong.

Eve tries again, this time using the next divisor in the list, which is 8 (Table 3.2). Without going through all the details this time, it turns out that she gets a variance of about 0.462, compared to an expected variance of $.381 \times (1 - .381) \times 8 \approx 1.887$ for random sets of 8 letters. Now there's a good possibility that Eve has found the right number of columns. And the rows do look to the naked eye as if they might be scrambled plaintext.

TABLE 3.2.
Starting a second try for the variance

I	II	III	IV	V	VI	VII	VIII
O	I	O	N	W	W	H	K
H	T	W	T	S	A	A	I
I	M	T	S	A	H	T	U
V	B	N	E	Y	M	E	E
R	E	A	M	I	T	H	U
S	T	T	A	A	G	M	H
V	M	H	T	P	E	E	E
A	N	I	A	T	D	I	S
H	O	G	D	N	E	E	L
T	E	N	A	A	R	L	M
B	I	S	A	E	H	B	T
L	O	D	E	O	F	O	K
R	I	E	R	W	D	H	S
H	T	N	A	B	D	I	E
L	E	U	R	C	T	U	C
H	T	P	I	D	E	S	R
L	K	B	A	R	A	I	E
B	R	L	O	E	E	E	P

3.7 ANAGRAMMING

The next step in cryptanalyzing a permutation cipher or a keyed columnar transposition cipher is to find the permutation that serves as the key. We do this by **anagramming**, which is pretty much what it sounds like. In ordinary conversation, an anagram is when you rearrange the letters of one word or phrase to get another. In cryptanalysis, anagramming is rearranging the letters of ciphertext to get plaintext. What makes this feasible is that we are rearranging not just individual letters, but entire columns. For example, it's pretty unlikely that column II follows column I. HT on the second line is an unlikely combination, but possible, especially if the H is the end of one word and the T is the beginning of the next. However, VB on the fourth line is almost impossible, since V hardly ever ends a word or even a syllable in English; the same is true for VM on the seventh line. In fact, columns VII and VIII are the only ones that are really likely to follow column I (Table 3.3).

TABLE 3.3.
The contact method

I	VII	Frequency	I	VIII	Frequency
O	H	.0005	O	K	—
H	A	.0130	H	I	.0060
I	T	.0100	I	U	—
V	E	.0080	V	E	.0080
R	H	.0010	R	U	.0015
S	M	.0005	S	H	.0050
V	E	.0080	V	E	.0080
A	I	.0010	A	S	.0080
H	E	.0165	H	L	.0005
T	L	.0015	T	M	.0005
B	B	—	B	T	.0005
L	O	.0020	L	K	—
R	H	.0010	R	S	.0045
H	I	.0060	H	E	.0165
L	U	.0015	L	C	.0020
H	S	—	H	R	.0010
L	I	.0045	L	E	.0090
B	E	.0055	B	P	—

Which column is a better fit, VII or VIII? We can put I next to VII and I next to VIII and see which looks like a better collection of digraphs. If it's not obvious to the eye, we can put the frequency of each digraph along with it. (This is sometimes called the **contact method**.) The dashes indicate digraphs whose frequencies are negligible, although not necessarily zero.

As a rough way of scoring the two options, Friedman suggests adding the frequencies in each case. That's easy and usually works, but it's wrong mathematically, as Friedman goes on to point out. After all, if you want to know what the probability is that the first row starts with OH and the second row starts with HA, you don't add the probabilities—you multiply them. Multiplying all these small numbers is such a pain, however, that Friedman suggests using logarithms, which is a common trick for making multiplication of a large number of things into a much easier addition. The relevant property is that

$$\log(x \times y) = \log x + \log y,$$

and so instead of doing the calculation

$$.0005 \times .0130 \times .0100 \times .0080 \times \cdots$$

for the first column, we can do

$$\log .0005 + \log .0130 + \log .0100 + \log .0080 + \cdots.$$

Even better, we can look up the logarithms of the frequencies in a table just as easily as we can look up the original frequencies, so we never actually have to compute the logarithms. In the case of a negligible frequency, we will use $\log 0.0001$, since 0.0001 is quite small compared to the other numbers in our table. The numbers we arrive at for each combination, which are sometimes called the **log weights**, are approximately -49 for I with VII and -51 for I with VIII. They are negative because the logarithm of a number between 0 and 1, such as a probability, is negative. The closer a log weight is to 0, the larger the probability is that the column is correct plaintext. Thus we suspect column VII should follow column I. Continuing along these lines, we can either consider likely digraphs starting with column VII or perhaps likely trigraphs starting with columns I and VII. Trigraph frequencies are even more imprecise than digraph frequencies, but we might note that the TL on line 10 is almost certainly followed by an E, which appears only in column II. Tentatively trying column II next, we have Table 3.4.

If you would like to finish the deciphering, you might note next that ITM on the third line is almost certainly followed by a vowel, and HST on the third-to-last line is probably followed by either a vowel or an R. At that point you might be ready to make a guess for a word that includes the letters at the beginning of the fifth line, and that should pretty much do the trick. When you are done, the numbers at the tops of the columns will be the same numbers Alice used when she encrypted, from which you could find the permutation she used or guess at the keyword.

TABLE 3.4.
Starting to anagram the ciphertext

I	VII	II	III	IV	V	VI	VIII
O	H	I	O	N	W	W	K
H	A	T	W	T	S	A	I
I	T	M	T	S	A	H	U
V	E	B	N	E	Y	M	E
R	H	E	A	M	I	T	U
S	M	T	T	A	A	G	H
V	E	M	H	T	P	E	E
A	I	N	I	A	T	D	S
H	E	O	G	D	N	E	L
T	L	E	N	A	A	R	M
B	B	I	S	A	E	H	T
L	O	O	D	E	O	F	K
R	H	I	E	R	W	D	S
H	I	T	N	A	B	D	E
L	U	E	U	R	C	T	C
H	S	T	P	I	D	E	R
L	I	K	B	A	R	A	E
B	E	R	L	O	E	E	P

■ ■ ■ SIDEBAR 3.2. BUT WHEN YOU TALK ABOUT DISRUPTION ■ ■ ■

One fairly easy way of making a columnar transposition more
complicated is to "disrupt" it by leaving some of the spaces in the
grid blank or, perhaps, taking them out of order. The easiest way
for Alice to make a **disrupted columnar transposition** is the
incompletely filled rectangle, which just means she leaves
spaces at the end of the grid blank instead of filling them with
nulls. This has the added bonus of making it more difficult for Eve
to guess the width of the rectangle, since there is no longer any
reason for it to be a divisor of the length of the message. On the
other hand, since Bob knows the width of the rectangle, all he
has to do is divide the length of the message by the width of the
rectangle. The quotient is the number of completely filled rows
and the remainder is the number of filled spaces in the last row.
Bob knows from the key which columns Alice filled in last, so he
know which columns are "short."

Eve, unlike Bob, has a problem. Suppose she correctly identifies the width of the rectangle—say, through our statistical technique—and therefore figures out how many blank spaces are in the last row. She still can't yet tell which columns correspond to the long ones and which correspond to the short ones. Therefore, she doesn't know exactly where each column begins and ends. This makes the contact method considerably more complicated, although not impossible. Alice and Bob can make life even harder for Eve by designating particular blank spaces in the middle of the grid. Too much of this and the contact method becomes almost impossible, although it also makes the system more inefficient. On the other hand, the technique of multiple anagramming we are about to see works just as well on disrupted transpositions as on any other sort.

If Eve is faced with a double transposition, or some system that uses a shape other than a rectangle, then her job is much harder. In the case of a single keyed columnar transposition, letters that are on the same row of the plaintext rectangle end up a fixed distance apart in the ciphertext, which gives her a way to test whether she has guessed the right shape of rectangle. In the case of a double transposition or irregular shape, this regularity won't be present, and it will be very difficult for Eve to start anagramming. However, if Eve has multiple messages in the same key, then she has a chance. In particular, she needs two or more messages that are the same length and encrypted with the same key. Then she can superimpose them in much the same way as she superimposed rows of the same message in Section 2.6. Since corresponding letters in the two messages are treated exactly the same way by the transposition, we can use the contact method to anagram the columns. This is called **multiple anagramming**.

For example, suppose Eve has 5 messages and believes that the first 12 letters of each message were encrypted in the same way, as shown in Table 3.5. The Js in column I in Table 3.5 are almost certainly both followed by vowels other than Y, so that means column III, X, or XI. LN

TABLE 3.5.
Superimposing multiple ciphertexts encrypted with the same key

I	II	III	IV	V	VI	VII	VIII	IX	X	XI	XII
S	E	U	I	S	M	D	M	N	A	A	S
J	Y	I	N	B	N	D	H	N	O	A	L
L	L	N	A	A	U	E	L	C	U	I	D
J	E	E	I	P	K	D	C	N	A	A	E
B	A	I	Y	R	D	B	D	D	U	N	G

TABLE 3.6.
Starting to anagram the multiple ciphertexts

I	X	II	III	IV	V	VI	VII	VIII	IX	XI	XII
S	A	E	U	I	S	M	D	M	N	A	S
J	O	Y	I	N	B	N	D	H	N	A	L
L	U	L	N	A	A	U	E	L	C	I	D
J	A	E	E	I	P	K	D	C	N	A	E
B	U	A	I	Y	R	D	B	D	D	N	G

on the third line is unlikely, and so is BN on the fifth line, so that leaves column X. Now we have Table 3.6.

What's next? Well, the Us on the third and fifth lines in Table 3.6 are probably followed by consonants, which means column VIII, IX, or XII. JAE on the fourth line doesn't look too good, so that rules out XII. Both VIII and IX look reasonable. We could try to distinguish them using digraph frequencies, but remember that frequency tests are about which option is most *probable*—they don't always give you the right answer. In the end, we might have to try them both. I'll let you finish this up if you want; as hints, I'll tell you that there are a bunch of names in the plaintext, and since we have only the first 12 letters of each message, they do cut off in the middle of a word.

3.8 LOOKING FORWARD

We will see in Chapter 4 that transpositions are extremely important building blocks of modern ciphers. All kinds are used, including columnar transpositions, geometric transpositions, permutations, expansion and compression functions, products of these, and many others. In most cases, however, fixed transpositions (no key) are used in conjunction with keyed substitutions. The reason is partly historical. Fixed transpositions were easy to implement in the early days of computers; you could implement them by running wires from one place to another. Keyed transpositions are harder. Since then, the cases where ciphers are implemented using wires or other hardware have been decreasing, and the cases where they are done with software programs have been increasing. So, the use of keyed transpositions is increasing, but it's still somewhat rare.

One partial exception to this rule is rotations, which are a certain simple type of permutation. A **rotation** on n letters with a key of k is a permutation of the form

$$\begin{pmatrix} 1 & 2 & \cdots & n-k & n-k+1 & n-k+2 & \cdots & n \\ k+1 & k+2 & \cdots & n & 1 & 2 & \cdots & k \end{pmatrix}.$$

In other words, the block of plaintext letters is rotated around into ciphertext positions without actually mixing it up. This would not be very secure by itself, but it's useful as a component of other ciphers. Furthermore, rotations, even with variable keys, can be implemented relatively easily in both hardware and software.

Rotations with keys have been used as a part of several modern ciphers, including Madryga, RC5, RC6, and Akelarre. Unfortunately, these ciphers have had mixed success. Madryga was published in 1984 as an alternative to existing ciphers that would be faster and easier to implement in software. It is now considered seriously flawed. RC5 was published in 1995 and was likewise intended to be fast in software as well as hardware. It was considered strong for its time, although some attacks on it have been discovered. Its lack of widespread adoption probably had more to do with licensing fees than security. RC6, published in 1998, was deliberately designed to improve on RC5. It is considered a strong cipher, although it is not especially common. The Advanced Encryption Standard, which we will meet in Chapter 4, is generally preferred, because it has roughly equal security, government endorsement, and no licensing fees.

Akelarre is an interesting case. Published in 1996, it was also based in part on RC5, in hopes of combining the strengths of RC5 with the security features of another cipher known as IDEA. Unfortunately, several attacks on it were quickly discovered, including one that essentially bypasses everything but the rotation. That attack shows that a combination of two ciphertexts can be expressed as a rotation of a combination of two plaintexts. If some plaintexts are known, or even if something is known about the frequencies at which different sorts of plaintexts occur, then a process very much like anagramming can be used to determine which rotation is correct. These experiences probably didn't increase confidence in ciphers with keyed rotations, but RC6 shows that when properly handled they can still be very useful in modern cipher design.

4

Ciphers and Computers

A distinction is sometimes made between polygraphic substitution ciphers and **polyliteral** substitution ciphers. Polygraphic ciphers, as we saw in Section 1.6, transform a block of plaintext letters into a block of ciphertext letters of the same size. Polyliteral ciphers, on the other hand, transform a single letter into a block of letters or symbols. For our first example, we go back to the ancient Greeks again. In the second century BCE, the Greek historian **Polybius** wrote a 40-volume history of ancient Greece and Rome with considerably many digressions, including one about cryptography and, in particular, what he refers to as "signaling by fire"—torches or beacon-fires. Polybius may be the first author to distinguish between codes and ciphers and gives several examples of using torches to send coded messages. However, it's the cipher which interests us here. In Polybius' words:

> It is as follows: We take the alphabet and divide it into five parts, each consisting of five letters. There is one letter less in the last division, but this makes no practical difference.* Each of the two parties who are about to signal to each other must now get ready five tablets and write one division of the alphabet on each tablet .... [T]he dispatcher of the message will now raise the first set of torches on the left side indicating which tablet is to be consulted, i.e. one torch if it is the first, two if it is the second, and so

---

*Polybius' Greek alphabet had 24 letters.

on. Next he will raise the second set on the right on the same principle to indicate what letter of the tablet the receiver should write down.

Modern descriptions of this usually use a 5-by-5 square instead of 5 tablets, and the system is generally called the **Polybius square**, or Polybius checkerboard. Also, since the modern English alphabet has 26 letters instead of 24, we will have to leave 1 out or put 2 of them in the same square—putting i and j together is traditional. The square then looks like this.

	1	2	3	4	5
1	a	b	c	d	e
2	f	g	h	ij	k
3	l	m	n	o	p
4	q	r	s	t	u
5	v	w	x	y	z

So if Alice wanted to encode "I fear the Greeks," it would look like this:

plaintext:	i	f	e	a	r	t	h	e	g	r	e	e	k	s
ciphertext:	24	21	15	11	42	44	23	15	22	42	15	15	25	43

Better still,

ciphertext:	24211	51142	44231	52242	15152	543

This is a biliteral cipher, since each plaintext letter becomes two ciphertext digits. Like most ciphers from the ancient world, this one doesn't have a key, although we could add one by scrambling the order of the letters in the square, the numbers along the top and side, or both.

A version a little more suited to the English alphabet might look like this:

	0	1	2	3	4	5	6	7	8
0		a	b	c	d	e	f	g	h
1	i	j	k	l	m	n	o	p	q
2	r	s	t	u	v	w	x	y	z

One suited to the 29-letter Danish and Norwegian alphabet might be[*]

	0	1	2	3	4	5	6	7	8	9
0		a	b	c	d	e	f	g	h	i
1	j	k	l	m	n	o	p	q	r	s
2	t	u	v	w	x	y	z	æ	ø	å

You might be thinking that the last example was gratuitous, but I want to call your attention to exactly what the transformation is in this instance.

plaintext:	a	b	c	d	e	f	g	h	i	j	k	l	⋯
ciphertext:	01	02	03	04	05	06	07	08	09	10	11	12	⋯

Aside from some leading zeros, all we have done is convert the letters to numbers! This is because, if we leave the upper-left square blank, the letter in row r and column c is the $(r \cdot 10 + c)$th letter of the alphabet. However, our normal system of writing numbers represents this number as rc, that is, the digit for r followed by the digit for c. For example, the letter in row 2 and column 3 is w, which is the $(2 \cdot 10 + 3)$rd = 23rd letter of the alphabet in English, Danish, or Norwegian.[†]

So what about the English version of the table? The transformation here is

plaintext:	a	b	c	d	e	f	g	h	i	j	k	l	⋯
ciphertext:	01	02	03	04	05	06	07	08	10	11	12	13	⋯

Now the letter in row r and column c is the $(r \cdot 9 + c)$th letter of the alphabet. For example, the letter in row 2 and column 7 is "y," which is

---

[*]I want to make it clear here that the 29-letter Swedish alphabet would have worked just as well for this example. Having grown up in Minnesota, I know how dangerous it is to appear to take sides between Norwegians and Swedes.

[†]Or Swedish.

the $(2 \cdot 9 + 7)$th $= 25$th letter of the alphabet. If we instead write this number as 27, then we are using a base 9, or nonary, **numeral system** instead of the usual base 10, or decimal, system.

If we use a small base, such as 3 (ternary), then 2 digits are not enough to represent all the plaintext letters. One way to get around this is to use multiple tables and put the digit of the table before the row and column:

Table 0	0	1	2
0		a	b
1	c	d	e
2	f	g	h

Table 1	0	1	2
0	i	j	k
1	l	m	n
2	o	p	q

Table 2	0	1	2
0	r	s	t
1	u	v	w
2	x	y	z

This gives

plaintext:	a	b	c	d	e	f	g	h	i	j	k	l	$\cdots$
ciphertext:	001	002	010	011	012	020	021	022	100	101	102	110	$\cdots$

Now we have a triliteral system. The letter in table t, row r, and column c is the $(t \cdot 3^2 + r \cdot 3 + c)$th letter of the alphabet. For example, the letter in table 2, row 0, and column 1 is "s," which is the $(2 \cdot 3^2 + 0 \cdot 3 + 1)$th $= 19$th letter of the alphabet. If multiple tables seem too cumbersome, it is also common to use one table, grouping multiple digits in the rows, columns, or both:

	00	01	02	10	11	12	20	21	22
0		a	b	c	d	e	f	g	h
1	i	j	k	l	m	n	o	p	q
2	r	s	t	u	v	w	x	y	z

Base 2 (**binary**) numerals are also very common in modern ciphers due to their use in computers. In this case we would need very many nested tables unless we group the digits, so we will (group them, that is):

	000	001	010	011	100	101	110	111
00		a	b	c	d	e	f	g
01	h	i	j	k	l	m	n	o
10	p	q	r	s	t	u	v	w
11	x	y	z					

This gives

plaintext:	a	b	c	d	e	· · ·
ciphertext:	00001	00010	00011	00100	00101	· · ·

Here, a numeral like 10010 represents the $(1 \cdot 2^4 + 0 \cdot 2^3 + 0 \cdot 2^2 + 1 \cdot 2 + 0)$th = 18th letter of the alphabet, or r. As you probably know, it's very convenient to use binary numerals in computers because the two digits can be represented by something, such as electrical current, being either on or off.

The first use of a binary numeral system in cryptography, however, was well before the advent of digital computers. **Sir Francis Bacon** alluded to this cipher in 1605 in his work *Of the Proficience and Advancement of Learning, Divine and Humane* and published it in 1623 in the enlarged Latin version *De Augmentis Scientarum*. It is actually a combination of a cipher with a steganographic system, as not only the meaning but the very existence of the message is hidden in an innocuous "covertext." As usual, we will give a modern English example.

Suppose we want to encrypt the word "not" into the covertext "I wrote Shakespeare." First convert the plaintext into binary numerals:

plaintext:	n	o	t
ciphertext:	01110	01111	10100

Then stick the digits together into a string:

$$011100111110100$$

Now we need what Bacon called a "biformed alphabet," that is, one where each letter can have a "0-form" and a "1-form." We will use roman letters for our 0-form and italic for our 1-form. Then for each letter of the covertext, if the corresponding digit in the ciphertext is 0, use the 0-form, and if the digit is 1 use the 1-form:

0   11100   111110100xx

I   *wrote*   *Shake*speare.

Any leftover letters can be ignored, and we leave in spaces and punctuation to make the covertext look more realistic. Of course, it still looks odd with two different typefaces—Bacon's examples were more subtle, although it's a tricky business to get two alphabets that are similar enough to fool the casual observer but distinct enough to allow for accurate decryption.

Ciphers with binary numerals were reinvented many years later for use with the telegraph and then the printing telegraph, or teletypewriter. The first of these were technically not cryptographic since they were intended for convenience rather than secrecy. We could call them **nonsecret ciphers**, although for historical reasons they are usually called codes or sometimes encodings. They are not codes in the cryptographic sense, since they work with letters or characters, not words. To try to avoid confusion as much as possible I am going to call them **nonsecret encodings**. The most well-known nonsecret encoding is probably the Morse code used for telegraphs and early radio, although Morse code does not use binary numerals. In 1833, Gauss, whom we met in Chapter 1, and the physicist Wilhelm Weber invented probably the first telegraph code, using essentially the same system of 5 binary digits as Bacon. **Jean-Maurice-Émile Baudot** used the same idea for his **Baudot code** when he invented his teletypewriter system in 1874. And the Baudot code is the one that **Gilbert S. Vernam** had in front of him in 1917 when his team at AT&T was asked to investigate the security of teletypewriter communications.

Vernam realized that he could take the string of binary digits produced by the Baudot code and encrypt it with a version of the tabula recta polyalphabetic substitution ciphers we looked at in Section 2.4. Each digit from the plaintext would be added to a corresponding digit from the key modulo 2 to produce the ciphertext. This process is sometimes called **noncarrying addition** because each digit is added separately without carrying over to the next digit. It is often designated by the symbol ⊕. For example, the digits 10010, which ordinarily represent 18, and the digits 01110, which ordinarily represent 14, would be added:

$$
\begin{array}{r}
1 \quad 0 \quad 0 \quad 1 \quad 0 \\
\oplus \quad 0 \quad 1 \quad 1 \quad 1 \quad 0 \\
\hline
1 \quad 1 \quad 1 \quad 0 \quad 0
\end{array}
$$

This gives 11100, which ordinarily represents 28—not the usual sum of 18 and 14. The big difference between this and the tabula recta polyalphabetic substitution ciphers we looked at earlier is that the addition is modulo 2 instead of modulo 26.

Some of the systems that AT&T was using were equipped to automatically send messages using a paper tape, which could be punched with holes in 5 columns—a hole indicated a 1 in the Baudot code and no hole indicated a 0. Vernam configured the teletypewriter to add (modulo 2) each digit represented by the plaintext tape to the corresponding digit from a second tape punched with key characters. The resulting ciphertext is sent over the telegraph lines as usual.

At the other end, Bob feeds an identical copy of the tape through the same circuitry. Since $1 \equiv -1$ modulo 2, subtraction modulo 2 is the same as addition. Thus the same operation at Bob's end subtracts the key, and the teletypewriter can print the plaintext. Vernam's invention and its further developments became extremely important in modern cryptography; we will see these ideas again in Sections 4.3 and 5.2.

In general, polyliteral ciphers trade off the advantage of fewer distinct symbols for the disadvantage of longer messages. In some situations, such as holding torches, digital computers, steganography, and telegraphs, the utility of this trade-off is obvious. In other situations it may not be as clear why one would be willing to put up with the longer messages. In the next section, however, we will see one big advantage that can be gained from polyliteral encryption.

4.2 FRACTIONATING CIPHERS

The polyliteral ciphers we have been looking at are not very different from simple substitution ciphers. The only trick to attacking them is figuring out how many symbols correspond to each letter, which can be done fairly easily by just guessing and starting the attack. With the right guess, Eve can do a frequency analysis just like simple substitution ciphers. In order to make these ciphers secure, something fancier needs to be added. One possibility is simply to scatter some null symbols throughout the message to disrupt the orderly division of the ciphertext into groups. The symbols can be placed in prearranged positions, or, if they are not otherwise used in the cipher, they can be distributed at Alice's whim and Bob will simply ignore them. This last technique works well with polyliteral ciphers, since they use a smaller set of symbols than the original message anyway.

Another possibility is to make different groups different lengths. An example of this is the **straddling checkerboard**:

	0	1	2	3	4	5	6	7	8	9
			a	b	c	d	e	f	g
1	h	i	j	k	l	m	n	o	p	q
2	r	s	t	u	v	w	x	y	z

The letters in the first row are encrypted with a single digit; the letters in the other rows are encrypted with 2 digits each. Since there are no 2-digit ciphertext equivalents that start with 3 through 9, it is unambiguous when Bob should pick a letter in the first row as opposed to one of the others.

Yet another possibility is to combine a polyliteral cipher with another cipher in such a way as to split up the ciphertext groups. This is often referred to as **fractionation**. The simplest thing to do is to perform a transposition after the polyliteral cipher, so that the symbols corresponding to a single plaintext letter are no longer adjacent. Perhaps the "most interesting and practical" of these fractionating product ciphers was the cipher invented by the German officer Lieutenant (later Colonel) **Fritz Nebel** and used by the German army during World War I. The Germans called this cipher **GedeFu 18**, short for *Geheimschrift der Funker 1918*, or Radio Operators' Cipher 1918. The French, seeing a bunch of cryptograms containing only the letters A, D, F, G, V, and X, called it the **ADFGVX cipher**. This system starts out with a 6 × 6 version of the Polybius square, containing both letters and digits in a scrambled order and labeled on the top and side with the letters ADFGVX. For example:

	A	D	F	G	V	X
A	b	5	x	q	j	c
D	6	y	r	k	d	7
F	z	s	l	e	8	1
G	t	m	f	9	2	u
V	n	g	0	3	v	o
X	h	a	4	w	p	i

So to encrypt the name Zimmermann, Alice would write

plaintext:	z	i	m	m	e	r	m	a	n	n
ciphertext:	FA	XX	GD	GD	FG	DF	GD	XD	VA	VA

This was followed by a keyed columnar transposition. For instance, if the keyword for this part of the cipher was GERMANY, then we would have

first ciphertext	second ciphertext
3 2 6 4 1 5 7	
G E R M A N Y	
F A X X G D G	G A F X D X G
D F G D F G D	F F D D G G D
X D V A V A X	V D X A A V X

and the final ciphertext would be

GFVAF DFDXX DADGA XGVGD X

A general method of breaking this cipher was not discovered until after the war was over, and an outline was first published in 1925, although during the war Allied cryptanalysts were able to break the cipher when they could compare ciphertexts that came from plaintexts with identical beginnings or endings or if the division into columns was easily guessed. That makes the ADFGVX cipher easily one of the most successful ciphers of World War I and one of the most difficult to break of any practical cipher not requiring a machine of some sort.

The reason that this cipher is so difficult to solve and the reason it is in the chapter on computer ciphers is because it embodies one of the two principles at the heart of all modern ciphers. These are now known as **diffusion** and **confusion**, and they were given their modern meanings by **Claude Shannon**. Shannon was an engineer and mathematician who is considered the founder of the field of information theory. In defining these principles, he was concerned about statistical attacks on a cipher. In a paper written in 1945 and declassified and published in 1949, Shannon defined diffusion as the idea that statistical structures in the plaintext that depend on looking at only a few letters at a time, such as letter frequency or digraph frequency, should be "diffused" in the ciphertext into statistics that require looking at long strings of letters. Shannon's definition of confusion, on the other hand, was the idea that given simple statistics of the ciphertext, it should be very complicated to find the key. In particular, the cipher should be resistant to known-plaintext attacks, since information about the frequency of letters and words in English (or any other human language) always allows a certain amount of successful guessing of plaintext.

The ADFGVX cipher does a reasonably good job of diffusion. The columnar transposition generally sends the two ciphertext letters corresponding to each plaintext letter to widely different parts of the final ciphertext, and thus any attempt to use letter frequency information to solve the biliteral substitution cannot succeed until substantial amounts of ciphertext rearranging is done. On the other hand, the usual techniques of solving transposition ciphers from Sections 3.6 and 3.7 require information about the original plaintext letters, such as whether they are vowels or consonants and which ones fit into high-frequency digraphs. This information is difficult to obtain before the substitution is solved.

On the other hand, Shannon's aim is not completely satisfied, since the plaintext letters are not truly diffused into a large number of ciphertext letters but merely into two widely separated ones. The ADGFVX cipher also exhibits rather good confusion, especially if the Polybius square part of the key is chosen carefully. If care is taken to avoid high-frequency letters clustering in the square, known-plaintext attacks become very difficult.

4.3 HOW TO DESIGN A DIGITAL CIPHER:
SP-NETWORKS AND FEISTEL NETWORKS

Shannon himself did not invent any ciphers, but in the same paper where he defined diffusion and confusion, he did outline a technique for building ciphers that might possess these characteristics. In order to talk about his ideas and, in general, about the sort of ciphers designed for use in computers, it will be convenient to talk a little first about how mathematicians think about functions. We already saw some glimpses of this in Section 3.3.

You probably already think you know what a function is; you are no doubt familiar with functions like $f(x) = x^2$ and $f(x) = \sin x$. The first thing that pops into your head when you see one of these functions is probably the graph of the function, as in Figure 4.1.

You are not alone: in elementary algebra and calculus—and, in fact, in almost every mathematical field developed between the seventeenth and nineteenth centuries—the study of functions is closely tied to the study of curves in the plane or, sometimes, surfaces in 3 or more dimensions. Toward the end of the nineteenth century, however, mathematicians began to think of functions in a somewhat more general way. A function is simply something that "takes in" objects of one type and "spits out" other objects, possibly of the same type and possibly not, according to some definite rule. The rule may be a formula, a set of instructions, a table, or even a picture, as long as it is unambiguous and always produces the same output from the same input.

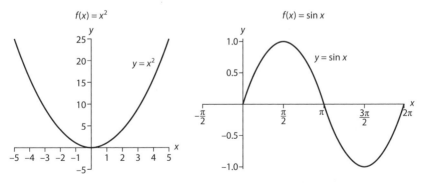

Figure 4.1. The graphs of $f(x) = x^2$ and $f(x) = \sin x$.

For example, $f(x) = x^2$ is a function that takes in real numbers and spits out real numbers according to a formula. The Caesar cipher, $f(P) =$ "the letter P shifted 3 letters down the alphabet, wrapping around" is a function that takes in letters and spits out letters, according to a set of instructions. The Baudot code is a function that takes in letters and spits out strings of binary digits according to a table. The permutation

$$\begin{pmatrix} 1 & 2 & 3 & 4 \\ 2 & 4 & 3 & 1 \end{pmatrix}$$

can be thought of as a function that takes in numbers between 1 and 4 (representing ciphertext positions) and spits out numbers between 1 and 4 (representing the plaintext letters that go in those positions), according to a table, and so on.

Shannon proposed using a **mixing function** in ciphers to provide diffusion and confusion. This is an idea that Shannon admits cannot be defined precisely for ciphers. "Speaking loosely, however," he says "we can think of a mixing transformation as one which distributes any reasonably cohesive region in the space fairly uniformly over the entire space. If the first region could be described in simple terms, the second would require very complex ones." If this were a simple substitution cipher, for instance, we would want plaintext letters near the beginning of the alphabet to give ciphertext letters scattered in a complex way through the alphabet. We would also want plaintext letters with high probabilities to be scattered in a complex way, and so on. To achieve diffusion, on the other hand, we want to operate on larger blocks of letters. Shannon suggests a function of the form

$$F(P_1 P_2 \cdots P_n) = H(S(H(S(H(T(P_1 P_2 \cdots P_n)))))),$$

as shown in Figure 4.2, where T is some transposition acting on groups of n letters, H is some Hill cipher on n-letter blocks that is not too complicated, and S is a simple substitution cipher applied to each letter of the block. Each step is simple, but it is perfectly believable that the combination and repetition would provide good mixing properties.

We have not yet discussed the role of the key. In Shannon's conception, F is not secret and does not involve a key, so there is no security yet. However, this makes F particularly easy to perform using a computer or other machine, which is important because F will be the most

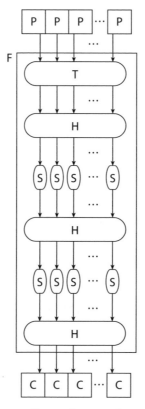

Figure 4.2. Shannon's mixing function *F*.

complicated part of the cipher. Shannon goes on to say that while a good mixing function provides good diffusion, we can add confusion by extending the function to something like

$$V_k(F(U_k(P_1 P_2 \cdots P_n))),$$

as shown in Figure 4.3, where U_k and V_k are two relatively uncompli-cated ciphers, say simple substitutions, which depend on a key k. The idea is that some key information is applied immediately, which is "mixed around" by the mixing function, adding confusion as well as diffusion, and then more key information is applied. This last step does not provide confusion, but it is necessary to keep Eve from immediately undoing whatever operations were in the nonsecret function *F*. If she could "unmix" the ciphertext in this way, then she would be left with a

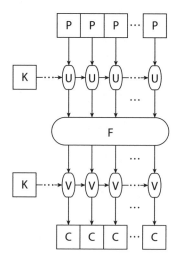

Figure 4.3. A Shannon cipher.

very easy cipher to solve. For even more security, this could be extended with more repetition to

$$W_k(F_2(V_k(F_1(U_k(P_1P_2\cdots P_n)))))\text{,}$$

as shown in Figure 4.4, and so on.

Shannon was considerably ahead of his time; cipher designers didn't really start thinking about his principles in a systematic way until the 1970s. Around this time people began thinking about applications for computers outside the military and the government, and among these was **Horst Feistel**. Feistel was born in Germany but fled the Nazi military conscription and came to the United States in 1934. In 1944 he became a US citizen and started working for the US Air Force Cambridge Research Center on Identification: Friend or Foe (IFF) systems, which were not precisely cryptographic but were closely related. After that Feistel held a couple of jobs doing defense contract work at nonprofit research centers, and in 1967 he joined IBM's Watson Research Center. Throughout this period Feistel continued to think about ciphers for computers, but (perhaps because of NSA pressure) he was not able to work on them until he reached IBM, which had a contract with Lloyds Bank in Great Britain to provide them with some of the

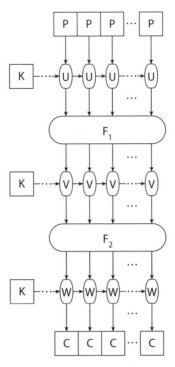

Figure 4.4. An even more secure Shannon cipher.

earliest automated teller machines. Obviously, it was necessary to encrypt the communications between the ATMs and the central bank to
prevent unauthorized transactions from taking place. Feistel's group
eventually came up with two different schemes for creating a secure
computer cipher, both variations on Shannon's scheme and both still
used today.

The one that resembles Shannon's ideas most closely is now called
a **substitution-permutation network**, or **SP-network**. Like Shannon's
scheme, it has a pattern of substitutions and transpositions (which are
really the same as permutations, as we mentioned in Section 3.3). Unlike
Shannon, instead of alternating simple polygraphic substitutions (Hill
ciphers) and more general monographic substitutions, with one larger
transposition, SP-networks alternate large transpositions with smaller
but still complicated polygraphic substitutions and usually throw in
something like a polyalphabetic substitution besides. Also, since these

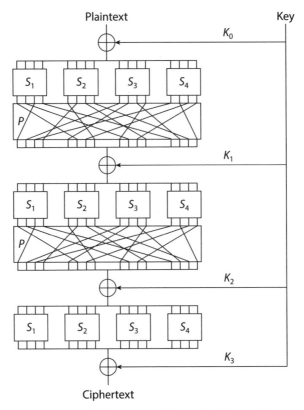

Figure 4.5. The structure of a sample SP-network.

ciphers are designed for computers, they all act on patterns of binary digits, or **bits**, rather than letters.

The easiest way to see what's going on in a modern computer cipher is usually with a diagram. A "standard" sort of modern SP-network might look something like Figure 4.5. A fairly typical block size for a modern cipher is 128 bits, so we'll assume that there are 128 bits of plaintext going in. The key might also be 128 bits or it might be larger. This key is split up into parts according to some **key schedule**, which could be as simple as just taking the first 128 bits, and the then the next 128, and so on, or it could be much more complicated. Some bits may be used more than once and some might be added together or otherwise transformed before use. At any rate, a series of 128-bit **round keys** K_0, K_1, K_2, ... is eventually arrived at.

■ ■ ■ SIDEBAR 4.1. DIGITIZING PLAINTEXT ■ ■ ■

You might be wondering just how the bits going into a digital computer cipher represent plaintext. One could do this with the 5-bit Baudot code we mentioned in Section 4.1. A more modern way is to use the **American Standard Code for Information Interchange**, or **ASCII**, a 7-bit nonsecret encoding developed in the 1960s. Seven bits instead of 5 gives $2^7 = 128$ possibilities instead of $2^5 = 32$, which allows ASCII code to represent both upper- and lowercase letters as well as digits, punctuation marks, and other symbols. In addition, there are some combinations of bits designated as "control characters"; they were originally designed to make a computer do something, such as force a line break or ring a bell, rather than to print on the screen. The printable characters in ASCII are shown in Table 4.1.

Seven bits is considered unwieldy by modern designers of computer hardware and software, who prefer powers of 2, like 2, 4, 8, 16, 32, and so on. Therefore, an extra bit is usually tacked onto the beginning of an ASCII representation to make an even 8 bits. Sometimes this bit is used for error checking, sometimes it indicates that the characters should be displayed a special way, and sometimes it is just set to zero. Thus, if a cipher with a block size of 128 bits is operating on plaintext represented with ASCII, each block will ordinarily hold 16 characters.

As I write this, ASCII is in the process of being superseded by encodings using 16 or even 32 bits, with the goal of eventually encoding all the letters and symbols used by all the living and dead languages of the world. Ciphers are also likely to move to larger block sizes as computers get more powerful. More importantly, modern ciphers are used in ways that make the number of characters per block less important, as we will see in Section 5.3.

Table 4.1.
The printable ASCII characters

Decimal Numeral	Binary Encoding	Character	Decimal Numeral	Binary Encoding	Character	Decimal Numeral	Binary Encoding	Character
32	0100000	[space]	64	1000000	@	96	1100000	`
33	0100001	!	65	1000001	A	97	1100001	a
34	0100010	"	66	1000010	B	98	1100010	b
35	0100011	#	67	1000011	C	99	1100011	c
36	0100100	$	68	1000100	D	100	1100100	d
37	0100101	%	69	1000101	E	101	1100101	e
38	0100110	&	70	1000110	F	102	1100110	f
39	0100111	'	71	1000111	G	103	1100111	g
40	0101000	(	72	1001000	H	104	1101000	h
41	0101001	)	73	1001001	I	105	1101001	i
42	0101010	*	74	1001010	J	106	1101010	j
43	0101011	+	75	1001011	K	107	1101011	k
44	0101100	,	76	1001100	L	108	1101100	l
45	0101101	-	77	1001101	M	109	1101101	m
46	0101110	.	78	1001110	N	110	1101110	n
47	0101111	/	79	1001111	O	111	1101111	o
48	0110000	0	80	1010000	P	112	1110000	p
49	0110001	1	81	1010001	Q	113	1110001	q
50	0110010	2	82	1010010	R	114	1110010	r
51	0110011	3	83	1010011	S	115	1110011	s
52	0110100	4	84	1010100	T	116	1110100	t
53	0110101	5	85	1010101	U	117	1110101	u
54	0110110	6	86	1010110	V	118	1110110	v
55	0110111	7	87	1010111	W	119	1110111	w
56	0111000	8	88	1011000	X	120	1111000	x
57	0111001	9	89	1011001	Y	121	1111001	y
58	0111010	:	90	1011010	Z	122	1111010	z
59	0111011	;	91	1011011	[	123	1111011	{
60	0111100	<	92	1011100	\	124	1111100		
61	0111101	=	93	1011101	]	125	1111101	}
62	0111110	>	94	1011110	^	126	1111110	~
63	0111111	?	95	1011111	_			

The first step in the actual encryption is the polyalphabetic substitution that I mentioned earlier. It adds the bits of the plaintext modulo 2 with the bits of the first round key, using noncarrying addition, just like the teletypewriter system in Section 4.1. Then the bits are divided up into a large number of small groups—Feistel suggested 32 groups of 4. Each group of 4 bits is passed through a (nonsecret) **substitution box**, or **S-box**, which carries out a polygraphic substitution on the 4 bits that is as complicated to describe mathematically as possible—often the designers just give a table and leave it at that. The S-boxes may all be the same or they may be different, and sometimes the S-boxes used depend on the key in some way, but not necessarily. After the S-box, the bits are rejoined and pass through a (nonsecret) **permutation box**, or **P-box**, which performs some complicated permutation on the whole block. Note the similarities to the ADFGVX cipher, which also combines a smaller substitution with a large transposition.

Finally, the bits are added modulo 2 to the bits of the next part of the round key and the cycle repeats for many, many rounds—perhaps 10 or 20, depending on how secure and how fast the designer wants the cipher to be. As Shannon suggested, both the first and last operations should involve adding the key; otherwise Eve can just undo them, since everything else is essentially not secret. In order to decrypt, Bob merely runs each stage of the cipher backward. As we saw in Section 4.1, subtracting the round key modulo 2 is the same as adding it, and thus this step works the same way backward as it does forward. We will see an example of a substitution-permutation network when we look at the Advanced Encryption Standard in Section 4.5.

The idea of a substitution-permutation network is to use the S-boxes to provide confusion and diffusion 4 bits at a time and the P-boxes to "spread it around" to the whole 128 bits. The complicated mathematical relation expressed in the S-boxes provides the confusion. To ensure diffusion, the S-boxes are designed to produce what Feistel called an "avalanche effect"; if any one bit of the input is changed from 0 to 1, or vice versa, a large percentage of the output bits should be changed. Modern cryptologists have quantified this as the **strict avalanche criterion**: if any one input bit is changed and the others are held constant, each output bit will change for half the values of the other input bits and remain the same for the other half. Let's take a small, 3-bit example:

input	output
000	110
001	100
010	010
011	111
100	011
101	101
110	000
111	001

For example, consider the case where the middle bit of the input is 0 and we want to know what happens to the last bit of the output. There are 4 input cases: 000, 001, 100, and 101. The next table indicates what happens in each case, with the bits we are focusing on in bold:

input	changes to	output	changes to
0**0**0	0**1**0	11**0**	01**0**
0**0**1	0**1**1	10**0**	11**1**
1**0**0	1**1**0	01**1**	00**0**
1**0**1	1**1**1	10**1**	00**1**

As you can see, in two cases the output bit we care about changes and in the other two, it doesn't. You can check that the same thing will be true if we pick any other input bit and any other output bit.

If we could simply create a 128-bit S-box that was mathematically complicated and exhibited the strict avalanche criterion, then we would have confusion and diffusion and be all set. But that's not really practical, even with modern technology. Instead, once the S-boxes produce several changed bits out of 1, we use the P-boxes to scatter the changed bits throughout the whole block. Unlike S-boxes, large P-boxes are easy, especially on a computer chip—all we need to do is move the wires around. Then those changed bits go through more S-boxes and change more bits, and so on. If the strict avalanche criterion holds in every S-box, the P-boxes are carefully constructed, and enough rounds are used, then in the end, changing one bit of the input plaintext will have a 50% chance of changing every bit of the output ciphertext.

The other scheme which Feistel's group came up with is now called merely a **Feistel network**. Again, a diagram will probably be the most useful, as in Figure 4.6. As in an SP-network, a key schedule is used to determine a sequence of round keys. The plaintext bits are divided in half, and in each round the right (according to the diagram) half of the input bits are used to modify the left half. First, the right half of the input bits passes through a **round function** f, which typically involves one or more S-boxes and/or P-boxes and adding the bits of the round key modulo 2. The bits of the output of the round function are then added modulo 2 to the left half of the input bits. Then the 2 halves are swapped and the round is repeated a large number of times. Again, 10

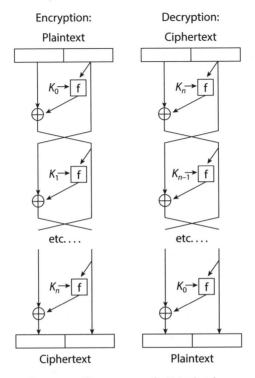

Figure 4.6. The structure of a Feistel cipher.

to 20 rounds are typical. There is no swap at the end of the last round—leaving it out doesn't affect the security and makes decryption more convenient.

One of the interesting things about Feistel networks is that decryption actually goes in the *same* direction as encryption, rather than backward (except for the key schedule). This is due to the fact that the bits of the output of the round function are added on modulo 2 rather than used directly, and addition modulo 2 is the same as subtraction modulo 2. You can see if you put the encryption diagram from Figure 4.6 on top of the decryption diagram that in the middle there will be two identical additions modulo 2 in a row, which thus cancel out. Once those are canceled out, the additions above and below them are effectively in a row, and those are also identical, so they cancel out, and so on. Because the S-boxes in the round function don't have to go backward, there is more freedom to choose them to have good properties, like avalanche.

In fact, it is possible for the S-boxes to have a different number of inputs and outputs. Likewise, the P-boxes don't have to be proper permutations; they can be expansion functions or compression functions like we saw in Section 3.3.

Just like for SP-networks, the confusion in a Feistel network is provided by the S-boxes in the round function. The P-boxes then provide diffusion through each half of the bits, and the addition modulo 2, swapping, and repeating spread the diffusion and confusion to the other half. The most famous example of a Feistel network cipher is the Data Encryption Standard, which we will meet in the next section.

4.4 THE DATA ENCRYPTION STANDARD

Feistel's research group and its successors at IBM produced at least 5 ciphers from 1971 to 1974, all of which very confusingly seem to have been called Lucifer. One of the early versions was used in the IBM 2984 automated teller machine. A later version became known internally as the DSD-1, and this was the one that the group was working on when the National Bureau of Standards (NBS, now the National Institute of Standards and Technology, or NIST) starting soliciting proposals for a new national standard encryption algorithm.

The DSD-1 was the only serious competitor for the new standard, which became known as the **Data Encryption Standard**, or **DES**, but some controversial changes were made along the way. The NSA had not wanted to design the new standard cipher, fearing that releasing any of its design work to the public would already give away too much information about what the agency knew and how it worked. However, when the NBS requested the NSA's help to evaluate the algorithm's security, the NSA was presumably happy to agree. Exactly what happened next is not clear, because the NSA insisted that everyone involved be sworn to secrecy. What is known is that the length of the key, which Feistel had intended to be 128 bits, was reduced to 64 and then to 56. In addition, the tables used in the S-boxes were changed from the original design. According to people in the IBM group, the reduction from 128 bits to 64 bits was done for purely practical reasons—the circuitry that implemented the DES algorithm was supposed to fit on a single chip and at the time, working with 128 bits on a chip would certainly have been difficult. Furthermore, even 64 bits meant there would be 2^{64} different keys. A single computer trying a million keys per second (which would have been quite fast in the 1970s) would still take around 300,000 years to complete a brute force attack on a 64-bit key.

The reduction from 64 bits to 56 bits was much more controversial. That cut the time to complete a brute-force attack down to around 1000 years on a single computer, or 1 year on a thousand computers, and that seemed within the range of possibility for an organization like the NSA. Even some people at IBM suspected that the NSA had insisted that the key size be reduced so that they might have a chance at cracking the cipher. The head of the product-development group insisted that the reason was, rather, an IBM internal specification that required that the 8 bits taken away from the key be used for an error-checking mechanism. According to a book published by the NSA's Center for Cryptologic History in 1995, the NSA had actually pushed for a 48-bit key and the 56-bit key size was a compromise. Whether this means that the NSA could have broken DES at the time may never be known.

As far as the S-boxes were concerned, the NSA seems to have made them more secure, rather than less. In 1990, two academic researchers announced that they had discovered a **differential attack** on DES, which compares ciphertexts that come from two or more closely related plaintexts—actually, this particular attack uses about 2^{47} plaintexts, which is an awful lot but still is potentially quicker than a brute-force search. They also discovered that DES seemed particularly resistant to differential cryptanalysis. After hearing about this announcement, one of the IBM researchers revealed that in 1974 the S-boxes in fact had been redesigned, with or without NSA help, to withstand exactly this attack. It is still not known for certain whether the NSA knew about the technique used in the differential attack before 1974 and, if so, whether they helped the IBM researchers discover it. At any rate, the NSA decided that the technique was too powerful to reveal to the world at large and made sure that both the technique and the design considerations used to make it difficult were kept secret until they were rediscovered almost 20 years later.

So what is the DES algorithm? It's a Feistel network, like in Section 4.3, with the small addition of a P-box at the very beginning of the cipher and its inverse at the very end. As we remarked earlier,

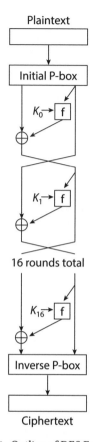

Figure 4.7. Outline of DES Encryption.

nonsecret transformations at the very beginning and end of a cipher don't affect the security, since Eve can just undo them. Apparently the P-boxes are merely there to make the data easier to handle on the original chip. The blocks are 64 bits long and there are 16 rounds. Figure 4.7 shows the cipher in broad outline.

The key schedule involves P-boxes that do rotations, which are a simple type of permutation that we mentioned in Section 3.8. It also uses a compression function, like we discussed in Section 3.3, which picks out some of the 56 bits and rearranges them to get one of the 16 round keys of 48 bits each, as shown in Figure 4.8. Each round key uses a different amount of rotation before the compression function, which makes each round key different.

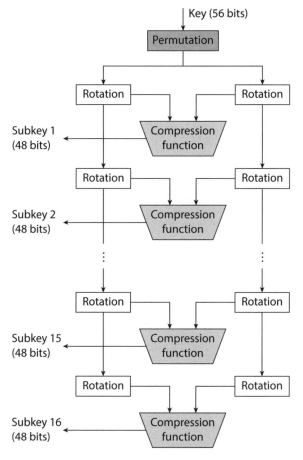

Figure 4.8. DES Key Schedule.

Finally, we need to look at the DES round function, shown in Figure 4.9. This involves an expansion function, also discussed in Section 3.3, which rearranges the 32 bits in the right half of the block and also repeats some of them to get 48 bits, which can be added to the bits of the round key. Then the bits are divided into 8 groups of 6 bits each, and each group is passed through one of the famous DES S-boxes— each of the 8 is different. This, as we have said, provides the confusion. We also mentioned in Section 4.3 that in a Feistel network the S-boxes can have a different number of output than inputs, and this is the case in DES—each S-boxes takes in 6 bits and puts out 4. This gives us 32 bits

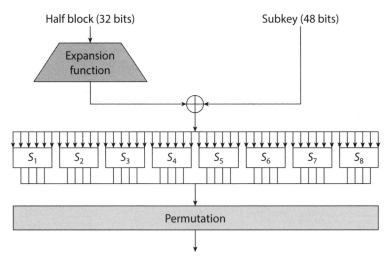

Figure 4.9. DES Round Function.

again, which then go through an ordinary P-box to provide diffusion and finish the round.

Despite the concerns about the key size, DES held up remarkably well as a cryptographic standard—it lasted more than 20 years before being definitively shown to be insecure. As we mentioned, the differential attack on DES was rediscovered in 1990, but the number of carefully arranged plaintext-ciphertext pairs it requires is not really considered practical. In 1993 another attack, called linear cryptanalysis, was discovered. This is also a known-plaintext attack, but the plaintext-ciphertext pairs don't have to be carefully chosen. However, it still requires an average of 2^{43} pairs and considerable computation. This attack does not seem to have been known to the DES designers. In 1997–98 the Electronic Frontier Foundation decided to see whether a concerted group with a reasonable budget could break DES by brute force. They designed and built a custom computer using 1728 custom chips. The whole process, including the design and manufacturing time, took 18 months and cost less than $250,000, plus the volunteer time of fewer than 10 part-time people on the core staff and a separate short volunteer project for the software. This machine took about 56 hours to crack a 56-bit DES key, although they were somewhat lucky—an average search would have taken about twice as long. Furthermore, the system was **scalable**, in that two machines of the same size and cost could be used to break DES in half the time. At this point it was generally acknowledged that DES was breakable.

4.5 THE ADVANCED ENCRYPTION STANDARD

On September 12, 1997, the National Institute of Standards and Technology announced a "Request for Candidate Algorithm Nominations for the Advanced Encryption Standard." The **Advanced Encryption Standard**, or **AES**, was intended to replace DES as the new government cipher standard. Almost everything about the process of choosing the AES cipher was different from the DES process. The key size and block size were specified up front. The algorithm was required to work with multiple key sizes (128, 192, and 256) in order to allow for future as well as present security needs. The criteria on which the submissions would be judged were specified: security, cost, flexibility, suitability for both hardware and software, and simplicity. Foreign nationals were allowed and even invited to participate, both as submitters and reviewers. And most importantly, the entire evaluation would be considered in full public view. A series of three public conferences would be held, at which the designers of the candidate ciphers would be invited to make presentations, NIST scientists and outside experts would present their analyses of the candidates, and the public at large was invited to watch, ask questions, and make comments.

By the due date of June 15, 1998, 21 ciphers had been submitted, of which 15 were judged to have met the minimum specified requirements. Ten of the 15 had been primarily developed outside the United States, and all but one had at least one non-US national on the design team. In August 1999, NIST narrowed the field to 5 finalists, and on October 2, 2000, NIST announced that the winner was a submission known as "Rijndael" by two Belgian cryptographers, **Joan Daemen** and **Vincent Rijmen**. The standard took effect on May 26, 2002.

As I mentioned in Section 4.3, AES is basically an SP-network. The blocks are 128 bits long and are generally thought of as being broken up into a 4×4 square of 8-bit groups, as in Figure 4.10. The key can be 128, 192, or 256 bits long, and the number of rounds depends on the key size—10 rounds for 128 bits, 12 for 192, and 14 for 256.

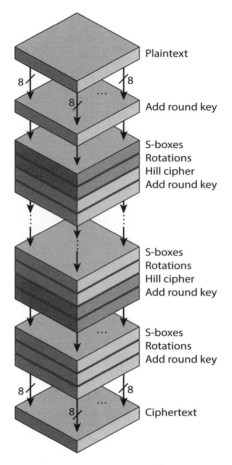

Figure 4.10. Overview of AES.

For simplicity, I'm going to describe only the key schedule for the version of AES with a 128-bit key, which is not only simplest, but also the most common as of this writing. The first round key is just the same as the original cipher key. From then on, to get the next round key, the previous round key is passed through a function that involves a rotation P-box, a set of identical S-boxes (about which more later), and a lot of addition modulo 2, including using a "round constant," which is just like it sounds—it depends on the round but nothing else. See Figure 4.11.

AES starts with adding the round key as usual for SP-networks. Then each 8-bit group passes through an identical S-box, for confusion.

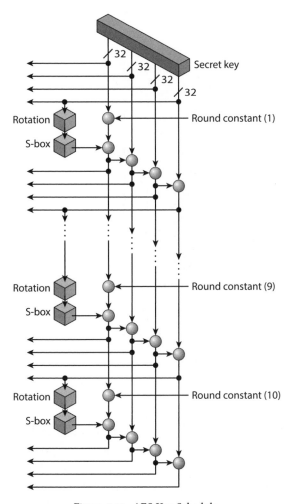

Figure 4.11. AES Key Schedule.

After the S-boxes comes the diffusion. The designers of AES, unlike Feistel, thought that one gigantic 128-bit P-box was too much. There-fore, they made the diffusion work in 2 steps. Remember that the 8-bit groups are thought of as arranged in a 4 × 4 square. The first diffusion step is a set of P-boxes that rotate each row by a different number of 8-bit groups. This takes care of what the designers called **dispersion**, which is a process of moving bits that start out near each other to posi-tions that are widely separated. The second diffusion step is not quite a

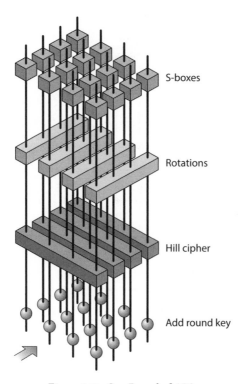

S-boxes

Rotations

Hill cipher

Add round key

Figure 4.12. One Round of AES.

permutation. It performs a Hill cipher encryption on each column using a fixed key and a special multiplication which I am going to talk about shortly. One advantage of using a Hill cipher, in which every bit has a chance to influence every other bit, is that it is possible to prove that every bit has to be influenced by a large number of S-boxes. This makes it difficult to carry out differential attacks and linear cryptanalysis. At the end of each round, the new round key is added.

Now we're ready to talk about the design of the AES S-box—there's only one, so it better be a good one! Remember that an S-box is usually given as a table. There are basically three ways of choosing the entries in the table: they can be randomly selected, "human made," or "math made." The DES S-boxes were human made: the designers thought hard about what criteria they wanted the tables to fit and then searched until they found entries that succeeded. The AES S-boxes, on the other hand, are basically math made. Remember that the S-box function is supposed to be complicated mathematically in order to provide confusion. The AES designers deliberately decided to choose a function that is very complicated when looked at from the level of bits but not so complicated when looked at in a different mathematical way.

The math that the AES S-boxes use is modular arithmetic, but it's modular arithmetic using polynomials. Remember that pretty much everything in AES works with 8-bit groups. We start by turning the 8-bit groups into polynomials:

$$
\begin{aligned}
01010111 \quad \rightarrow \quad & 0x^7 + 1x^6 + 0x^5 + 1x^4 + 0x^3 + 1x^2 + 1x + 1 \\
= \quad & x^6 + x^4 + x^2 + x + 1,
\end{aligned}
$$

$$
\begin{aligned}
10000011 \quad \rightarrow \quad & 1x^7 + 0x^6 + 0x^5 + 0x^4 + 0x^3 + 0x^2 + 1x + 1 \\
= \quad & x^7 + x + 1.
\end{aligned}
$$

To add the groups we just add the polynomials modulo 2, which turns out to be just what we would have done to add two groups of bits anyway:

$$
\begin{array}{rll}
01010111 & \leftrightarrow & x^6 + x^4 + x^2 + x + 1 \\
+ \quad 10000011 & \leftrightarrow & + \quad x^7 + x + 1 \\
\hline
11010100 & \leftrightarrow & x^7 + x^6 + x^4 + x^2 + 2x + 2 \\
& = & x^7 + x^6 + x^4 + x^2.
\end{array}
$$

To multiply the groups we multiply the polynomials modulo 2:

$$
\begin{array}{rll}
01010111 & \leftrightarrow & x^6 + x^4 + x^2 + x + 1 \\
\times \quad 10000011 & \leftrightarrow & \times \quad x^7 + x + 1 \\
\hline
???????? & \leftrightarrow & x^{13} + x^{11} + x^9 + x^8 + 2x^7 + x^6 + x^5 + x^4 \\
& & + x^3 + 2x^2 + 2x + 1 \\
& = & x^{13} + x^{11} + x^9 + x^8 + x^6 + x^5 + x^4 + x^3 + 1.
\end{array}
$$

But now we have a problem, because we have too many polynomial coefficients to turn back into an 8-bit group. This should remind you of a problem we've dealt with before, when we needed to turn a ciphertext number larger than 26 into a ciphertext letter. The solution there was to wrap around using the number 26. The solution here is the same, only we need to wrap around using a polynomial of degree 8—in other words, the term in the polynomial with the largest exponent will be x^8. If we divide our result by a polynomial of degree 8 and take the remainder, it will have degree 7 or less. Then we can turn it back into an 8-bit group; remember we need to make room for the constant coefficient also, so 8 bits means degree 7 or less.

Which polynomial should we choose? We could just use x^8, but we will see in a bit that we want to use a **prime polynomial**, also known as an **irreducible polynomial**. The polynomial x^8 isn't prime, since it's $x \times x \times x \times x \times x \times x \times x \times x$. The designers of AES consulted a published list of polynomials of degree 8 that are prime modulo 2 and took the first one, which turned out to be $x^8 + x^4 + x^3 + x + 1$.

The simplest way to reduce modulo a polynomial is still to divide by it and take the remainder, working modulo 2 the entire time, and so we can finish our example:

$$
\begin{array}{r}
x^5 \quad\;\; +x^3 \qquad\qquad\qquad\qquad \\
\hline
x^8 + x^4 + x^3 + x + 1 \,\big)\, x^{13} +x^{11} +x^9 +x^8 \quad\; +x^6 +x^5 +x^4 +x^3 +1 \\
-x^{13} \qquad\quad -x^9 -x^8 \quad\;\; -x^6 -x^5 \qquad\qquad\quad \\
\hline
x^{11} \qquad\qquad\qquad\qquad +x^4 +x^3 \qquad\; \\
-x^{11} \qquad\qquad -x^7 -x^6 \quad\; -x^4 -x^3 \qquad\; \\
\hline
-x^7 -x^6 \qquad\qquad\qquad\; +1
\end{array}
$$

The remainder is

$$-x^7 - x^6 + 1 \equiv x^7 + x^6 + 1 \quad \text{modulo 2,}$$

so

$$
\begin{array}{rcl}
01010111 & \leftrightarrow & x^6 + x^4 + x^2 + x + 1 \\
\times \quad 10000011 & \leftrightarrow & \times \quad x^7 + x + 1 \\
\hline
& & x^{13} + x^{11} + x^9 + x^8 + 2x^7 + x^6 + x^5 \\
& & + x^4 + x^3 + 2x^2 + 2x + 1 \\
& = & x^{13} + x^{11} + x^9 + x^8 + x^6 + x^5 + x^4 + x^3 + 1, \\
11000001 & \leftarrow \;\equiv & x^7 + x^6 + 1 \quad \text{modulo } x^8 + x^4 + x^3 + x + 1 \quad \text{modulo 2.}
\end{array}
$$

Now we know how to add and multiply modulo a polynomial modulo 2. Since we are working modulo 2, subtraction is the same as addition. That just leaves division. The reason we picked a prime polynomial for the modulus is the same as when we worked modulo a prime number: every nonzero polynomial has a multiplicative inverse, and we can compute it using the Euclidean algorithm just like for numbers.

For instance, to find the multiplicative inverse of $x^5 + x^4 + x^2 + x$, we compute the greatest common divisor of $x^5 + x^4 + x^2 + x$ and $x^8 + x^4 + x^3 + x + 1$, reducing modulo 2 whenever convenient. We confirm that the GCD is 1, and rewrite each line to eventually discover that

$$1 \equiv (x^3 + x + 1) \times (x^8 + x^4 + x^3 + x + 1)$$
$$+ (x^6 + x^5 + x^2 + x) \times (x^5 + x^4 + x^2 + x) \quad \text{modulo 2,}$$

so

$$1 \equiv (x^6 + x^5 + x^2 + x) \times (x^5 + x^4 + x^2 + x)$$
$$\text{modulo } x^8 + x^4 + x^3 + x + 1 \quad \text{modulo 2,}$$

giving

$$\overline{x^5 + x^4 + x^2 + x} = x^6 + x^5 + x^2 + x$$
$$\text{modulo } x^8 + x^4 + x^3 + x + 1 \quad \text{modulo 2,}$$

or in terms of bits,

$$\overline{00110110} = 01100110.$$

AES uses this polynomial arithmetic in two places. The first is to do the multiplications in the Hill cipher step. The second is in the design of the S-box. The S-box function basically has just two steps. The first is to take the multiplicative inverse of the 8-bit group using the procedure we just went through. (If all 8 bits are 0, then we get the zero polynomial, which doesn't have an inverse, so we just leave it alone.) The second step is to separate the 8 bits and perform a specified 8-by-8 affine Hill encryption on them. And that's it.

As I've mentioned before, in practice the S-box function is usually specified by a table that can be built into the cryptographic hardware and software so that the computations don't have to be done over and over—especially the multiplicative inverse, which, as you can see, is somewhat time consuming. Nevertheless, the fact that the mathematical structure is there makes it easier to analyze the resistance of AES to certain types of attacks. The polynomial structure makes it difficult to use differential and linear attacks against AES, as I mentioned earlier, because these attacks only "know about" bits and not polynomials. On the other hand, the affine Hill step of the S-box is intended to make it hard to attack AES using polynomial techniques, since this step acts on the individual bits and doesn't respect the polynomial structure. Ever since the original AES competition, there has been some concern that this step is not enough to protect against attacks using the "high-level" structure of AES. In 2002 an attack on AES called XSL (eXtended Sparse Linearization) was announced that uses, among other things, this structure. Although the consensus now seems to be that the XSL attack is not better than brute force, it is still possible that attacks based on the polynomial structure of AES may become important in the future.

In 2009 and 2010 several papers appeared describing possible known-key and related-key attacks on AES. Since these require knowing some or all of the key before the attack starts, they are not directly applicable against a well-designed implementation of AES as it was designed to be used. However, history shows that ciphers do not always get used in the way they were intended. In addition, these attacks might (or might not) be signs of weaknesses that could be exploited in more conventional attacks.

In 2011, a more standard attack on AES was announced, which is agreed to be better than brute force, although not by very much. It is the equivalent of reducing a 128-bit key to a 126-bit key, which would still take an unreasonably long time to break using any computers we know of. Also, the attack requires knowing the plaintexts corresponding to 2^{88} carefully chosen ciphertexts, which seems difficult to arrange in practice. Nevertheless, this seems to be the most serious attack on AES known as of this writing.

4.6 LOOKING FORWARD

This is actually about as far forward as we are going to pursue ciphers based on SP-networks, Feistel networks, and similar ideas in this book. There is certainly still active research going on in these areas, though. Presumably, AES will eventually need to be replaced, and cryptographers are already thinking about ciphers that might succeed it. The original expectations were that the new standard should last at least 30 years, but NIST is supposed to reevaluate AES every 5 years to see if it is still acceptable. So far, no serious problems with AES have surfaced, but everyone wants to be ready just in case.

Cryptographers are also interested in finding encryption methods that have special properties, such as making some aspects of the data secret but not others. In **format-preserving encryption**, the goal is for data of a certain type to still look like the same type after encryption. For example, an encrypted sound file should still be playable as a sound file on a computer. It will probably sound like noise, but it won't cause an error message. Similarly, if a database was originally set up to hold names in some fields and credit-card numbers in other fields, then after encryption the encrypted names should still fit in the name fields and the encrypted numbers should still fit in the number fields. This idea goes back at least as far as a National Bureau of Standards document from 1981, but early methods were very inefficient. In 2013, after several proposals from researchers, NIST published a draft proposal for a national standard incorporating three of the most efficient suggested methods. Unfortunately, a report from April 2015 indicates that one of the three methods was not as secure as previously thought.

An even more exciting idea is **homomorphic encryption**. The goal here is for Alice to be able to encrypt her data and store it on Bob's computer. She can ask Bob to manipulate the encrypted data and return an encrypted answer without having to give Bob the ability to decrypt the data. For example, Alice could ask Bob to add up a column of numbers in an encrypted spreadsheet without him ever finding out what either the numbers or the total were. Or, she could ask Bob to find all the names in a list that begin with A without having to worry about the fact that diffusion will cause A to be encrypted differently, depending on the other letters in the name. This would have great implications for storing financial and other sensitive data in the cloud, as well as for other applications like electronic vote counting.

Like format-preserving encryption, the idea of homomorphic encryption goes back to the early days of computer ciphers, at least as far back as 1978. The first actual fully homomorphic system, one that could perform arbitrary operations on data, was not invented until 2009. Early systems were much too slow to be practical, but there have been major advances in both speed and technique since then. Two government research agencies have committed more than $20 million to researching practical solutions to the problem. And as of 2013, at least one company was hoping to have a solution ready to market by 2015, although it has not yet appeared as of this writing.

Efforts to find new attacks on AES are continuing, both in public and in secret government projects. In 2013, the NSA contractor Edward Snowden released to journalists a large number of classified documents that he had taken from the NSA computer system. Some of these dealt with NSA attempts to read encrypted communications, although most of their techniques had to do with finding ways around the encryption rather than breaking it outright. One excerpt published in *Der Spiegel* caused some fuss by noting that

> Electronic codebooks, such as the Advanced Encryption Standard, are both widely used and difficult to attack cryptanalytically. The NSA has only a handful of in-house techniques. The TUNDRA project investigated a potentially new technique—the Tau statistic—to determine its usefulness in codebook analysis.

A closer examination of the full document reveals that this was a summer project for undergraduate students interested in future work at the NSA, so it is not clear how much of a threat this project really is. Journalists are still sorting through the Snowden documents, so it is possible that more information about the "handful of in-house techniques" may come to light in the future.

.......5.......

Stream Ciphers

5.1 RUNNING-KEY CIPHERS

The ciphers we have discussed so far, classical and modern, are all basically **block ciphers**. They break the plaintext up into relatively large blocks, which might be one or more letters or one or more bits. Then what happens to each block is independent of what happens to the blocks before or after. The alternative is a **stream cipher**, where letters, bits, or small blocks are encrypted one at a time, and the result of each encryption might depend on what happened during previous encryptions. This has several possible advantages. For one thing, it is very convenient in situations where you do not have any idea ahead of time whether the plaintext is going to be long or short and you don't want to worry about transmitting padding. Digital wireless communication is a good example of this. For another, diffusion is almost automatic, and because operations that contribute to confusion can pile up over the course of the encryption, it is possible to achieve good confusion using simple, fast operations.

The first steps on the road to stream ciphers came out of the keyed polyalphabetic ciphers that we looked at in Section 2.4. It became clear early on that the shorter the keyword or keyphrase was, the easier it was to break the cipher. Most early cryptographers seem to have been under the opinion that there was a "sweet spot" of no more than a sentence or so, beyond which making the key longer caused more trouble than it was worth. As cryptanalytic techniques improved, it became clear that any repetition of the key could be exploited to break the cipher, as we saw in Chapter 2. By the end of the nineteenth century, it had become common to suggest the use of a **keytext** that could be made as long as the plaintext—for example, the text starting on an agreed-on page of a common book. Using the tabula recta cipher from Section 2.4, we could have

keytext:	D	O	R	O	T	H	Y	L	I	V	E	D	I	N	T	H	E
plaintext:	a	s	l	o	w	s	o	r	t	o	f	c	o	u	n	t	r
ciphertext:	E	H	D	D	Q	A	N	D	C	K	K	G	X	I	H	B	W

keytext:	M	I	D	S	T	O	F	T	H	E	G	R	E	A	T	K	A
plaintext:	y	s	a	i	d	t	h	e	q	u	e	e	n	n	o	w	h
ciphertext:	L	B	E	B	X	I	N	Y	Y	Z	L	W	S	O	I	H	I

keytext:	N	S	A	S	P	R	A	I	R	I	E	S	W	I	T	H	U
plaintext:	e	r	e	y	o	u	s	e	e	i	t	t	a	k	e	s	a
ciphertext:	S	K	F	R	E	M	T	N	W	R	Y	M	X	T	Y	A	V

keytext:	N	C	L	E	H	E	N	R	Y	W	H	O	W	A	S	A	F
plaintext:	l	l	t	h	e	r	u	n	n	i	n	g	y	o	u	c	a
ciphertext:	Z	O	F	M	M	W	I	F	M	F	V	V	V	P	N	D	G

keytext:	A	R	M	E	R	A	N	D	A	U	N	T	E	M	W	H	O
plaintext:	n	d	o	t	o	k	e	e	p	i	n	t	h	e	s	a	m
ciphertext:	O	V	B	Y	G	L	S	I	Q	D	B	N	M	R	P	I	B

keytext:	W	A	S	T	H	E	F	A	R	M	E	R	S	W	I	F	E
plaintext:	e	p	l	a	c	e											
ciphertext:	B	Q	E	U	K	J											

Even polyalphabetic ciphers with repeating keys have caused many a cryptanalyst to throw up their hands and look for something easier. These **running-key ciphers**, where the key does not repeat, are even more difficult—but not impossible—to break.

There are two basic situations. One is if Eve has several messages encrypted with the same running key. The other, harder, situation is if she has only a single message. If Eve has reason to suspect that several messages were encrypted with the same key, she can test this using the kappa test from Section 2.5. If it comes up negative, Eve should consider the possibility that the texts were encrypted using the same key text but starting in different places:

keytext 1:	D	O	R	O	T	H	Y	L	I	V	E	D	I	N	T	H	E	M	I
plaintext 1:	a	s	l	o	w	s	o	r	t	o	f	c	o	u	n	t	r	y	s
ciphertext 1:	E	H	D	D	Q	A	N	D	C	K	K	G	X	I	H	B	W	L	B
keytext 1:								L	I	V	E	D	I	N	T	H	E	M	I
plaintext 2:								m	o	w	g	l	i	w	a	s	f	a	r
ciphertext 2:								Y	X	S	L	P	R	K	U	A	K	N	A

keytext 1:	D	S	T	O	F	T	H	E	G	R	E	A	T	K	A	N	S	A	S
plaintext 1:	a	i	d	t	h	e	q	u	e	e	n	n	o	w	h	e	r	e	y
ciphertext 1:	E	B	X	I	N	Y	Y	Z	L	W	S	O	I	H	I	S	K	F	R
keytext 1:	D	S	T	O	F	T	H	E	G	R	E	A	T	K	A	N	S	A	S
plaintext 2:	a	n	d	f	a	r	t	h	r	o	u	g	h	t	h	e	f	o	r
ciphertext 2:	E	G	X	U	G	L	B	M	Y	G	Z	H	B	E	I	S	Y	P	K

keytext 1:	P	R	A	I	R	I	E	S	W	I	T	H	U	N	C	L	E	H	E
plaintext 1:	o	u	s	e	e	i	t	t	a	k	e	s	a	l	l	t	h	e	r
ciphertext 1:	E	M	T	N	W	R	Y	M	X	T	Y	A	V	Z	O	F	M	M	W
keytext 1:	P	R	A	I	R	I	E	S	W	I	T	H	U	N	C	L	E	H	E
plaintext 2:	e	s	t	r	u	n	n	i	n	g	h	a	r	d	a	n	d	h	i
ciphertext 2:	U	K	U	A	M	W	S	B	K	P	B	I	M	R	D	Z	I	P	N

keytext 1:	N	R	Y	W	H	O	W	A	S	A	F	A	R	M	E	R	A	N	D
plaintext 1:	u	n	n	i	n	g	y	o	u	c	a	n	d	o	t	o	k	e	e
ciphertext 1:	I	F	M	F	V	V	V	P	N	D	G	O	V	B	Y	G	L	S	I
keytext 1:	N	R	Y	W	H	O	W	A	S	A	F	A	R	M	E	R	A	N	D
plaintext 2:	s	h	e	a	r	t	w	a	s	h	o	t	i	n	h	i	m	h	e
ciphertext 2:	G	Z	D	X	Z	I	T	B	L	I	U	U	A	A	M	A	N	V	I

keytext 1:	A	U	N	T	E	M	W	H	O	W	A	S	T	H	E	F	A	R	M
plaintext 1:	p	i	n	t	h	e	s	a	m	e	p	l	a	c	e				
ciphertext 1:	Q	D	B	N	M	R	P	I	B	B	Q	E	U	K	J				
keytext 1:	A	U	N	T	E	M	W	H	O	W	A	S	T	H	E	F	A	R	M
plaintext 2:	c	a	m	e	t	o	t	h	e	c	a	v	e	a	s	t	h	e	e
ciphertext 2:	D	V	A	Y	Y	B	Q	P	T	Z	B	O	Y	I	X	Z	I	W	R

keytext 1:	E	R	S	W	I	F	E
plaintext 1:							
ciphertext 1:							
keytext 1:	E	R	S	W	I	F	E
plaintext 2:	v	e	n	i	n	g	m
ciphertext 2:	A	W	G	F	W	M	R

As in Section 2.5, when the ciphertexts are correctly lined up then the kappa test index of coincidence should be closest to 6.6%.

Once Eve has identified a series of ciphertexts that were enciphered by the same key, she can superimpose them (correctly lined up, of course), as in Section 2.6. Unfortunately for her, she cannot take advantage of the repetition of the key to make the columns longer. If she has many messages enciphered with the same key, then this is not necessarily a problem, and a letter frequency analysis can often be made to work. Alternatively, if there are fewer messages but Eve knows which types of ciphers the columns are encrypted with, she could just do a brute-force attack as in Section 2.6. If there are too few messages, however, there may not be enough data in each column. Now in a long message, there should be columns that can be combined, and in fact lots of them—in our example, there were 11 columns that were encrypted with the key letter A, and all of those columns could be combined, if Eve knew where they were. Luckily for her, there is one more of Friedman and Kullback's index of coincidence tests that will help, namely, the **chi test**, or cross-product sum test.

You might recall that the phi test checks to see if one ciphertext was encrypted monoalphabetically, and the kappa test checks to see if two ciphertexts were encrypted with the same polyalphabetic key. The chi test checks to see if two ciphertexts were encrypted with the same *monoalphabetic* key. We first use the phi test to make sure that each ciphertext was encrypted monoalphabetically. Then the basic idea is that if we lump the two ciphertexts together, the result should still look like a monoalphabetic encryption, and we can check this using the same technique as in the phi test. Alternatively, Kullback showed algebraically that once you had done the phi test on each ciphertext individually, using the phi test on the combined ciphertexts was equivalent to calculating the **cross-product sum** of the two ciphertexts. This is computed by taking the chance that you will pick an A out of the first ciphertext times the chance that you will pick an A out of the second ciphertext, plus the same thing for B, plus the same thing for C, and so on. If this sum is approximately equal to 6.6%, then the index of coincidence of the two ciphertexts together will also be approximately 6.6%, so they were encrypted with the same key.

For example, suppose Eve has the ciphertexts in Table 5.1.

TABLE 5.1.
A series of ciphertexts encrypted with the some running key

column:	I	II	III	IV	V	VI	VII
ciphertext 1:	Z	Q	K	I	Q	I	G
ciphertext 2:	G	C	Z	B	J	F	R
ciphertext 3:	H	N	T	V	T	B	P
ciphertext 4:	J	X	M	U	U	U	S
ciphertext 5:	G	W	J	X	N	X	O
ciphertext 6:	Z	Q	K	V	Q	F	Q
ciphertext 7:	Y	Y	U	N	Y	M	S
ciphertext 8:	Y	N	G	W	M	J	G
ciphertext 9:	Z	Q	K	F	F	X	H
ciphertext 10:	U	O	Z	B	J	G	Z
ciphertext 11:	S	J	T	N	M	J	Q
ciphertext 12:	V	J	V	Y	W	X	W
ciphertext 13:	M	X	Z	I	G	W	W
ciphertext 14:	G	C	Z	B	J	X	W
ciphertext 15:	U	O	Z	B	J	X	D
ciphertext 16:	V	X	C	X	J	W	O
ciphertext 17:	G	A	S	M	Y	M	S
ciphertext 18:	B	X	E	U	L	J	K
ciphertext 19:	O	Q	K	U	W	I	W
ciphertext 20:	Z	Q	K	U	U	U	Z
ciphertext 21:	H	J	X	L	J	Q	Q
ciphertext 22:	U	O	C	U	W	M	C
ciphertext 23:	Z	Q	K	M	M	N	D
ciphertext 24:	C	J	Y	U	G	F	B
ciphertext 25:	Z	Q	K	D	T	Q	Z
ciphertext 26:	K	W	J	I	K	Y	V
ciphertext 27:	Y	R	R	P	J	W	G
ciphertext 28:	O	B	Z	L	N	P	S
ciphertext 29:	V	R	K	W	J	X	C
ciphertext 30:	G	W	J	F	F	X	H
Most common letter:	Z	Q	K	U	J	X	S
Corresponding key:	U	L	F	P	E	S	N
Chi-test value with column I:		0.018	0.072	0.037	0.038	0.023	0.058

If Eve takes the most common ciphertext letter from each column and assumes that it corresponds to plaintext e, then she gets the key letters shown. The chi-test values indicate that columns I and III are most likely encrypted with the same key. Decrypting the two columns with various keys gives the following:

column	key	result	frequency sum
I	U	elmoleddezxarlzalgtemzehepdtal	0.060
I	F	tabdatsstompgaopavitbotwtesipa	0.061
III	U	peyropzlpeyaeeehxjppchpdpowepo	0.053
III	F	etngdeoaetnptttwmyeerwesedlted	0.076

As you can see, F gives a marginally better result for column I but a considerably better result for column III. Given that the chi test suggests that they have the same key, that is strong evidence that the key for both is F. Proceeding in this way, Eve can see that columns V and VI are also encrypted with the same key, and that the final running key is "fifteen."

■ ■ ■ SIDEBAR 5.1. WE HAVE ALL BEEN HERE BEFORE ■ ■ ■

When you got to the part of this book about the contact method, in Section 3.4, you might have wondered about something. We said there that it was more accurate to add the logarithms of the digraph frequencies than the frequencies themselves, but earlier, when we reduced a polyalphabetic ciphertext to monoalphabetic terms in Section 2.6, we just went ahead and added the terms. Is there something that makes the two situations different?

It turns out that there is. In the case of reduction to monoalphabetic terms, we have some extra information. We know the phi-test index of coincidence of the set of letters we are working with, which should be about 0.066 if we are doing the technique correctly. We also know the index of coincidence that we are looking for, namely, 0.066 for English plaintext. That should remind you of the situation we have when using the chi test from this section, and in fact the frequency sum from Section 2.6 is equivalent to the chi test.

To see why, it will be easiest to set up some equations. Suppose we have a set of n suspected plaintext letters with n_a a's, n_b b's, and so on. Let f_a, f_b, and so on, be the frequencies of these letters in English text. Then, when we add up the frequencies, we will be adding f_a n_a times, f_b n_b times, and so on. So the sum of the frequencies of the set we have will be

$$n_a f_a + n_b f_b + \cdots + n_z f_z.$$

Now imagine doing a chi test on the suspected plaintext letters and a large number of real plaintext letters. The chance of picking the letter a (for instance) out of the suspected plaintext is n_a/f_a, and the chance of picking the letter a out of the real plaintext is f_a. Continuing like this, we see that the chi-test index of coincidence for these two texts will be

$$\frac{n_a}{n} f_a + \frac{n_b}{n} f_b + \cdots + \frac{n_z}{n} f_z,$$

which is the same as the sum of the frequencies divided by n. If the number of letters is the same, comparing the frequency sums gives us the same result as comparing the chi-test values.

But does this test do what we want? Remember that the chi test checks to see if two texts are encrypted with the same monoalphabetic cipher. We know what cipher the hypothetical real plaintext was encrypted with—it's the trivial cipher! So, if the chi-test value is good, that means that our suspected plaintext was also encrypted with the trivial cipher; in other words, it's also plaintext. And that's what we wanted to test.

These techniques so far work only if Eve has multiple messages encrypted with the same key. What if Eve has only a single message? This is the second basic situation I mentioned earlier. It would seem that there is not enough frequency information even to get started. However, there is one set of frequency information that we have not used, and that is the frequency of letters in the key. Since we have been picking key texts from common books, we should expect the letter frequency distribution in those texts to be more or less the same as in our plaintexts.

Just for variety, let's suppose now that Eve knows that Alice and Bob are using a tabula aversa cipher table instead of the tabula recta that we have been assuming. In this table, the ciphertext numbers are the key numbers minus the plaintext numbers modulo 26, so we have $C \equiv k - P \equiv 25P + k$ modulo 26. The table looks like this:

	a	b	c	d	e	f	g	h	$\cdots$	s	t	u	v	w	x	y	z
z	Y	X	W	V	U	T	S	R	$\cdots$	G	F	E	D	C	B	A	Z
A	Z	Y	X	W	V	U	T	S	$\cdots$	H	G	F	E	D	C	B	A
B	A	Z	Y	X	W	V	U	T	$\cdots$	I	H	G	F	E	D	C	B
C	B	A	Z	Y	X	W	V	U	$\cdots$	J	I	H	G	F	E	D	C
$\vdots$									$\vdots$								
x	W	V	U	T	S	R	Q	P	$\cdots$	E	D	C	B	A	Z	Y	X
Y	X	W	V	U	T	S	R	Q	$\cdots$	F	E	D	C	B	A	Z	Y

Suppose that Eve has an O in her ciphertext. According to the table, the plaintext could be k and the key could be z, but that's not especially likely if the key came from an average book. Or the plaintext could be l and the key could be A, which is more likely. Or, there are 24 other combinations, which are mostly in between. Assuming the keytext and the plaintext are chosen independently, which is likely, we can find the probability of each combination by multiplying the probabilities for the plaintext letter and the key letter. For example, the probability of l and A is about $.040 \times .082 \approx .0033$, while the probability of k and Z is about $.0077 \times .00074 \approx .0000057$. Note that we're going to need to use more accurate numbers for the low-frequency letters than appear in Table 2.2.

We can use this to set up a table with each ciphertext letter and the most likely plaintext letters. For instance, suppose Eve has the ciphertext

OFKOP QZHUL XSFTJ JRAHY

Then we can observe the following:

ciphertext:	O	F	K	O	P	Q	Z	H	U	L	X	S	F	T	J	J	R	A	H	Y
plaintext
Most likely:	e	n	t	e	o	n	e	a	t	s	t	a	n	t	e	e	i	s	a	t
Second:	t	i	i	t	s	a	t	s	n	o	e	l	i	o	i	i	a	d	s	e
Third:	s	h	h	s	d	r	a	l	s	h	n	h	h	n	t	t	t	n	l	o
keytext
Most likely:	T	T	E	T	E	E	E	I	O	E	R	T	T	N	O	O	A	T	I	S
Second:	I	O	T	I	I	R	T	A	I	A	C	E	O	I	S	S	S	E	A	D
Third:	H	N	S	H	T	I	A	T	N	T	L	A	N	H	D	D	L	O	T	N

Now Eve looks for high-frequency combinations of letters, or common words, keeping in mind that they could be scattered through the three lines of plaintext and the three lines of keytext, but for any given ciphertext letter the plaintext and keytext lines have to correspond. For example, the first word of the plaintext might be "this," which would correspond to "INTH," which is probably followed by "E" to make "IN THE," giving "this o." After that it gets harder, and Eve might have to do some trial and error and maybe add more lines to the table, but it's quite possible to finish given enough time and patience.

Another technique that often works well in this situation is the **probable word** method. Instead of looking for common words among the most likely plaintext and keytext possibilities, Eve could just select a very common word, such as *the*, or a word she has reason to think is in the plaintext for other reasons. Then she can try the word as plaintext in each possible place and see if she gets high-frequency letters or parts of words as keytext. Or, she can try a common word as keytext and see if she gets likely plaintext. The probable word method is also useful in several other situations we have looked at, so I thought I ought to mention it. There's not very much mathematics in it, though, so I'm going to leave it at that.

5.2 ONE-TIME PADS

If a repeating-key cipher can be broken by using the repetitions and a running-key cipher can be broken by using either multiple messages or the frequency of letters and words in the key, is there a cipher that cannot be broken either way? There is such a "perfectly secure" system, and it seems to have been independently discovered more than once in the late nineteenth and early twentieth centuries. The first we know of is from 1882, when a California banker named Frank Miller published a code and cipher combination system for use in telegraphs. Sadly, his system seems to have been ignored and forgotten. There is some disagreement about exactly who next put all the pieces necessary for a perfectly secure system together, but it seems to have been either Gilbert Vernam, from Section 4.1, or Major **Joseph O. Mauborgne**, or both, possibly with help from their colleagues. In 1928, Mauborgne was head of the Army Signal Corps' research and engineering division when AT&T reported the success of Vernam's device for encrypting teletypewriter communications. Since the Signal Corps was responsible for communications security in the Army, Mauborgne was sent to see a demonstration. He loved the system—but there was a problem with the key. The AT&T engineers had originally put the random key onto a loop of tape, which would cycle through the machine over and over again. They quickly realized that this was a form of repeating-key cipher and could be broken by the same techniques we used in Chapter 2. Two solutions were suggested. One was to make two shorter, looped keytapes of different lengths and encrypt with both of them. As in Section 2.7, the length of the resulting key is the least common multiple of the lengths of the two tapes. Even this system is vulnerable, however, as we saw in Section 2.7, especially under heavy traffic.

The other solution was to have a key as long as the cipher, like a running key, but a purely random key with no frequency information or probable words to let the cryptanalyst get started. Furthermore, the key could never be reused. If it was reused several times, the techniques of superimposition of multiple messages from Section 5.1 could be used to break the message. Even worse, the message could be broken even if the key was used only twice. This is easiest to see in the form of equations: suppose Eve has acquired two ciphertexts C_1 and C_2, such that

$$C_1 \equiv P_1 + k \quad \text{modulo } 2$$

and

$$C_2 \equiv P_2 + k \quad \text{modulo } 2.$$

She adds them to get

$$C_1 + C_2 \equiv P_1 + P_2 + 2k \quad \text{modulo } 2.$$

But $2 \equiv 0$ modulo 2, so Eve has

$$C_1 + C_2 \equiv P_1 + P_2 \quad \text{modulo 2,}$$

which is the same result as if one plaintext had been enciphered by a running-key. Thus she can use frequency or probable word information from both texts to uncover the plaintexts, and then use those to get the key if she wishes.

If the key is used only once, on the other hand, this system holds up to even known-plaintext attacks. If Eve has a matching plaintext and ciphertext, it is trivial for her to recover the key. But if the key is chosen at random and never used again, knowing it does Eve no good. Due to the importance of using keys only once, this system has come to be known as the **one-time system**, **one-time tape**, or, most commonly, **one-time pad**. The pad aspect requires a little explanation: around the same time as Vernam and Mauborgne were working on their system, three cryptologists in the German Foreign Office also realized that an unbreakable system requires a one-time random key as long as the plaintext. They were using plaintexts made up of decimal digits rather than binary, adding modulo 10 rather than 2, and most importantly for our story, working on paper rather than teletypewriter tape. Their system, instituted for German diplomats in the early 1920s, used pads of 50 sheets, each of legal-size paper, filled with random digits. Exactly two matching pads were made for each sequence of digits, and the sheets were torn off and destroyed after each message.

Although it was generally acknowledged that the one-time pad was unbreakable, it wasn't until the 1940s that Claude Shannon gave a rigorous proof. He had to start, in fact, with a rigorous definition of "unbreakable." Shannon said that a cipher has **perfect security** if, given a ciphertext, it is just as likely to come from any plaintext as any other. Thus, Eve can't do any better at recovering the plaintext than random guessing. Shannon went on to show some consequences of this. One is that there must be as many keys as there are possible plaintexts—practically speaking this means that the key must be as long as the plaintext. Another consequence is that every key must be equally likely to be used, which means that the characters or digits must be chosen at random and never taken from a previous key. Both the teletypewriter system and the German diplomatic system fit these criteria, as does a tabula recta cipher with a random key as long as the plaintext.

We can use this last cipher to illustrate exactly why the one-time pad is unbreakable. Suppose Eve intercepts the ciphertext message

WUTPQGONIMM

which she has reason to believe says either "meet me at two" or "meet me at ten." She can try both possibilities; if the plaintext is "meet me at two," then she can find a key that works:

keytext:	J	P	O	V	D	B	N	T	O	P	X
plaintext:	m	e	e	t	m	e	a	t	t	w	o
ciphertext:	W	U	T	P	Q	G	O	N	I	M	M

"Aha!" says Eve. But wait—if she tries "meet me at ten," she also finds a key that works:

keytext:	J	P	O	V	D	B	N	T	O	H	Y
plaintext:	m	e	e	t	m	e	a	t	t	e	n
ciphertext:	W	U	T	P	Q	G	O	N	I	M	M

In fact, there's a key that works for every possible plaintext; for example,

keytext:	N	I	K	E	L	N	N	B	V	X	Y
plaintext:	i	l	i	k	e	s	a	l	m	o	n
ciphertext:	W	U	T	P	Q	G	O	N	I	M	M

If every key is just as likely as every other, then there's no way Eve can possibly tell which is the right one. So she can't identify the right plaintext.

Despite the attraction of perfect security, there's one big problem with the one-time pad: it requires a lot of random key material, and Alice and Bob have to figure out how to exchange it. During the first large-scale trial of the teletypewriter system, Vernam and Mauborgne ran out of key tape and had to fall back on the system with two looped key tapes—and that was a system that communicated between stationary participants, on a trial basis with plenty of warning, during peaceful circumstances. In practice, it's very seldom that conditions are suitable for using a one-time pad. Diplomatic communications are one such case. The key material can be sent by diplomatic courier at regularly scheduled times, and then used to communicate over insecure telephone or computer networks. For example, the "red phone" line linking the White House and the Kremlin was encrypted, at least originally, with a one-time system. One-time pads (on paper) were also used during the Cold War by the Soviet Union for most of its top-level spy communications. The pads can be made extremely small so they are easy to hide and easy to get rid of in an emergency. They were also made extremely flammable for the latter reason. Presumably the difficulty in transmitting new key material was dealt with by merely being conservative about how many messages are sent.

5.3 BABY, YOU CAN DRIVE MY CAR: AUTOKEY CIPHERS

There are systems that do not involve repeating keys and yet do not require a long keytext because sections of previous plaintext, ciphertext, or keytext are used to generate new keytext. These systems, called **autokey ciphers**, do not provide perfect security, but they can be more difficult to break than repeating-key ciphers and more convenient than running-key ciphers and one-time pads. The autokey idea was first thought of, or at least first described, around the same time as the early polyalphabetic ciphers by **Girolamo Cardano**. His idea was to use the plaintext itself as a "key" to encrypt the ciphertext, starting it over with each word. The modern term for this key would be a **keystream**. As we will see, it's not a proper key at all.

For example, Alice might have encrypted the title of Cardano's book on gambling as

keystream:	O	N	O	N	C	A	S	T	I	O	N	C	A	S	T
plaintext:	o	n	c	a	s	t	i	n	g	t	h	e	d	i	e
ciphertext:	D	B	R	O	V	U	B	H	P	I	V	H	E	B	Y

In order to decrypt, Bob needs to decrypt the first word (more on that later), and then he can use it to decrypt the next two letters:

keystream:	O	N	O	N										
ciphertext:	D	B	R	O	V	U	B	H	P	I	V	H	E	B	Y
plaintext:	o	n	c	a											

This gives him the key for the next two, and so on.

There are three big problems with Cardano's system. The first might be more obvious when we look at the first word of our example in terms of modular arithmetic:

keystream:	O	N
numbers:	15	14
plaintext:	o	n
numbers:	15	14
ciphertext:	D	B
numbers:	4	2

All we are doing to the first word is adding it to itself, or multiplying it by 2. But we know two is a bad key, so there is more than one possible way to decrypt the first word; it could also have been

keystream:	B	A
numbers:	2	1
plaintext:	b	a
numbers:	2	1
ciphertext:	D	B
numbers:	4	2

This might not be fatal. It applies only to the first word, and not only will the incorrect decipherment probably be gibberish, but it will make the rest of the text gibberish also. The second problem is more important. Cardano's cipher doesn't have a key that Alice and Bob can change at will, and thus it violates Kerckhoffs' principle. The third problem is common to many autokey ciphers. If Bob makes a mistake in deciphering the cipher early on, it is pretty much impossible to recover from it, since the early plaintext is needed to decipher the rest. These three problems pretty much doomed Cardano's cipher. Bellaso improved the situation by combining the autokey idea with a progressive substitution cipher, but the system never caught on, probably due to its complexity.

Unlike in the case of the Vigenère cipher from Section 2.4, this time it really was **Blaise de Vigenère** who made the big breakthrough, which he described in his 1586 book, *Treatise on Ciphers or Secret Manners of Writing*. He avoided Cardano's first problem by having Alice use a "priming key" to encrypt the first letter and by using the plaintext starting with the first letter as the key starting with the second letter:

keystream:	V	A	W	O	R	T	H	L	E	S	S	C	R	A	C	K	I	N
plaintext:	a	w	o	r	t	h	l	e	s	s	c	r	a	c	k	i	n	g
ciphertext:	W	X	L	G	L	B	T	Q	X	L	V	U	S	D	N	T	W	U

The modern term for this priming letter is an **initialization vector**. A **vector** is a list of things of a fixed length, and this is a list of starting letters of length 1. It's easy to see how it could be any other length that Alice and Bob agree on.

This doesn't fix Cardano's second problem, since the priming key isn't really much of a key—it affects only the encryption of the first letter. Vigenère also included a provision that fixed this by adding an extra step that alters the plaintext before using it as the keystream. The real key to the cipher is then the method of alteration. For example, we could apply the $25P + 1$, or atbash, transformation to each ciphertext letter before using it:

shifted plaintext:	v	a	w	o	r	t	h	l	e	s	s	c	r	a	c	k	i	n
keystream:	E	Z	D	L	I	G	S	O	V	H	H	X	I	Z	X	P	R	M
plaintext:	a	w	o	r	t	h	l	e	s	s	c	r	a	c	k	i	n	g
ciphertext:	F	W	S	D	C	O	E	T	O	A	K	P	J	C	I	Y	F	T

This idea eventually came to be called a **plaintext autokey cipher**, with the key being the $25P + 1$ cipher. Cardano's third problem is still an issue. If Bob makes a mistake anywhere in the deciphering or if there is an error in transmitting the ciphertext, all deciphering from then on is doomed. This problem is inherent in plaintext autokey ciphers and is generally given as the reason for their scarcity.

Vigenère did propose another option that solves this problem, known as **error propagation**. Instead of a plaintext autokey cipher, Alice could use a **ciphertext autokey cipher**. In this case, the ciphertext is shifted one or more letters, after an initialization vector, and becomes the keystream. For example,

keystream:	I	F	G	Z	T	Y	Z	L	X	W	L	G	Y	N	W
plaintext:	w	a	s	t	e	a	l	l	y	o	u	r	o	i	l
ciphertext:	F	G	Z	T	Y	Z	L	X	W	L	G	Y	N	W	I

This time the initialization vector percolates through and influences all the ciphertext, but on the other hand, almost all the keystream is sitting out in plain sight. Kerckhoff's principle suggests that Eve might know Alice and Bob are using a ciphertext autokey cipher, so giving her the keystream is a bad idea. All she has to do is try it in different positions until it works and then figure out the initialization vector.

As with the plaintext autokey cipher, the security is improved by applying another transformation. A ciphertext autokey cipher using the $25P + 1$ transformation, for example, would look like this:

shifted ciphertext:	I	O	M	G	N	R	J	C	J	P	Z	V	W	S	Q
keystream:	R	L	N	T	M	I	Q	X	Q	K	A	E	D	H	J
plaintext:	w	a	s	t	e	a	l	l	y	o	u	r	o	i	l
ciphertext:	O	M	G	N	R	J	C	J	P	Z	V	W	S	Q	V

Once again, the $25P + 1$ transformation would be considered the key here.

Since the keystream depends only on the ciphertext, not the decrypted plaintext, ciphertext autokey ciphers don't suffer from the same problem, in which deciphering errors propagate through the deciphering process. Instead, they have a related problem if Alice makes an *enciphering* error. Since changes in the ciphertext affect all the rest of the enciphering process, any mistake Alice makes will make Bob's deciphered message gibberish from that point on. Ciphertext autokey ciphers were rather rare before modern computers came around, but a sort of encryption that is really the same idea has become fairly common since then.

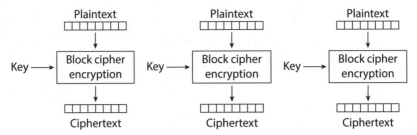

Figure 5.1. ECB encryption.

```
^(@@@)^(@@@)^        (*&&&!(*&&&!(
(@@@@@@@@@@)         *&&&&&&&&&&&!
^(@@@@@@@@)^         (*&&&&&&&&&&((
^^^(@@@@@)^^^        (((*&&&&&!(((
^^^^^(@)^^^^^        (((((*&!(((((

Plaintext            ECB encryption


&*)((&&*)((&&        &!^!^($@()#)&
*)(((((((((&         !^!^!^!^!^!^(
&*)(((((((((&&       $@()#@()#@(*!
%%%*)((((&&%%        )&*$%*&$%^%@#
%%%%*)&&%%%%         ^%@#^*&@#^%@#

Plaintext autokey    Ciphertext autokey
encryption           encryption
```

Figure 5.2. The effect of different modes of encryption on a picture.

In fact, each of the types of precomputer autokey ciphers corresponds to a **mode of operation** that can be used with a modern block cipher. The way we showed block ciphers being used in Chapter 4, where each block is enciphered separately, is technically known as **electronic codebook mode**, or ECB (see Figure 5.1). The drawback of this mode is that the same block of plaintext always encrypts to the same block of ciphertext. If blocks are 128 bits long, as in the case of AES, that's 16 text characters, which is a pretty long repeat. However, in the case of other types of data, such as high-resolution pictures or high-quality music, such a repeat could be quite common. This could leak a lot of information, so ECB is not considered secure.

I have tried to illustrate this in Figure 5.2 by using a cipher with a very small block size and a very low-resolution picture. The first picture

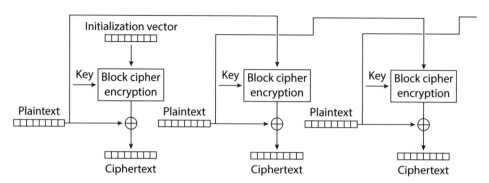

Figure 5.3. PFB encryption.

shown is the plaintext. This was encrypted by converting the symbols to numbers according to their positions on a standard US keyboard:

symbol:	!	@	#	$	%	^	&	*	(	)
number:	1	2	3	4	5	6	7	8	9	0

In the second picture, each digit was encrypted separately using the transformation $3P + 1$ modulo 10 and converted back into a symbol. The general shape of the picture is still easy to make out. In the third picture, the string of digits is encrypted using a plaintext autokey cipher with the transformation $3P + 1$ modulo 10 and an initialization vector of 0. The picture is somewhat harder to make out, although the long strings of identical plaintext symbols still allow too much information to leak through. In the fourth picture, the string of digits is encrypted using a ciphertext autokey cipher with the transformation $3P + 1$ modulo 10 and an initialization vector of 0. The final result is then considerably more difficult, although not impossible, to relate back to the original picture.

Suppose we take the idea of the plaintext autokey cipher but use a modern block cipher like AES instead of $25P + 1$ and add bits modulo 2, as in Section 4.1, instead of adding letters modulo 26. Then we have **plaintext feedback mode**, or PFB, as in Figure 5.3. A variation on this is to combine each plaintext block with the next *before* encrypting the combined block. This is called **plaintext block chaining**, or PBC, and is shown in Figure 5.4. PFB and PBC still suffer from the issue of error

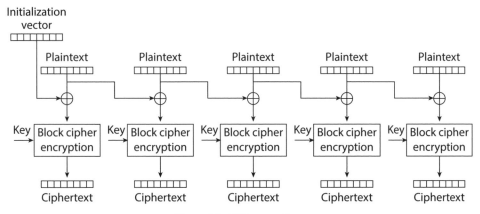

Figure 5.4. PBC encryption.

propagation we mentioned above. With modern computers, errors in encryption and decryption are much rarer than they used to be. However, errors in transmission can still be a very serious problem, so these modes are rarely used. They also leak some frequency information, as we saw in Figure 5.2. If some plaintext blocks occur substantially more often than others, it will be relatively easy to tell this by looking at the ciphertext.

We can do the same thing with ciphertext autokey ciphers: Vigenère's ciphertext autokey cipher becomes **ciphertext feedback mode**, or CFB, as shown in Figure 5.5. Or, if we combine each ciphertext block with the next plaintext block before encryption, we get **ciphertext block chaining**, or CBC, as shown in Figure 5.6. These modes don't suffer from error propagation caused by mistakes in transmission, and, as we said, the errors in encryption that could cause propagation are rare with modern computers. Therefore, these are considered very useful modes of operation and are quite common.

The third major type of autokey cipher is the **key autokey cipher**. This apparently didn't occur to Vigenère, probably because the idea of copying the keystream to the keystream doesn't seem like it would get you anywhere. However, if we add an extra transformation in the way we've been doing, then interesting things start to happen. If Alice adds 1 to the keystream letter each time, she gets Trithemius' progressive cipher:

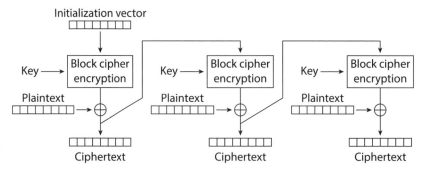

Figure 5.5. CFB encryption.

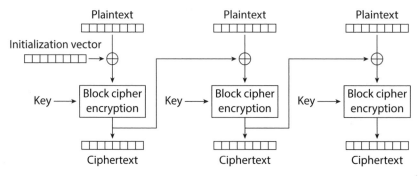

Figure 5.6. CBC encryption.

shifted keystream:	Z	A	B	C	D	E	F	G	H	I	J
keystream:	A	B	C	D	E	F	G	H	I	J	K
plaintext:	t	h	e	o	p	p	o	s	i	t	e
ciphertext:	U	J	H	S	U	V	V	A	R	D	P

shifted keystream:	K	L	M	N	O	P	Q	R	S	T
keystream:	L	M	N	O	P	Q	R	S	T	U
plaintext:	o	f	p	r	o	g	r	e	s	s
ciphertext:	A	S	D	G	E	X	J	X	M	N

Other transformations will give us other repeating-key ciphers, none of which are especially interesting or secure.

But there's nothing that says that the extra transformation has to work on only one letter or number at a time. It will be easier to think of the keystream as composed of numbers for this type of cipher, so suppose Alice starts out with an initialization vector of 5 decimal digits, such as 17742. Before she adds this to her plaintext, she adds each of the 5 digits to a new block of 5 digits, say 20243, using addition modulo 10.

shifted keystream:	1 7 7 4 2	
keystream:	3 7 9 8 5	
plaintext:	t u r n i	ngpointontheeasternfront
ciphertext:	W B A V N	

For the next block of 5 digits of the keystream, Alice adds 20243 to the first 5 digits, and so on.

shifted keystream:	1 7 7 4 2	3 7 9 8 5	5 7 1 2 8	7 7 3 6 1	9 7 5 0 4	1 7 7 4
keystream:	3 7 9 8 5	5 7 1 2 8	7 7 3 6 1	9 7 5 0 4	1 7 7 4 7	3 7 9 8
plaintext:	t u r n i	n g p o i	n t o n t	h e e a s	t e r n f	r o n t
ciphertext:	W B A V N	S N Q Q Q	U A R T U	Q L J A W	U L Y R M	U V W B

This is a slightly simplified version of a cipher used by Soviet troops during World War II to encipher numerical code groups, but as you can see, it is possible to use it on letters as well. Technically, this is still a repeating-key cipher, but the period has increased to 50. That's not unbreakable and there are some relations between the blocks that Eve can use, but it's pretty good for only 5 key digits—and, of course, we could use larger blocks. Furthermore, we could use a more interesting block cipher to transform our keystream, such as a transposition or a Hill cipher, which could make the period quite large. Key autokey ciphers are extremely flexible, although they can be extremely complicated to use. But then that's what computers are for.

The modern block cipher mode of operation that corresponds to the key autokey cipher is called **output feedback mode**, or OFB. Once again, the basic idea is to operate on the previous keystream block with a block cipher and then add the bits of the result modulo 2 to the bits of the plaintext. Figure 5.7 shows how that looks.

Output feedback mode uses a single previous key block to produce the next one and therefore in some sense is still a progressive cipher.

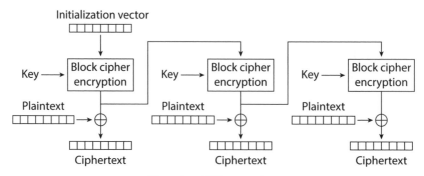

Figure 5.7. OFB encryption.

However, since there are a lot of different possible blocks, the period is likely to be very long. Also, there are supposed to be a lot of possible keys for the block cipher; therefore, it's pretty unlikely that Eve would be able to attack a good cipher in OFB mode using repeating-key techniques. It is possible, however, and for that reason some experts suggest not using it. Even so, it's still pretty common.

There's one more common mode of operation for block ciphers, and it corresponds to a key autokey cipher of a sort, although not one that was ever likely to have been used before computers. Instead of taking the previous keystream block and encrypting it to get the new keystream block, we could start with the initialization vector, alter it a little bit for each new block, and encrypt that. The most common alteration is just to add 1 each time before encrypting, so this is generally called **counter mode**, or CTR. One reason this doesn't make a good by-hand cipher is that the block cipher needs to have good diffusion properties. We'll use a 2×2 Hill cipher (modulo 10) for our example. The Hill cipher generally has good diffusion, as we saw in Section 4.5.

Suppose Alice picks 17 as her initialization vector and 1, 2, 3, 5 as the key to the Hill cipher. Then her encryption would look like this:

counter:	1 7	1 8	1 9	2 0	2 1	2 2	2 3	2 4
keystream:	5 8	7 3	9 8	2 6	4 1	6 6	8 1	0 6
plaintext:	y o	u c	a n	c o	u n	t o	n m	e x
ciphertext:	D W	B F	J V	E U	Y O	Z U	V N	E D

For a computer cipher, the setup looks like Figure 5.8.

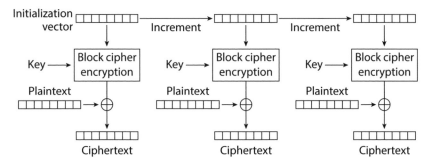

Figure 5.8. Counter mode encryption.

As with output feedback mode, the initialization vector for counter mode doesn't necessarily have to be secret, but it does need to be different for every message that uses a particular key. Otherwise, the keystream is the same for two messages and Eve can use the superimposition techniques of Sections 5.1 and 5.2. Also as with output feedback mode, the cipher will eventually repeat. In this case, it's easy to see how long the period is: it's whenever the counter wraps around back to where it started. Therefore, all Alice and Bob need to do is change the key before that happens. One last interesting feature of counter mode is that unlike stream ciphers generally, you can easily start encrypting or decrypting a message in the middle just by setting your counter to the appropriate number. This makes counter mode useful for encrypting data files that store information that might need to be changed piece by piece.

5.4 LINEAR FEEDBACK SHIFT REGISTERS

We noted in the previous section that making the block size of a block cipher larger can improve the security of several of its modes of operation. An alternative is to use very small blocks, even 1 letter or 1 bit, but make each new key block depend on more than 1 previous block. Let's again start out with an initialization vector of 5 decimal digits. Let the sixth keystream number be equal to the sum of the first 2 keystream digits modulo 10. The seventh keystream number is the sum of the second and third digits modulo 10, and so on. For example, if the initialization vector is $(1, 2, 0, 2, 9)$, we get

$$1, 2, 0, 2, 9, 3, 2, 2, 1, 2, 5, 4, 3, 3, 7, 9, 7, 6, 0, \ldots$$

This sequence will eventually repeat, but not for 16401 steps!

The process we are using to generate this keystream is called **chain addition** in older sources or a **lagged Fibonacci generator** (modulo 10) in newer ones. Fibonacci refers to the famous **Fibonacci sequence**, which you get if you start with an initialization vector of $(1, 1)$ and do not use modular arithmetic:

$$1, 1, 2, 3, 5, 8, 13, 21, 34, 55, 89, 144, 233, 377, 610, \ldots$$

This sequence was well known to ancient Indian mathematicians but was introduced to Western Europe by **Leonardo "Fibonacci (of the Bonacci family)" of Pisa**. Lagged refers to the fact that unlike the Fibonacci sequence, we are not adding the two terms on the end of the current stream, but two that are farther back.

Alice's encryption with our lagged Fibonacci keystream looks like this:

keystream:	1	2	0	2	9	3	2	2	1	2	5	4	3	3	7	9	7	6	0
plaintext:	m	u	l	t	i	p	l	y	l	i	k	e	r	a	b	b	i	t	s
ciphertext:	N	W	L	V	R	S	N	A	M	K	P	I	U	D	I	K	P	Z	S

This cipher looks like something Vigenère might have come up with, if not Fibonacci himself. Actually, it was invented in 1969 as a challenge to the American Cryptogram Association and is called the **Gromark cipher**. The technique of chain addition modulo 10 seems to have first appeared in unclassified literature after the 1957 trial of the Soviet spy Rudolf Ivanovich Abel. At the trial, former Soviet spy Reino Hayhanen, who had defected to the United States, described the use of chain addition to generate key numbers for a complicated cipher. Unlike the Gromark, it used the key numbers in a straddling checkerboard and a complicated transposition rather than a polyalphabetic cipher. Hayhanen's cipher is generally known as the VIC cipher, after Hayhanen's code name.

For a classical chain addition system, the initialization vector is generally used as the key. It's easy to change and very strongly influences the rest of the keystream, and it can also be easily used as the indication of how far back in the keystream to go to get the numbers that are added together. However, a lagged Fibonacci system can also be varied by using two numbers from the previous keystream that are not next to each other or by finding a different rule to combine the numbers—it could be a modulus other than 10, it could be multiplication instead of addition, or it could be something even more complicated.

We could change the system even more by using more than two numbers from the previous keystream. Suppose we write the formula for the key number at position n in the Gromark cipher as

$$k_n \equiv k_{n-5} + k_{n-4} \quad \text{modulo 10.}$$

For a lagged Fibonacci system, more generally we would have

$$k_n \equiv k_{n-i} + k_{n-j} \quad \text{modulo } m,$$

where i and j tell us how back in the keystream to go and m is the modulus. Now suppose we use the formula

$$k_n \equiv c_1 k_{n-j} + c_2 k_{n-j+1} + \cdots + c_{j-1} k_{n-2} + c_j k_{n-1} \quad \text{modulo } m.$$

The coefficients $c_1, c_2, \ldots, c_j$ can be part of the key or they can be considered fixed as part of the cipher method, but either way they are the same throughout the message.

For example, let's take $m = 2$, $j = 4$, $c_1 = c_3 = 1$, and $c_2 = c_4 = 0$, so we have

$$k_n \equiv 1k_{n-4} + 0k_{n-3} + 1k_{n-2} + 0k_{n-1} \quad \text{modulo 2.}$$

If we start with an initialization vector of $k_1 = k_2 = k_3 = k_4 = 1$, then we have

$$k_5 \equiv 1 \times 1 + 0 \times 1 + 1 \times 1 + 0 \times 1 \equiv 0 \quad \text{modulo 2,}$$

$$k_6 \equiv 1 \times 1 + 0 \times 1 + 1 \times 1 + 0 \times 0 \equiv 0 \quad \text{modulo 2,}$$

$$k_7 \equiv 1 \times 1 + 0 \times 1 + 1 \times 0 + 0 \times 0 \equiv 1 \quad \text{modulo 2,}$$

$$k_8 \equiv 1 \times 1 + 0 \times 0 + 1 \times 0 + 0 \times 1 \equiv 1 \quad \text{modulo 2,}$$

and so on.

A real or simulated machine that produces a keystream using a formula of this type is called a **linear feedback shift register**, or **LFSR**. Linear refers to the type of formula. Equations where one set of variables is multiplied by another set and then added to together are called linear

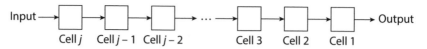

Figure 5.9. A shift register.

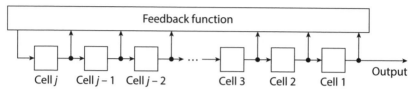

Figure 5.10. A feedback shift register.

because the most famous such equation is the equation $y = mx + b$ for a line in the two-dimensional plane. Feedback refers to the fact that previous values are used to produce new values. And shift register refers to a particular type of electronic circuit that was used early on to build these machines. A shift register, as shown in Figure 5.9, is a sequence of storage cells, each of which holds a number. The shift register is controlled by a clock so that at each tick of the clock, a new input number goes into cell j, the contents of cell j move to cell $j - 1$, and so on. The contents of cell 1 become the output of the shift register. If the shift register starts with k_1 in cell 1 through k_j in cell j, as it continues to run it will output the original $k_1, k_2, \ldots, k_j$, and then new numbers k_{j+1}, k_{j+2}, and so on, based on the input.

A feedback shift register uses the contents of the various cells in some way to produce the new input to the first cell, as in Figure 5.10; the procedure used to do this is called the **feedback function**. And a linear feedback shift register produces the new input by a linear feedback function. Figure 5.11 gives you an idea how you might build such a thing, with the circles labeled c_1 to c_j indicating to multiply the thing going in by that number modulo m and the circles labeled with a plus sign indicating to add the two things going in together modulo m.

The most common modulus for LFSRs is 2, in which case all the numbers can be taken to be either 0 or 1 and we can think of them as bits. Multiplying a number by either 0 or 1 and then adding means either doing nothing or adding the number, so we can also think of the multiplication circles in Figure 5.11 as switches that either let the bit

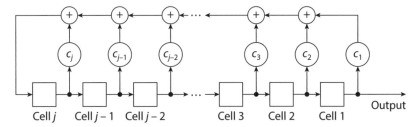

Figure 5.11. A linear feedback shift register.

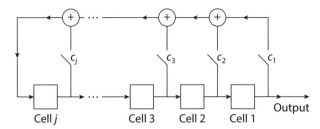

Figure 5.12. A linear feedback shift register represented with switches.

be added or not. Figure 5.12 shows an LFSR represented in this form. As you can probably imagine, this setup is very easy to implement in a specialized piece of digital hardware, and a variation that produces the same results is easy in software as well, although not as fast. The use of digital LFSRs in cryptography goes back at least as far as 1952, when the brand-new NSA started designing the KW-26 for their own use and that of the US military.

Going back to our figures, if we know what c_1 through c_j are, yet another way to represent them is by either drawing a line or not. So, the example we just gave can be drawn as in Figure 5.13. If we start it with 1, 1, 1, 1 in the cells, it will output

$$1, 1, 1, 1, 0, 0, 1, 1, \ldots$$

just like we calculated. I encourage you to try this yourself and check.

Now that we know what an LFSR is, how do we use it for encryption? The LFSRs we are going to use always output numbers modulo 2, or bits, so it would make sense to represent the plaintext in bits as well. We will use the ASCII representation explained in Sidebar 4.1. If Alice wants to encrypt a message, she first converts it to ASCII:

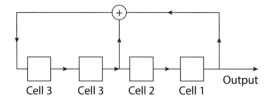

Figure 5.13. A particular example of a linear feedback shift register.

plaintext:	S	e	n	d
ASCII:	1010011	1100101	1101110	1100100

plaintext:		$	.	
ASCII:	0100000	0100100	0101110	

Then she generates a keystream using her LFSR:

plaintext:	S	e	n	d
ASCII:	1010011	1100101	1101110	1100100
keystream:	1111001	1110011	1100111	1001111

plaintext:		$	.	
ASCII:	0100000	0100100	0101110	
keystream:	0011110	0111100	1111001	

Next, she adds the corresponding bits modulo 2:

plaintext:	S	e	n	d
ASCII:	1010011	1100101	1101110	1100100
keystream:	1111001	1110011	1100111	1001111
ciphertext bits:	0101010	0010110	0001001	0101011
decimal numerals:	42	22	9	43

plaintext:		$	.	
ASCII:	0100000	0100100	0101110	
keystream:	0011110	0111100	1111001	
ciphertext bits:	0111110	0011000	1010111	
decimal numerals:	62	24	87	

If Alice and Bob are computers, then Alice is done when she has the ciphertext bits. If Alice is a person, she might want to put them in a more compact form, such as the decimal equivalent.

You might have noticed that the output of this LFSR repeats with a fairly short period of only 6 bits. You might be asking whether every LFSR is going to repeat and, if so, whether the periods are always this short. The answers are yes and no, respectively. The output of an LFSR has to repeat because it depends only on the numbers in the cells of the shift register modulo m, and there are only so many options for those numbers. Once the same set of numbers repeats, then the output repeats from then on. How many possible sets of numbers are there? In the example we gave with four cells and a modulus of 2, each cell will hold 0 or 1; thus there will be $2 \times 2 \times 2 \times 2 = 2^4 = 16$ possibilities. If all the cells are zero, then it should be pretty clear that the output is always also going to be zero. So we should avoid that possibility. There are 15 others that we could go through before the output repeats, so there might be a modulo-2 LFSR with 4 cells and a period of 15, and in fact there is. In general, an LFSR with j cells and a modulus of m can have a period of at most $m^j - 1$. If m is a prime number, multiple LFSRs with this period exist and there are well-known ways of finding them.

This all seems like an excellent situation if we want a stream cipher. We have a fast way of producing a keystream and a reliable way of guaranteeing a period as long as we want. Unfortunately, the equations describing LFSRs, like those in the Hill cipher, are linear. That means that LFSRs, like Hill ciphers, are extremely vulnerable to known-plaintext attacks.

Suppose Eve knows that she is looking at an LFSR with j cells, and she has $2j$ pairs of plaintext bits and corresponding ciphertext bits. There might be s bits we don't have at the beginning, so call the plaintext bits we do have P_{s+1} through P_{s+2j} and the ciphertext bits C_{s+1} through C_{s+2j}. Since the encrypting is done using

$$C_n \equiv P_n + k_n \quad \text{modulo } 2,$$

Eve can easily recover the keystream bits using

$$k_n \equiv C_n - P_n \quad \text{modulo } 2.$$

Remember that Eve's goal is to recover the key, which for an LFSR is usually considered to be the initialization vector k_1 through k_j and, sometimes, the coefficients c_1 through c_j as well. At this point Eve could always do a known-plaintext version of a brute force attack by using every possible key to see if it generates the right keystream. However, there is a much better way.

Without knowing k_1 through k_j or even c_1 through c_j, Eve can set up a system of equations using the keystream bits she does know:

$$
\begin{aligned}
k_{s+j+1} &\equiv c_1 k_{s+1} + c_2 k_{s+2} + \cdots + c_{j-1}k_{s+j-1} + c_j k_{s+j} &&\text{modulo 2}\\
k_{s+j+2} &\equiv c_1 k_{s+2} + c_2 k_{s+3} + \cdots + c_{j-1}k_{s+j} + c_j k_{s+j+1} &&\text{modulo 2}\\
&\ \ \vdots\\
k_{s+2j} &\equiv c_1 k_{s+j} + c_2 k_{s+j+1} + \cdots + c_{j-1}k_{s+2j-2} + c_j k_{s+2j-1} &&\text{modulo 2}
\end{aligned}
$$

That's just a system of j equations in the j unknowns c_1 through c_j, and it can be solved using the same techniques we saw Eve use for a Hill cipher in Section 1.6. This will give her the coefficients c_1 through c_j.

If Eve wants to know the initialization vector k_1 through k_j, she can now set up the system

$$
\begin{aligned}
k_{j+1} &\equiv c_1 k_1 + c_2 k_2 + \cdots + c_{j-1}k_{j-1} + c_j k_j &&\text{modulo 2}\\
k_{j+2} &\equiv c_1 k_2 + c_2 k_3 + \cdots + c_{j-1}k_j + c_j k_{j+1} &&\text{modulo 2}\\
&\ \ \vdots\\
k_s &\equiv c_1 k_{s-j} + c_2 k_{s-j+1} + \cdots + c_{j-1}k_{s-2} + c_j k_{s-1} &&\text{modulo 2}\\
k_{s+1} &\equiv c_1 k_{s-j+1} + c_2 k_{s-j+2} + \cdots + c_{j-1}k_{s-1} + c_j k_s &&\text{modulo 2}\\
&\ \ \vdots\\
k_{s+j} &\equiv c_1 k_s + c_2 k_{s+1} + \cdots + c_{j-1}k_{s+j-2} + c_j k_{s+j-1} &&\text{modulo 2}
\end{aligned}
$$

Since Eve now knows c_1 through c_j as well as k_{s+1} through k_{s+j}, this is a system of s equations in the s unknowns k_1 through k_s. Thus she can solve it to get the entire keystream up to the point she knew already, including the initialization vector.

5.5 ADDING NONLINEARITY TO LFSRS

If linear feedback shift registers are not secure because of the linearity, what can we do to improve the situation? One option would be to use a nonlinear feedback function, but those are slower, and their strengths and weaknesses are harder to analyze. Another option is to take the values of more than one cell of an LFSR or the output values of more than one LFSR and combine these values in a nonlinear way. One pitfall of this approach is possible vulnerability to a **correlation attack**—if the nonlinear function is not well chosen, then it may be possible to make an educated guess about the values in one or more of the LFSRs from the values of the output. A third option is to alter the clock that controls when the bits in the LFSR shift. There can be multiple LFSRs shifting at different times, or the output of one LFSR can control the shifting of another or even itself. And these ideas can be combined.

As I'm writing this book, the most used and most studied stream cipher based on LSFRs is probably the A5/1 cipher used in the first generation of GSM digital cell phones. The details of the development of A5/1 are extremely unclear, and the deliberations that led to it are apparently classified. According to anonymous sources cited by one researcher, there was a disagreement among the intelligence agencies of the Western European countries originally involved in the development of GSM in the 1980s. In particular, West German intelligence wanted strong encryption, presumably to protect against eavesdropping by Soviet-bloc countries. The other countries' agencies preferred weaker encryption, possibly to make it easier for them to conduct their own surveillance. The cipher chosen in the end seems to have been one of the weaker ones. On the other hand, the final choice is particularly efficient in terms of speed, number of components, and power consumption. These features may have also played a role in the decision.

The details of the cipher were developed in 1987 and 1988, and the cipher was first officially used in 1991. At this point the cipher was kept as a trade secret. Sometime around early 1994, a British telephone company gave documents describing the cipher to a researcher at a British university, apparently without requiring him to sign a nondisclosure agreement. By mid-1994 an almost-complete description of the cipher was posted on the Internet. In 1999 the complete design was reverse-engineered from an actual cell phone and again posted on the Internet. The GSM Association eventually confirmed that this description was correct.

The A5/1 cipher starts with three LFSRs, one with 19 cells, one with 22 cells, and one with 23 cells, as shown in Figure 5.14. Each has

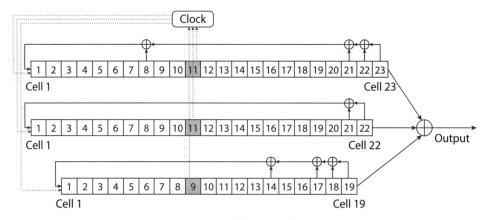

Figure 5.14. The A5/1 cipher.

maximum period, that is, $2^{19} - 1 = 524,287$, $2^{22} - 1 = 4,194,303$, and $2^{23} - 1 = 8,388,607$. This makes 64 cells total, which are initialized with the bits of a 64-bit key. The outputs from the three LFSRs are all added together modulo 2 and then added modulo 2 to the plaintext bits. So far this is a lot like multiple encryption by repeating-key ciphers, as in Section 2.7. Since each period has a GCD of 1 with each other period, if we left the cipher here it would have a period of $(2^{19} - 1) \times (2^{22} - 1) \times (2^{23} - 1) \approx 18 \times 10^{18}$, which is certainly very long. But we have combined three linear things in a linear way, so this cipher would still be linear and still vulnerable to a known-plaintext attack.

The nonlinearity comes in by the third idea mentioned previously, making the clock system more complicated. Not all the 3 LFSRs are shifted at every tick of the clock. Notice that a bit near the middle of each LFSR is highlighted in Figure 5.14; these are the **clock control bits**. Every time the clock ticks, these 3 bits "vote" for 0 or 1, and the majority wins. Then each register shifts if its clock control bit voted with the majority, and otherwise it stays put. Table 5.2 might make this clearer. As you can see from the table, each LFSR shifts $\frac{3}{4}$ of the time and at least 2 LFSRs shift on each clock tick. Experiments suggest that this change cuts down on the period considerably, but with careful use this particular disadvantage would be tolerable compared to the added security against known-plaintext attacks.

Alas, that added security seems to have turned out to be not as great as the GSM researchers originally thought. The first hints of this

TABLE 5.2.
The A5/1 clock control system

Clock control bit			Shift?		
LFSR-19	LFSR-22	LFSR-23	LFSR-19	LFSR-22	LFSR-23
0	0	0	yes	yes	yes
0	0	1	yes	yes	no
0	1	0	yes	no	yes
0	1	1	no	yes	yes
1	0	0	no	yes	yes
1	0	1	yes	no	yes
1	1	0	yes	yes	no
1	1	1	yes	yes	yes

came as early as 1994, when a known-plaintext attack was proposed. It involved guessing some bits of the initialization vectors of the LFSRs and working out what the other bits had to be. The details of this were worked out in 1997 in a paper that also proposed another attack: a **precomputation attack**, which streamlines a known-plaintext brute-force attack by letting Eve compute and store part of the information before she gets her hands on the plaintext-ciphertext pairs. A third type of attack developed against A5/1 was a variation on the correlation attack. As described before, Eve makes an educated guess about the input to a nonlinear function, based on the output and the function. Here, she makes an educated guess about previous values in the three LFSRs based on a later value of the keystream and an estimate of the number of times that each LFSR has been shifted. Such an attack was first applied to A5/1 in 2001. Both the precomputation attacks and the correlation attacks on A5/1 have been considerably refined since they were first proposed.

Known-plaintext attacks are somewhat difficult to mount against actual cell phones, for various logistical reasons. However, several researchers have discovered various peculiarities of the way A5/1 is used in actual GSM phones. These allow Eve to identify the equivalent of probable words or otherwise take advantage of known-plaintext-type attacks with ciphertext-only data. In 2006 it was estimated that one correlation-type attack could be mounted with 4 minutes of cell phone communication and less than 10 minutes average computation time on a personal computer and would succeed more than 90% of the time. A precomputation attack was described in 2003 that required 140 personal computers running for 1 year to develop the precomputed tables and twenty-two 200-gigabyte hard drives to store them. That's a lot of precomputation, but a single personal computer using the tables could decrypt a cell phone communication as fast as it could intercept it. A project to create these tables as proof that the attack is feasible was started and showed some partial cryptanalysis successes. The GSM Association downplayed the significance of these attacks but pointed out that a new cipher, which does not involve LFSRs at all, is "being phased in to replace A5/1."

The new cipher, known as A5/3, is standard on 3G and 4G networks, but progress on retrofitting the older networks was slow until 2013. In that year, the internal NSA documents obtained by Edward Snowden revealed that the NSA can "process" encrypted A5/1 without the key. This was generally taken to refer to an attack similar to the attacks described before. In response, a number of major wireless carriers announced that they were either moving their older GSM networks to the A5/3 cipher or simply replacing them with 3G or higher technology.

If LFSRs by themselves are insecure—and the GSM Association seems to have given up on using LFSRs to secure cell phone communications—you might be forgiven for asking if there are any ciphers based on LFSRs still considered secure. As it happens, cipher designers are still using LFSRs in their plans. The United States doesn't have a standard for stream ciphers, but in 2004 the European Network of Excellence for Cryptology (ECRYPT), a research initiative funded by the European Union, started the eSTREAM project to "identify new stream ciphers suitable for widespread adoption." Out of 34 ciphers submitted for consideration, 7 were eventually determined to be secure enough, efficient enough, and useful enough for the eSTREAM portfolio. Out of those 7, 3 use LFSRs in some fashion. The answer, it seems, is simply that one must be very careful in how one adds nonlinearity to an LFSR-based cipher.

5.6 LOOKING FORWARD

Like Chapter 4, this chapter takes us about as far forward with this type of cipher as we are going to go in this book. Both the development of new stream ciphers and of new modes of operation for block ciphers are still areas of active research. The NIST Web site currently lists 12 approved modes of operation, with many more proposed for further consideration. Five of the approved modes, ECB, CFB, CBC, OFB, and CTR, are included in those we have discussed earlier. Another, XTS, is related to counter mode but is designed specifically to encrypt information stored on hard drives. The last six modes involve **authentication** of messages rather than, or in addition to, encryption.

One problem with many of the modes and stream ciphers we have discussed is that they don't protect against the possibility that Eve can change the message that Alice sends. I mentioned that it is desirable for modes to avoid the propagation of errors in encryption or transmission. This also means that Eve might be able to make a small change to one part of a message without making the whole thing unreadable. This is especially important if Eve might know that a particular part of a message contains numbers or computer data rather than text. For instance, she might be able to change the amount of money in an electronic transaction or corrupt a critical part of a computer program to crash Bob's computer. Even if Eve can't tell exactly what she is changing the message to, she can still cause a lot of trouble.

The goal of an authentication mode is to use a key to produce a **message authentication code**, or **MAC**, to go with the message. This is a short piece of information that should change unpredictably if even 1 bit of the message is changed. Eve might be able to change Alice's message, but without the key she won't be able to change Alice's MAC to match it. Bob uses the key to verify the MAC and make sure that the message hasn't been changed. Alice can employ a MAC whether she encrypts her message or not—maybe she doesn't care who knows the message as long as Eve can't change it.

One of the earliest and simplest MACs was CBC-MAC, a version of which was made a US government standard in 1985. Essentially one encrypts the message in CBC mode, using a special key, and throws away everything except the last block of ciphertext. This needs some tweaks to be really secure, but the CMAC authentication mode approved by NIST is a close relative of CBC-MAC.

One problem with CBC-MAC and CMAC is that if Alice wants to both encrypt and authenticate her message, she has to go through the encryption process twice with two different keys. Authenticated encryption modes produce the MAC and the ciphertext at the same

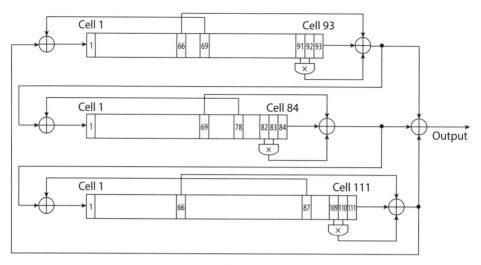

Figure 5.15. The Trivium cipher.

time. How to do this both securely and efficiently is an important area of current cryptographic interest.

As I said, cryptographers are also still working on completely new stream ciphers. I mentioned that three of the eStream portfolio ciphers use LFSRs. Two of those use both LFSRs and **nonlinear feedback shift registers** (NLFSRs). NLFSRs put the nonlinear function directly into the feedback function instead of using it to combine LFSRs. I mentioned earlier that they were slower and harder to analyze, which is one reason to use both NLSFRs and LFSRs as a sort of backup. One of the eStream ciphers has attracted a lot of attention for using only NLFSRs but keeping the nonlinearity to the minimum that seems necessary for security. This is the **Trivium** cipher, shown in Figure 5.15. The only nonlinear operations are the three places where two keystream bits are multiplied instead of added. It's too early to tell whether this cipher will see widespread use, but it looks promising.

That leaves three eStream ciphers that don't use shift registers at all. These are primarily designed to be implemented in software rather than directly in circuitry. Such ciphers have more design flexibility and use a wide variety of techniques. These include looking up values in constantly changing tables and ideas of confusion and diffusion taken from block cipher design.

Will any major government ever standardize on a single stream cipher the way that the United States standardized AES? It doesn't look very likely. Stream ciphers that are not directly based on block cipher modes are generally used in specific situations where block ciphers are not suitable. Sometimes this is for reasons of speed; other times it might be because of limited processing ability, such as in cell phones or smart cards. Or it might be for reduced power consumption, limited bandwidth, to make the procedure easier to parallelize, for specific error correction properties, or .... Since there are a variety of such situations calling for ciphers with different strengths and weaknesses, it doesn't seem like anyone could pick a single "best" stream cipher.

6

Ciphers Involving Exponentiation

6.1 ENCRYPTING USING EXPONENTIATION

We'd like our next cipher to be a simple mathematical cipher resistant to both ciphertext-only and known-plaintext attacks, as explained in Section 1.7. For the first, we'll make it a polygraphic cipher, although the way we construct the blocks is just a little bit different from what we did in Section 1.6. Once again, we'll take a block size of 2 in our example and divide up the plaintext into 2-letter blocks.

po   we   rt   ot   he   pe   op   le

This time, we'll convert each 2-letter block into a number by just jamming the numbers from the 2 letters together, putting in 0s where appropriate.

plaintext:	po	we	rt	ot	he	pe	op	le
numbers:	16, 15	23, 5	18, 20	15, 20	8, 5	16, 5	15, 16	12, 5
"jammed together":	1615	2305	1820	1520	805	1605	1516	1205

We will also need to pick a modulus for the cipher. A modulus of 26 is no longer going to do it, since our blocks can be as large as 2626. It will be convenient to pick a modulus that is a prime number, although later in this chapter, we will see that we can get around that. For the moment, 2819 will be a good choice, since it is prime and larger than 2626.

We've tried addition, multiplication, and various combinations of them. A mathematician's next idea might be to try exponentiation, or raising a number to a power. Remember that raising a number to a power means multiplying it by itself repeatedly. For example, $2^3 = 2 \times 2 \times 2 = 8$. In particular, we will use

$$C \equiv P^e \quad \text{modulo } 2819.$$

The key for this cipher is traditionally called e, for **encryption exponent**. Note that e has nothing to do here with the number $2.71828\ldots$, which is the base of the natural logarithm. The encryption exponent is a number between 1 and 2818, with some restrictions, which we will explore in more detail shortly. For the moment, let's take $e = 769$.

plaintext:	po	we	rt	ot	he	pe	op	le
numbers:	16, 15	23, 5	18, 20	15, 20	8, 5	16, 5	15, 16	12, 5
together:	1615	2305	1820	1520	805	1605	1516	1205
to the 769th power:	1592	783	2264	924	211	44	1220	1548

What we are doing here is raising 1615 to the 769th power, wrapping around every time we get to 2819, which means really a lot of multiplications and wraparounds. You need a computer, or at least a very good calculator, to have any hope of doing this. We can't change all these blocks back into letters, but that's okay. Alice can just send Bob the numbers.

How is Bob going to decrypt this? Just as the opposite of addition is subtraction and the opposite of multiplication is division, the opposite of taking a power is taking a root. For example, if $8 = 2^3$, then $2 = \sqrt[3]{8}$, and if $C = P^e$, then $P = \sqrt[e]{C}$. But if you thought doing division and making sure you get a whole number was problematic, taking roots is even worse. For instance, in our example the first ciphertext block was 1592, and the 769th root of 1592 is approximately 1.0096, which is pretty useless for our purposes.

6.2 FERMAT'S LITTLE THEOREM

In order to help Bob, we're going to have to go a little bit deeper into number theory than we have so far. Up until now, we've basically been using one big mathematical idea, namely, modular arithmetic, as formalized by Gauss. Now we need a second big idea, which is generally credited to **Pierre de Fermat**. Fermat was a seventeenth-century Frenchman who was a lawyer by profession and a mathematician by avocation. Possibly because of this, he had a bit of a mathematical chip on his shoulder. He had a habit of writing letters to his colleagues in which he announced that he had proven something. Instead of giving the proof, he challenged the recipient to come up with the proof himself. He also claimed he had proved some things that turned out to be false, and at least one, now known as Fermat's last theorem, that turned out to be true but probably a lot harder to prove than Fermat thought.

The mathematical fact, or **theorem**, that we need here is definitely true, and Fermat may very well have come up with a proof, although as usual he didn't write it down. It's now called **Fermat's little theorem**, even though it has big implications. We don't know how Fermat discovered it, but here's how you might have discovered it using the ideas we've already explored.

Suppose you are working with a multiplicative cipher with a very small alphabet that has a prime number of letters. The 13-letter Hawaiian alphabet would work. With a key of 3, the table for this alphabet looks like this:

plaintext	number	times 3	ciphertext
a	1	3	I
e	2	6	H
i	3	9	M
o	4	12	W
u	5	2	E
h	6	5	U
k	7	8	L
l	8	11	P
m	9	1	A
n	10	4	O
p	11	7	K
w	12	10	N
`	13	13	`

The important thing here is that since 13 is prime, 3 is a good key, and so is every other number from 1 to 12. Thus the column of numbers on the left-hand side is the same as the column of the numbers on the right-hand side, except in a different order. If you were playing around with this, you might have tried adding each column. You would get the same answer modulo 13, since they are the same numbers modulo 13:

$$1 + 2 + 3 + \cdots + 13 \equiv (1 \times 3) + (2 \times 3) + (3 \times 3) + \cdots + (13 \times 3) \qquad \text{modulo 13.}$$

Collect like terms on the right:

$$1 + 2 + 3 + \cdots + 13 \equiv (1 + 2 + 3 + \cdots + 13) \times 3 \qquad \text{modulo 13,}$$

or

$$91 \equiv 91 \times 3 \qquad \text{modulo 13,}$$

or

$$\equiv 0 \times 3 \qquad \text{modulo 13.}$$

That wasn't that interesting. Instead of adding up each column, you could try multiplying it instead. Then you would get

$$1 \times 2 \times 3 \times \cdots \times 13 \equiv (1 \times 3) \times (2 \times 3) \times (3 \times 3) \times \cdots \times (13 \times 3) \qquad \text{modulo 13,}$$

$$1 \times 2 \times 3 \times \cdots \times 0 \equiv (1 \times 3) \times (2 \times 3) \times (3 \times 3) \times \cdots \times (0 \times 3) \qquad \text{modulo 13,}$$

$$0 \equiv 0 \qquad \text{modulo 13.}$$

That's even less interesting, but clearly the problem is the 13 at the end of each column. You could try just leaving that out.

$$1 \times 2 \times 3 \times \cdots \times 12 \equiv (1 \times 3) \times (2 \times 3) \times (3 \times 3) \times \cdots \times (12 \times 3) \qquad \text{modulo 13.}$$

Now you could pull out all the 3s on the right, which came from the key.

$$1 \times 2 \times 3 \times \cdots \times 12 \equiv (1 \times 2 \times 3 \times \cdots \times 12) \times 3^{12} \qquad \text{modulo 13.}$$

Cancel $1 \times 2 \times 3 \times \cdots \times 12$:

$$1 \equiv 3^{12} \qquad \text{modulo 13.}$$

And that, I hope you agree, is interesting.

Notice that the choices of 13 and 3 weren't important. Any prime modulus p and any good key number k will do. So Fermat's little theorem tells us the following.

Theorem (Fermat's Little Theorem) *For any prime p and any k between* 1 *and* $p - 1$,

$$k^{p-1} \equiv 1 \quad \text{modulo } p.$$

6.3 DECRYPTING USING EXPONENTIATION

Now would probably be a good time to drop back and try to remember our goal. We wanted to undo the equation

$$C \equiv P^e \quad \text{modulo } 2819.$$

Remember from Section 1.3 that in modular situations we should be able to go forward to go backward. So it should be reasonable to look for a number $\bar{e}$ such that

$$C^{\bar{e}} \equiv P \quad \text{modulo } 2819.$$

Since $C \equiv P^e$ modulo 2819, this is the same as saying

$$(P^e)^{\bar{e}} \equiv P \quad \text{modulo } 2819,$$

or, using the laws of exponents,

$$P^{e\bar{e}} \equiv P \quad \text{modulo } 2819.$$

If we look at Fermat's little theorem closely here, we see that it says

$$P^{2818} \equiv 1 \quad \text{modulo } 2819,$$

but we could also write it as

$$P^{2818} \equiv P^0 \quad \text{modulo } 2819.$$

We are working modulo 2819, which means 2819 is the same as 0 if we are looking at the whole equation. But if we are looking at the *exponent*, then Fermat's little theorem says 2818 is the same as 0. In general, if we are looking at an equation modulo a prime p, then we can treat the exponent as if we were working modulo $p - 1$. Therefore, the number $\bar{e}$ that we are looking for should be the inverse of e modulo 2818. For future reference, it is important to note that exponents work quite this way only for primes. We will see the equivalent for other numbers in Section 6.6.

So we'll use the Euclidean algorithm on e (which was 769) and 2818 like we did in Section 1.3. I'll put in a little less detail than I did there, but feel free to fill in the gaps.

$$2818 = 769 \times 3 + 511 \qquad 511 = 2818 - (769 \times 3)$$

$$769 = 511 \times 1 + 258 \qquad 258 = 769 - (511 \times 1)$$
$$= (769 \times 4) - (2818 \times 1)$$

$$511 = 258 \times 1 + 253 \qquad 253 = 511 - (258 \times 1)$$
$$= (2818 \times 2) - (769 \times 7)$$

$$258 = 253 \times 1 + 5 \qquad 5 = 258 - (253 \times 1)$$
$$= (769 \times 11) - (2818 \times 3)$$

$$253 = 5 \times 50 + 3 \qquad 3 = 253 - (5 \times 50)$$
$$= (2818 \times 152) - (769 \times 557)$$

$$5 = 3 \times 1 + 2 \qquad 2 = 5 - (3 \times 1)$$
$$= (769 \times 568) - (2818 \times 155)$$

$$3 = 2 \times 1 + 1 \qquad 1 = 3 - (2 \times 1)$$
$$= (2818 \times 307) - (769 \times 1125)$$

so

$$1 = (2818 \times 307) + (769 \times -1125)$$

and

$$1 \equiv 769 \times -1125 \quad \text{modulo } 2818 \equiv 769 \times 1693 \quad \text{modulo } 2818.$$

This tells us that the inverse of 769 modulo 2818 is 1693, so we get, for the first plaintext block,

$$P \equiv C^{1693} \equiv 1592^{1693} \equiv 1615 \quad \text{modulo } 2819.$$

Aha! The number 1615 corresponds to the plaintext "po." Bob's complete decryption goes as follows:

ciphertext:	1592	783	2264	924	211	44	1220	1548
to the 1693rd power:	1615	2305	1820	1520	805	1605	1516	1205
split apart:	16, 15	23, 5	18, 20	15, 20	8, 5	16, 5	15, 16	12, 5
plaintext:	po	we	rt	ot	he	pe	op	le

The number $\bar{e}$ that Bob needs to decrypt is traditionally called d, for **decryption exponent**. So, to summarize, Alice and Bob need to pick a prime p larger than the largest possible plaintext number. They also need a key e such that the GCD of e and $p - 1$ is 1, so that e has an inverse modulo $p - 1$. Then Bob needs to calculate the number d that is the inverse of e modulo $p - 1$. Alice encrypts using the formula

$$C \equiv P^e \quad \text{modulo } p$$

and Bob decrypts using the formula

$$P \equiv C^d \quad \text{modulo } p.$$

This cipher is called the **Pohlig-Hellman exponentiation cipher**. It was invented by **Stephen Pohlig** and **Martin Hellman** in 1976 while they were working on the first public-key cryptography systems, which we shall explore in Chapter 7.

6.4 THE DISCRETE LOGARITHM PROBLEM

Now we can encrypt and decrypt using the Pohlig-Hellman cipher. What about Eve's methods of attack? The way to measure resistance to brute-force attacks is to see how many keys there are. The good keys are the numbers between 1 and $p - 1$ that don't share any factors with $p - 1$. If $p = 2819$, then $p - 1 = 2818 = 2 \times 1409$, and 1409 is prime. So e can be any number between 1 and 2818 that doesn't have a factor of 2 or 1409, which means any odd number except 1409. There are 1408 such numbers, so there are 1408 good keys. That's not a huge number, but all we have to do to get more is choose a larger modulus, which also lets us use a larger block size. So brute-force attacks aren't a big problem, and ciphertext-only frequency attacks can be defeated by using a large block size.

What about known-plaintext attacks? For us to consider a cipher to be resistant to known-plaintext attacks, it needs to be clearly harder for Eve to recover the key than it is for Alice to encrypt or Bob to decrypt. If it weren't for the modular arithmetic, recovering the key would be easy. In order to find the exponent of an exponential expression when you know the base, you take a logarithm. If $C = P^e$, then $e = \log_P C$. In this case, Eve would see that the plaintext is 1615 and the ciphertext is 1592. So she knows $1615^e = 1592$ and $e = \log_{1615} 1592$. However, $\log_{1615} 1592$ is approximately 0.9981, and once again the modular arithmetic has messed things up. The problem of finding a whole number e such that $C \equiv P^e$ modulo p is called the **discrete logarithm problem**, and this is what Eve needs to solve.

It's not clear that solving the discrete logarithm problem is in fact harder than encryption or decryption—if Eve has some examples of P and C, her first step is to guess p, which she can do fairly easily by looking at the largest ciphertext number in the message. Then she can multiply P by itself repeatedly modulo p until she gets C, keeping track of how many times it takes, and that will be e.

That seems remarkably like what Alice does to encrypt, right? The issue is that multiplying P by itself e times is actually *not* the best way for Alice to encrypt. Here's a better way.

Consider $e = 769$. I reminded you in Section 4.1 that 769 really means $7 \times 10 \times 10 + 6 \times 10 + 9$. So

$$P^{769} = P^{7 \times 10 \times 10 + 6 \times 10 + 9} = \left(\left(P^{10} \right)^{10} \right)^7 \left(P^{10} \right)^6 P^9.$$

If you count this out, you'll see that Alice needs only 46 multiplications, not 768. On the other hand, Eve will need all 768, since she doesn't know e beforehand, so she can't split it up this way. As of 2016, people have been working hard for more than 35 years to find a fast way to solve the discrete logarithm problem, and so far Eve is not even close to being able to keep up with Alice and Bob. On the other hand, no one has been able to prove that she can't, either. Like several other problems we shall see in the next few chapters, the discrete logarithm problem is one that we think is hard, but no one knows for sure. We will talk more about this problem in Section 7.2.

6.5 COMPOSITE MODULI

You might think it's kind of annoying to have to use a prime number as the modulus in the Pohlig-Hellman cipher. Round numbers are easier to work with, so maybe you'd rather use 3000 as the modulus when the block size is 2. Alternatively, maybe the extra numbers in between the largest block and the modulus were bothering you and you'd rather use a modulus of exactly 2626. These are **composite numbers**, because they are made up of more than one prime multiplied together.

Encryption using exponentiation is no problem with a composite modulus. For example, if Alice wants to send Bob a message using a modulus of 2626 and the same key, $e = 769$, as before, she converts the plaintext to numbers and raises them to the 769th power as before.

plaintext:	de	co	mp	os	in
numbers:	4, 5	3, 15	13, 16	15, 19	9, 14
together:	405	315	1316	1519	914
to the 769th power:	405	1667	1992	817	1148

plaintext:	gc	om	po	se	rs
numbers:	7, 3	15, 13	16, 15	19, 5	18, 19
together:	703	1513	1615	1905	1819
to the 769th power:	1405	603	1615	137	1819

Decryption, once again, is the problem, and this time Fermat's little theorem is not going to come to our rescue. We can see the problem if we try to go through an example similar to the one in Section 6.2. Instead of the 13-letter Hawaiian alphabet, we will use the 15-letter Maori alphabet. Note that 13 is prime, but $15 = 3 \times 5$ is composite. Since 15 is not prime, not every number between 1 and 14 will be a good key. The number 2 will be, though, since the GCD of 15 and 2 is 1.

plaintext	number	times 2	ciphertext
a	1	2	E
e	2	4	I
h	3	6	M
i	4	8	O
k	5	10	R
m	6	12	U
n	7	14	NG
o	8	1	A
p	9	3	H
r	10	5	K
t	11	7	N
u	12	9	P
w	13	11	T
ng	14	13	W
wh	15	15	WH

For the prime case, we multiplied all the numbers in the left column together and all the numbers in right column together, leaving out the number at the end of each column because it reduces to zero. If we do that here, we get

$$1 \times 2 \times 3 \times \cdots \times 14 \equiv (1 \times 2) \times (2 \times 2) \times (3 \times 2) \times \cdots \times (14 \times 2) \qquad \text{modulo } 15,$$

$$1 \times 2 \times 3 \times \cdots \times 14 \equiv (1 \times 2 \times 3 \times \cdots \times 14) \times 2^{14} \qquad \text{modulo } 15.$$

Now we want to cancel out $1 \times 2 \times 3 \times \cdots \times 14$ from each side, but unfortunately not all of those numbers have multiplicative inverses. Only the ones that have a GCD of 1 with 15 have inverses, and those are the only ones we can cancel out.

This is just like the problem with the bad keys. Since $15 = 3 \times 5$, we need to start over, leaving out the numbers that are multiples of 3, or 5, or both.

plaintext	number	times 2	ciphertext
a	1	2	E
e	2	4	I
i	4	8	O
n	7	14	NG
o	8	1	A
t	11	7	N
w	13	11	T
ng	14	13	W

The numbers on the left-hand side are still the same as the numbers on the right-hand side, but in a different order. This kind of makes sense, since if a number on the left was a multiple of 3 or 5, we would expect 2 times it to be one also. So we crossed out the same numbers from each side.

If we try multiplying the columns again, we get

$$1 \times 2 \times 4 \times 7 \times 8 \times 11 \times 13 \times 14$$

$$\equiv (1 \times 2) \times (2 \times 2) \times (4 \times 2) \times \cdots \times (14 \times 2) \qquad \text{modulo } 15$$

$$1 \times 2 \times 4 \times 7 \times 8 \times 11 \times 13 \times 14$$

$$\equiv (1 \times 2 \times 4 \times 7 \times 8 \times 11 \times 13 \times 14) \times 2^8 \qquad \text{modulo } 15.$$

And now we *can* cancel $1 \times 2 \times 4 \times 7 \times 8 \times 11 \times 13 \times 14$, for instance, by multiplying by the inverse of each of them, so finally we get

$$1 \equiv 2^8 \qquad \text{modulo } 15.$$

Once again, the choice of 2 isn't important; any good key will do. But the choice of 15 clearly does make a difference—the 15 in the modulus produced an 8 in the exponent, and if we figure out how that happened, we'll be well on our way to figuring out how Bob can decrypt his message.

6.6 THE EULER PHI FUNCTION

Let's take a closer look at where the 8 came from in the last example. We listed all of the numbers from 1 to 15,

1, 2, 3, 4, 5, 6, 7, 8, 9, 10, 11, 12, 13, 14, 15,

and we got rid of all those that did *not* have a GCD of 1 with 15:

1, 2, ~~3~~, 4, ~~5~~, ~~6~~, 7, 8, ~~9~~, ~~10~~, 11, ~~12~~, 13, 14, ~~15~~.

This leaves 8 numbers behind. In other words, 8 is the number of whole numbers less than or equal to 15 that have a GCD of 1 with 15.

In general, we can define $\phi(n)$ (that's the Greek letter phi) to be the number of positive whole numbers less than or equal to n that have a GCD of 1 with n. For example, we have

n	$\phi(n)$	n	$\phi(n)$
1	1	11	10
2	1	12	4
3	2	13	12
4	2	14	6
5	4	15	8
6	2	16	8
7	6	17	16
8	4	18	6
9	6	19	18
10	4	20	8

We already know what $\phi(n)$ should be if n is prime, since every whole number will be counted except the number itself. Other than that, the function seems pretty mysterious.

The person who figured out the pattern was the great genius mathematician of the eighteenth century, in the way Gauss was for the nineteenth and Fermat was in the seventeenth. His name was **Leonhard Euler**, and while he was born in Switzerland, he did most of his work at prestigious scientific academies in Russia and Prussia. In 1736 he was the first one to publish a proof of Fermat's little theorem, and he later published several more. In one of these papers, in 1763, he introduced the function we now write as $\phi(n)$ and call the **Euler phi function**. And, he uses this function to prove what we now call the **Euler-Fermat theorem**.

> **Theorem** (The Euler-Fermat Theorem) *For any positive whole number n and any k between 1 and n such that the GCD of n and k is 1,*
>
> $$k^{\phi(n)} \equiv 1 \quad \text{modulo } n.$$

If n is a prime number, then $\phi(n)$ will be $n - 1$, and we have Fermat's little theorem again. And if n is 15, then $\phi(n)$ is 8 and we have our example. Now we know what the Euler phi function is and we have some idea what it's good for. But if we have to calculate $\phi(n)$ by checking a GCD for every number between 1 and n, that's going to be a very slow process.

Luckily, there's an easier way. Let's go back to our example and watch a little more closely as we cross out the "bad keys." The divisors of 15 are 1, 3, 5, and 15, so we know we have to cross out the numbers that are multiples of 3:

$$
\begin{array}{ccc}
1 & 2 & \cancel{3} \\
4 & 5 & \cancel{6} \\
7 & 8 & \cancel{9} \\
10 & 11 & \cancel{12} \\
13 & 14 & \cancel{15}
\end{array}
$$

Since we are crossing out every third number, there are $15/3 = 5$ crossed-out numbers. We also have to cross out multiples of 5:

$$
\begin{array}{ccccc}
1 & 2 & 3 & 4 & \cancel{5} \\
6 & 7 & 8 & 9 & \cancel{10} \\
11 & 12 & 13 & 14 & \cancel{15}
\end{array}
$$

This time we have crossed out every fifth number, and there are $15/5 = 3$ crossed-out numbers. We don't have to cross out multiples of 15 because any multiple of 15 is a multiple of 3 (and of 5), so it's already been crossed out.

So how many numbers are not crossed out? It should be $15 - 3 - 5 = 7$, but when we did it before, there were 8. Do you see why? It's because we crossed out 15, which is a multiple of both 3 and 5, twice. So we have to add it back in, giving us $15 - 3 - 5 + 1 = 8$ numbers not crossed out. In general, we have this formula if p and q are two different prime numbers:

$$\phi(pq) = pq - p - q + 1.$$

With a little bit of algebra, we can rearrange that into the more common form:

$$\phi(pq) = (p-1)(q-1).$$

Now what about Bob and our cipher? In that case we had $n = 2626 = 2 \times 13 \times 101$, and if you work through all the crossings out, you will see that

$$\frac{2626}{2} + \frac{2626}{13} + \frac{2626}{101} = 13 \times 101 + 2 \times 101 + 2 \times 13$$

numbers get crossed out, but

$$\frac{2626}{2 \times 13} + \frac{2626}{2 \times 101} + \frac{2626}{13 \times 101} = 101 + 13 + 2$$

got crossed out twice and have to be added back. However 1 number, namely, 2626, has now been crossed out 3 times and added back in 3 times, so it has to come out again. In other words:

$$\phi(2626) = 2626 - 2 \times 13 - 2 \times 101 - 13 \times 101$$
$$+ 2 + 13 + 101 - 1 = 1200.$$

In general, if p, q, and r are three different prime numbers,

$$\phi(pqr) = pqr - pq - pr - qr + p + q + r - 1$$
$$= (p-1)(q-1)(r-1).$$

And you can probably see the pattern for any product of different primes.

<h3 style="text-align:center">6.7 DECRYPTION WITH COMPOSITE MODULI</h3>

Now we should be able to figure out how to decrypt a message encrypted using the Pohlig-Hellman cipher and a composite modulus. Once we know $\phi(n)$, the Euler-Fermat theorem tells us that

$$P^{\phi(n)} \equiv 1 \equiv P^0 \quad \text{modulo } n.$$

This means that if we are looking at an equation modulo n, then we can treat the exponent as if we were working modulo $\phi(n)$. This is the equivalent of what we did with Fermat's little theorem earlier. In the case of $n = 2626$, we have

$$P^{1200} \equiv P^0 \quad \text{modulo } 2626.$$

If the encryption exponent is $e = 769$, the decryption exponent will be the inverse of e modulo 1200—assuming there is one. Remember that for e to have an inverse modulo 1200, the GCD of e and 1200 needs to be 1. Otherwise, e is a bad key, and Alice shouldn't have picked it in the first place.

So Bob's first step in decrypting the message is to use the Euclidean algorithm to find the inverse of $e = 769$ modulo 1200.

$$1200 = 769 \times 1 + 431 \qquad 431 = 1200 - (769 \times 1)$$

$$769 = 431 \times 1 + 338 \qquad 338 = 769 - (431 \times 1)$$
$$= (769 \times 2) - (1200 \times 1)$$

$$431 = 338 \times 1 + 93 \qquad 93 = 431 - (338 \times 1)$$
$$= (1200 \times 2) - (769 \times 3)$$

$$338 = 93 \times 3 + 59 \qquad 59 = 338 - (93 \times 3)$$
$$= (769 \times 11) - (1200 \times 7)$$

$$93 = 59 \times 1 + 34 \qquad 34 = 93 - (59 \times 1)$$
$$= (1200 \times 9) - (769 \times 14)$$

$$59 = 34 \times 1 + 25 \qquad 25 = 59 - (34 \times 1)$$
$$= (769 \times 25) - (1200 \times 16)$$

$$34 = 25 \times 1 + 9 \qquad 9 = 34 - (25 \times 1)$$
$$= (1200 \times 25) - (769 \times 39)$$

$$25 = 9 \times 2 + 7 \qquad 7 = 25 - (9 \times 2)$$
$$= (769 \times 103) - (1200 \times 66)$$

$$9 = 7 \times 1 + 2 \qquad 2 = 9 - (7 \times 1)$$
$$= (1200 \times 91) - (769 \times 142)$$

$$7 = 2 \times 3 + 1 \qquad 1 = 7 - (2 \times 3)$$
$$= (769 \times 529) - (1200 \times 339)$$

so

$$1 = (769 \times 529) + (1200 \times -339)$$

and

$$1 \equiv 769 \times 529 \quad \text{modulo } 1200.$$

The decryption exponent is $d = 529$, and the decryption goes as follows:

ciphertext:	405	1667	1992	817	1148
to the 529th power:	405	315	1316	1519	914
split apart:	4, 5	3, 15	13, 16	15, 19	9, 14
plaintext:	de	co	mp	os	in

ciphertext:	1405	603	1615	137	1819
to the 529th power:	703	1513	1615	1905	1819
split apart:	7, 3	15, 13	16, 15	19, 5	18, 19
plaintext:	gc	om	po	se	rs

Actually, I have cheated a bit. The Euler-Fermat theorem guarantees only that the exponents behave like we want if the GCD of P and n is 1. This isn't true for some of our plaintext blocks, such as 1316; in fact the GCD of 1316 and 2626 is 2. It turns out that if n is a product of *different* primes, then decryption *does* always work properly, but I'm not going to try to justify that in this book. If you want to see the proof, I've put some references in the endnotes.

■ ■ ■ SIDEBAR 6.1. FEE-FI-FO-FUM ■ ■ ■

If n is a product of primes that appear multiple times, then we can still find a formula for $\phi(n)$, even though we won't be able to easily use the Pohlig-Hellman cipher. Suppose that $n = 12 = 2^2 \times 3$. The divisors of 12 are 1, 2, 3, 4, 6, and 12. When we are crossing out bad keys, we need to cross out multiples of 2 and multiples of 3, and this will also eliminate multiples of 4, 6, and 12. First we cross out all the multiples of 2:

$$
\begin{array}{cc}
1 & \cancel{2} \\
3 & \cancel{4} \\
5 & \cancel{6} \\
7 & \cancel{8} \\
9 & \cancel{10} \\
11 & \cancel{12}
\end{array}
$$

There are $12/2 = 6$ of those. And then we cross out all the multiples of 3:

$$
\begin{array}{ccc}
1 & 2 & \cancel{3} \\
4 & 5 & \cancel{6} \\
7 & 8 & \cancel{9} \\
10 & 11 & \cancel{12}
\end{array}
$$

There are $12/3 = 4$ of those. But both 12 and 6 have been crossed out twice, since they are both divisible by 2 and by 3, so we have to add them back. Thus $\phi(n) = 12 - 6 - 4 + 2 = 4$. In general, we have this formula if p and q are different prime numbers:

$$
\phi(p^a q^b) = p^a q^b - \frac{p^a q^b}{p} - \frac{p^a q^b}{q} + \frac{p^a q^b}{pq}.
$$

And we can rearrange that into the more common form:

$$
\phi(p^a q^b) = \left(p^a - \frac{p^a}{p}\right)\left(q^b - \frac{q^b}{q}\right) = \left(p^a - p^{a-1}\right)\left(q^b - q^{b-1}\right).
$$

If $n = p^b q^b r^c$ is a product containing three different primes, then

$$
\phi(p^a q^b r^c) = \left(p^a - p^{a-1}\right)\left(q^b - q^{b-1}\right)\left(r^c - r^{c-1}\right),
$$

and so on.

For instance, if $n = 3000 = 2^3 \times 3 \times 5^3$, then

$$
\phi(3000) = (2^3 - 2^2) \times (3 - 1) \times (5^3 - 5^2) = 800.
$$

Alice can encrypt a message with $e = 769$ and $n = 3000$:

plaintext:	sy	st	em	er	ro	rx
numbers:	19, 25	19, 20	5, 13	5, 18	18, 15	18, 24
together:	1925	1920	513	518	1815	1824
to the 769th power:	125	0	2073	368	375	2424

If Bob uses the Euclidean algorithm, he will find that the inverse of 769 modulo 800 is 129, so he attempts to decrypt using $d = 129$:

	ciphertext:	125	0	2073	368	375	2424
to the 129th power:		125	0	513	2768	375	1824
split apart:		1, 25	0, 0	5, 13	27, 68	3, 75	18, 24
plaintext?:		ay	??	em	??	c?	rx

Remember that the Euler-Fermat theorem does not guarantee that decryption will work properly unless the GCD of P and n is 1. Two of the blocks come through all right: 513, which has a GCD of 1 with 3000, and 1824, which has a GCD of $24 = 2^3 \times 3$ with 3000 but works anyway. However, most of the blocks come through with incorrect letters or numbers that do not correspond to letters at all. You might hope that reducing the individual 2-digit numbers modulo 26 would help, but it doesn't. If the system was working correctly, Bob would get the same numbers that Alice started with. The general formula for $\phi(n)$ is useful for other situations but not really for the Pohlig-Hellman cipher.

6.8 LOOKING FORWARD

So, you may ask, are exponentiation ciphers the state of the art in modern ciphers? As it happens, they aren't actually used that much. Ciphers such as AES appear to have just as good resistance to attacks and work much faster, even with the trick that we have noted for speeding up exponentiation. Instead, we shall see in Chapters 7 and 8 that the ideas used in this cipher, and especially the hardness of the discrete logarithm problem, turn out to be very important to a very exciting idea known as public-key cryptography.

When Pohlig and Hellman were developing their cipher, they briefly considered using composite moduli but rejected it on the grounds that the convenience wasn't worth the complication. They missed a bet, because exponentiation with composite moduli is a key ingredient in the very important system that we will see in Section 7.4.

On the other hand, Pohlig and Hellman also figured out how to use their cipher with the sort of finite field arithmetic we saw in Section 4.5. This eventually turned out to be another important idea because finite-field arithmetic modulo 2 is a convenient way for computers to manipulate bits, as we also saw in that section.

7

Public-Key Ciphers

Throughout our discussion so far, we have tacitly assumed a few things about Alice, Bob, and Eve. One is that before Alice and Bob start sending messages, they need to get together somewhere where Eve can't overhear them—or otherwise find a method of communication that Eve can't listen in on—in order to agree on the key they will use. This seems both reasonable and necessary to such a degree that for more than 2000 years, no one seriously questioned it. It might be inconvenient for Alice and Bob to meet securely, but they have control over when it happens and it doesn't really have to last very long, so in most situations it is feasible. Occasionally in history there have been cases where Alice and Bob didn't have anything prearranged; in an emergency Alice sent a secret message anyway, in the hopes that Bob was smart enough to figure it out and Eve wasn't. That's a big risk though and hardly the basis of a good secure system.

In the fall of 1974, **Ralph Merkle** was finishing his last semester as an undergraduate at the University of California, Berkeley, and taking a class in computer security. There was a little bit of cryptography in the class, but DES had not yet been officially announced, and there wasn't an awful lot of cryptography to discuss. What there was caught Merkle's attention, however, and he began wondering if there was a way around the assumption that everyone had always made. Was it possible for Alice to send Bob a message *without* having them agree on a key beforehand? Obviously there would have to be a key, but maybe Alice and Bob could agree on it through some process that Eve couldn't understand, even if she could overhear it. Merkle submitted a preliminary idea, which he later described as "simple, but inefficient," as one of two term-project proposals for the class. The idea really took only a page and a half, but Merkle used four and a half more pages trying to justify the importance of the problem, explain the difficulties, and improve on the initial concept. And he couldn't cite any sources, since apparently no one had ever thought about the idea before. It's not that surprising that his professor was hopelessly confused and suggested that Merkle work on his second-choice project. Instead Merkle dropped the class, but he continued to work on the project.

Merkle's idea, which is now commonly known as **Merkle's puzzles**, went through several revisions, but here is the version that was finally published. Alice starts by creating a large number of encrypted messages (the puzzles) and sends them to Bob, as in Figure 7.1. The encryption function should be chosen so that breaking each puzzle by brute force is "tedious, but quite possible." Merkle suggested using a cipher with a 128-bit key, specifying only a small fraction of all possible keys that would be used. We will use an additive cipher in a very small example:

VGPVY	QUGXG	PVYGP	VAQPG	UKZVG
GPUGX	GPVGG	PBTPU	XSNHT	JZFEB
GJBAV	ARSVI	RFRIR	AGRRA	GJRYI
RFRIR	AGRRA	VTDHC	BMABD	QMPUP
AFSPO	JOFUF	FOUFO	TFWFO	UXFOU
ZGJWF	TFWFO	UFFOI	RCXJQ	EHHZF
JIZJI	ZNDSO	RZIOT	ADAOZ	ZINZQ
ZIOZZ	IWOPL	KDWJH	SEXRJ	IKAVV
YBJSY	DSNSJ	YJJSY	BJSYD	KNAJX
JAJSK	TZWXJ	AJSYJ	JSFNY	UZAKM
QCTCL	RFPCC	RUCLR	WDMSP	RCCLD
GDRCC	LQCTC	LRCCL	JLXUW	HAYDT
ADLUA	FMVBY	ALUVU	LVULZ	LCLUZ
LCLUA	LLUGE	AMPWB	PSEQG	IKDSV
JXHUU	VYLUJ	XHUUJ	UDDYD	UIULU
DJUUD	AUTRC	SGBOD	ALQUS	ERDWN
RDUDM	SDDMS	VDMSX	RDUDM	SDDMM
HMDSD	DMRHW	SDDMR	DUDMS	DDMAW
BEMTD	MBEMV	BGBPZ	MMMQO	PBMMV
AMDMV	NQDMA	MDMVB	MMVUR	YCEZC

Alice   Bob

Makes puzzles

Puzzles, check number →

Figure 7.1. The beginning of Merkle's puzzles.

Alice explains to Bob that each puzzle consists of three sets of numbers that Alice chose at random, all encrypted with the same key. The first number is an ID number to identify the puzzle. The second set of numbers is a secret key from a secure cipher, which Alice and Bob could actually use to communicate. Merkle suggested a 128-bit cipher again, this time allowing all possible keys. We will use a 2 × 2 Hill cipher for our example. The last number is the same for all puzzles and is a check so that Bob can make sure he has solved the puzzle correctly. In our example, the check number is seventeen. Finally, the puzzles are padded with random nulls so that they are all the same length.

Bob picks one of the puzzles at random and solves it by a brute-force search, using the check number to make sure he did it correctly. He then sends Alice the ID number encrypted in the puzzle, as shown in Figure 7.2. For example, if Bob's solution to one of the puzzles is

twent   ynine   teent   wenty   fives
evenf   ourse   vente   enait   puvfh

then he knows the ID number is twenty and the secret key is 19, 25, 7, 4. He sends Alice twenty.

Alice

Bob

Makes puzzles

Puzzles, check number →

Picks a puzzle
↓
"Aha! I have the ID number
and the secret key!"

← ID number

Figure 7.2. Bob solves the puzzle.

Alice has a list of the plaintexts of the puzzles, sorted by ID number:

ID	secret key				check
zero	nineteen	ten	seven	twentyfive	seventeen
one	one	six	twenty	fifteen	seventeen
two	nine	five	seventeen	twelve	seventeen
three	five	three	ten	nine	seventeen
seven	three	twenty	fourteen	fifteen	seventeen
ten	two	seven	twentyone	sixteen	seventeen
twelve	twentythree	eighteen	seven	five	seventeen
seventeen	twenty	seventeen	nineteen	sixteen	seventeen
twenty	nineteen	twentyfive	seven	four	seventeen
twentyfour	ten	one	one	seven	seventeen

So, she can also look up the secret key and find that it is $19, 25, 7, 4$. Now Alice and Bob both know a secret key to a secure cipher (Figure 7.3), and they can start sending encrypted messages.

Can Eve figure out the secret key? She's been eavesdropping on Alice and Bob's conversation as usual, so let's see what she has overheard. As Figure 7.4 shows, she has the ciphertexts of all the puzzles and the check number. She doesn't know which puzzle Bob picked, but she does know that the ID number was twenty. And she doesn't have

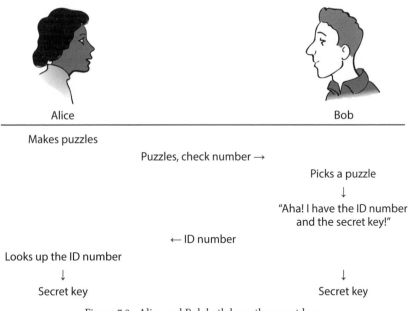

Alice   Bob

Makes puzzles

Puzzles, check number →

Picks a puzzle
↓
"Aha! I have the ID number
and the secret key!"

← ID number

Looks up the ID number
↓   ↓
Secret key   Secret key

Figure 7.3. Alice and Bob both have the secret key.

Alice's list of plaintexts. It looks like she has to solve *all* the puzzles before she can figure out which one Bob picked and get the secret key. This, of course, is possible, but it will take her a lot longer than the procedure took Alice or Bob. Alice had to encrypt 10 puzzles. Bob had to decrypt 1 puzzle (at worst) 25 times in his brute-force search. But Eve has to decrypt (at worst) 10 puzzles 25 times each, or 250 total decryptions. A modern (2016) desktop computer can do something very roughly in the range of 10 million puzzle encryptions or decryptions per second. If Alice generates 100 million puzzles, each with 100 million possible keys, then it will take her computer and Bob's computer less than a minute each for what they need to do. Eve, on the other hand, will need to do 10,000 million million decryptions, which will take her computer about 1000 million seconds, or about 32 years. If Alice and Bob are concerned that Eve has a faster computer, they just need to make more puzzles with more possible keys.

The study of systems that allow Alice and Bob to communicate securely without an initial secure meeting is now known as **public-key cryptography**. Merkle's puzzles is a public-key system, but it is not

Alice	Eve	Bob

Makes puzzles

Puzzles, check number →

Picks a puzzle

↓

"Aha! I have the ID number
and the secret key!"

"Which puzzle do I solve?"

← ID number

"That doesn't help?"

Looks up the ID number

↓

Secret key

↓

Secret key

"I still don't know!"

Figure 7.4. Eve can't keep up.

itself a code or cipher. Neither Alice nor Bob can predict what the final secret key is going to be, so they can't use it as a secret message by itself. Rather, what we have is a **key-agreement** system. Key-agreement systems are one major category of public-key systems, although we will see others, including some that actually are ciphers.

Merkle recognized from the beginning that his scheme was less than ideal. The puzzles take Alice a significant amount of time to set up, an even more significant amount of space to store, and a still more significant amount of time and/or data transfer capacity to transmit. Likewise Bob needs a significant amount of time to solve the puzzles, and Alice's and Bob's time commitment goes up at about the same rate that Eve's does. If Alice and Bob want to force Eve to spend twice as long, one of them has to spend twice as long. Merkle knew that if a key agreement system could be developed that forced Eve's required time to grow at a much larger rate compared to Alice and Bob, it would be much more useful in practice.

7.2 DIFFIE-HELLMAN KEY AGREEMENT

While Ralph Merkle was trying to get someone to take his idea seriously, two other people were also thinking about public-key ideas. In 1972, **Whitfield Diffie** was a researcher at the Stanford Artificial Intelligence Lab when his girlfriend, also a researcher at the lab, started working on a project related to stream ciphers. Diffie became interested in cryptography and then obsessed with it. Between 1972 and 1974 he drove back and forth across the United States looking for the few experts on cryptography who were not working for the NSA and would talk to him about the subject. In 1974 Diffie heard that someone back at Stanford was thinking about the same sorts of questions that he was. This turned out to be Martin Hellman, whom we met briefly in Section 6.3. Hellman was a former researcher at IBM who had gotten interested in cryptography there and at MIT. In 1971 he became an assistant professor at Stanford, which is where Diffie connected with him in 1974.

As Diffie later put it, his and Hellman's discovery was the result of "two problems and a misunderstanding." The first problem was the same one that Merkle was considering: how can two people who have never met before carry on a secure conversation? The second problem was that of authentication, or "digital signatures": how could the recipient of a digital message assure himself or herself and others that the sender was who the message said it was? We will postpone the solution until Section 8.4, but it's worth noting that this is not really an issue with traditional cryptography. The mere possession of the key to a cipher or MAC serves as some assurance that the sender of the message is a trusted member of your organization. The misunderstanding was this: Diffie and Hellman assumed that users of the cryptographic system would not want to have to trust any third party in order to complete their connection. In Diffie's later words:

> What good would it do to develop impenetrable cryptosystems, I reasoned, if their users were forced to share their keys with a key distribution center that could be compromised by either burglary or subpoena.

Perhaps it's not surprising that for thousands of years everyone had assumed that public-key cryptography was not possible, and all of a sudden in the early 1970s these three people independently started thinking about it. The microcomputer revolution was about to begin. People in the know were already thinking that someday ordinary people were going to be using them for communication, commerce, and who knew what else. At the same time, the United States was in the midst of the counterculture movement and the Watergate scandal. Distrust of government and other large organizations was strong, and privacy and self-reliance were on the minds of many, definitely including Whitfield Diffie and Martin Hellman.

Diffie and Hellman wrote a paper explaining, among other things, how useful public-key cryptography could be and some possible ways that one might be able to go about it. But they admitted that they didn't really know how to make it work. In early 1976 a draft copy of Diffie and Hellman's paper managed to find its way into the hands of Ralph Merkle. Merkle, excited to find someone else who understood what he was working on, sent Diffie and Hellman a copy of the paper he was writing on Merkle's puzzles and expressed an interest in working together to improve his scheme. Diffie, Hellman, and Merkle exchanged letters through the summer of 1976, and Diffie and Hellman started thinking about key-agreement systems as a particular way of implementing their ideas.

One idea that both Diffie and Merkle had been thinking about for several years was the use of things called **one-way functions**, which are easy to compute in one direction and hard to compute in the other. In fact, we've already seen in Section 6.4 that the exponentiation function, which takes e to P^e modulo p is easy to calculate, but it's hard to find e even if you know P^e, P, and p—this is the discrete logarithm problem from Section 6.4. So this function is an example of a one-way function. Bob's part of Merkle's puzzles can also be thought of as a one-way function: taking a ciphertext and extracting the ID number is (relatively) easy, but Eve's job of taking the ID number and trying to figure out which ciphertext it corresponds to is hard. Diffie, Hellman, and Merkle knew of these examples and a few others, and one day in the summer of 1976 Hellman managed to put it all together and make the exponentiation function into the system that is now called **Diffie-Hellman key agreement**. The paper announcing the new system has the memorable title "New Directions in Cryptography" and starts, only a little

Alice   Bob

Picks secret a Picks secret b

Figure 7.5. The beginning of Diffie-Hellman.

melodramatically, with the sentence, "We stand today on the brink of a revolution in cryptography."

Like Merkle's puzzles, the Diffie-Hellman system starts out with Alice and Bob setting out some ground rules. They need to pick a very large prime number p. As of 2015, experts were recommending 600 digits or more for acceptable security. Otherwise the discrete logarithm problem is not hard enough. They also need to find a **generator** modulo p, which is a number g between 1 and $p-1$ such that the numbers g, g^2, g^3, ..., g^{p-1}, taken modulo p, cover all of the possible numbers between 1 and $p-1$. For example, 3 is a generator modulo 7 because the numbers

$$3^1 = 3,\ 3^2 = 9,\ 3^3 = 27,\ 3^4 = 81,\ 3^5 = 243,\ 3^6 = 729$$

turn out to be

$$3,\ 2,\ 6,\ 4,\ 5,\ 1 \qquad \text{modulo } 7,$$

which is all the possibilities. Conveniently, it turns out that every prime has at least one of these generators, and they are not especially hard to find. Furthermore, p and g do not have to be kept secret, so it's fine to just look them up in a table.

Now, as in Merkle's puzzles, Alice picks some secret information. In this case, it's a number a between 1 and $p-1$. Unlike in Merkle's puzzles, Bob also picks a secret number b between 1 and $p-1$. This gives us the situation shown in Figure 7.5.

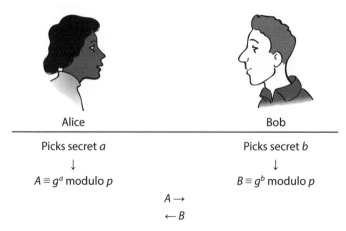

Figure 7.6. Alice and Bob exchange public information.

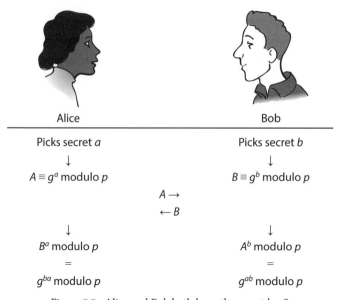

Figure 7.7. Alice and Bob both have the secret key?

Then Alice computes $A \equiv g^a$ modulo p. Bob computes $B \equiv g^b$ modulo p. Alice sends A to Bob, and Bob sends B to Alice, as shown in Figure 7.6.

Finally, Alice computes B^a modulo p and Bob computes A^b modulo p, as shown in Figure 7.7. Alice has $B^a \equiv (g^b)^a \equiv g^{ba}$ modulo p and Bob

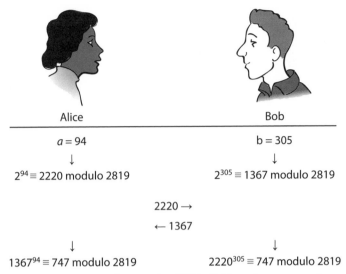

Alice	Bob
$a = 94$	$b = 305$
↓	↓
$2^{94} \equiv 2220$ modulo 2819	$2^{305} \equiv 1367$ modulo 2819

$$2220 \rightarrow$$
$$\leftarrow 1367$$

↓	↓
$1367^{94} \equiv 747$ modulo 2819	$2220^{305} \equiv 747$ modulo 2819

Figure 7.8. A specific example of Diffie-Hellman key agreement.

has $A^b \equiv (g^a)^b \equiv g^{ab}$ modulo p. But $ab = ba$, so these are the same. Now Alice and Bob share a piece of secret information that they can use as a key for some secure cipher.

For example, suppose Alice and Bob want to agree on a secret key for the Pohlig-Hellman exponentiation cipher, as set out in Section 6.1. The key needs to be a number between 1 and 2818, so for this example they will use $p = 2819$ in the Diffie-Hellman system. It happens that 2 is a generator modulo 2819, so Alice and Bob decide to use that. Alice picks a secret number, say 94, and Bob picks a secret number, say 305. Then the system proceeds as shown in Figure 7.8.

Now Alice and Bob both know the secret key, 747, and they can use it for their exponentiation cipher. It's worth noting once again that Alice and Bob have no idea what the secret key will end up being. In particular, it might not even be a good key for the exponentiation cipher. If that is the case, they will both find out quickly. Then all they have to do is try again with new secret numbers until they get a good key.

How hard will it be for Eve to get the secret key? She knows g and p, because Alice and Bob agreed on them over an insecure communications line. She doesn't know a or b, but she knows g^a modulo

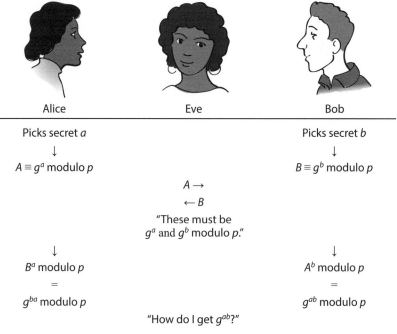

Alice	Eve	Bob
Picks secret a		Picks secret b
↓		↓
$A \equiv g^a$ modulo p		$B \equiv g^b$ modulo p
	$A \rightarrow$	
	$\leftarrow B$	
	"These must be g^a and g^b modulo p."	
↓		↓
B^a modulo p		A^b modulo p
=		=
g^{ba} modulo p		g^{ab} modulo p
	"How do I get g^{ab}?"	

Figure 7.9. Can Eve figure out the secret key?

p and g^b modulo p, as shown in Figure 7.9. The problem of getting the secret key g^{ab} modulo p from g^a and g^b modulo p is called the **Diffie-Hellman problem**. If Eve could figure out a or b, she could get the secret key, but that would require solving the discrete logarithm problem, and as we said in Section 6.4, that seems to be hard. Maybe there's another way to quickly solve the Diffie-Hellman problem—but once again, people have been trying for 35 years with no success. So far, the Diffie-Hellman problem is another one that we think is hard, but no one knows for sure.

The record as of June 2016 for finding a discrete logarithm modulo a large prime p was for the prime

$$p = \lfloor 2^{766}\pi \rfloor + 62762,$$

which is 232 digits, or 768 bits, long. The challenge was to take the logarithm of

$$y = \lfloor 2^{766}e \rfloor$$

with respect to the generator $g = 11$, and the result was announced on June 16, 2016, by a group of researchers at the University of Leipzig and the École Polytechnique Fédérale de Lausanne in Switzerland. The project took about 16 months to complete.

An important use of Diffie-Hellman is as part of the security system in one of the common types of virtual private networks. Virtual private networks, or VPNs, are systems designed to let members of an organization securely access the organization's network even from a location where you can't be sure someone isn't tapping the Internet connection. As I write this, a new version of the system which controls how data flows across the Internet is being slowly deployed. The new system, known as IPv6, is supposed to use that same Diffie-Hellman-based security much more extensively to protect the communications that control the Internet itself, as well as the messages of ordinary users.

7.3 ASYMMETRIC-KEY CRYPTOGRAPHY

Let's recap where we are historically. By 1976, Diffie and Hellman, with some help from Merkle, had come up with a practical public-key key agreement system. They were still working on another angle, however. One day in the summer of 1975, before Ralph Merkle had gotten them starting thinking about key agreement, Diffie had a crucial insight about another sort of system entirely. Traditional cryptography is symmetric in that Alice and Bob have essentially the same key information available. In many cases, the key Alice uses to encrypt and the key Bob uses to decrypt are the same, as in DES, AES, or shift register ciphers, among others. In other cases, Alice uses a key to encrypt and Bob uses some sort of inverse of the key to decrypt, as in the additive, multiplicative, and exponentiation ciphers, among others. There are two versions of the key, but if you know the encryption version, you can easily find its inverse, and vice versa. These systems are now called **symmetric-key systems**.

Diffie's new idea was to have an **asymmetric-key system**. Each party has two keys, an **encryption key** and a **decryption key**. This time, the relationship between the keys is such that even if you know the encryption key it is very difficult to find the decryption key. This is the asymmetry: if Alice knows only the encryption key and Bob knows

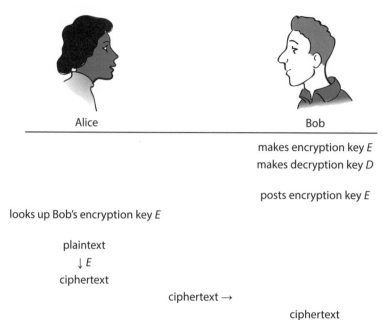

Figure 7.10. Asymmetric-key cryptography.

only the decryption key, then Alice can only encrypt and Bob can only decrypt. Of course, these keys don't appear by magic. At some point someone has to know both keys. Figure 7.10 shows how the system generally works in practice. Bob creates both an encryption key and a corresponding decryption key from some secret information. He posts the encryption key in a public place, such as a Web site, and keeps the decryption key secret. (For this reason the encryption key is often also called a **public key**, and the decryption key is also called a **private key**.)

When Alice wants to send Bob a message, she looks up his encryption key and uses it to encrypt the message. When she sends it to Bob, he can decrypt it using his decryption key, but no one else can decrypt it. Notice that Alice can't even decrypt her own message! If she loses the plaintext, she's out of luck, just the same as Eve is.

There have been lots of analogies proposed for asymmetric-key cryptography, going back to Diffie and Hellman themselves. My favorite

Figure 7.11. A nonmathematical asymmetric-key cryptographic system.

is a locked door with a mail slot in it, as shown in Figure 7.11. If Bob publishes the address of his door (the encryption key), then anyone can put a message through the slot. But only Bob has the key to the door (the decryption key), so only Bob can get the message and read it. Once Alice puts the message through the slot, even she can't get at it.

Unfortunately, Diffie and Hellman had only a vague idea how to actually make a system that would allow asymmetric encryption and decryption keys. They had one-way functions, which were easy to compute in one direction and hard in the other. But they needed more. They needed a function that would be easy for Alice to compute in one direction (encryption using the encryption key) and hard for Eve to compute in the other direction (cryptanalysis using the encryption key). *But the hard direction needed to be easy* for Bob to compute using an extra piece of secret information (decryption using the decryption key). This would be a **trap-door one-way function**—the extra piece of secret information would act like a trap door, letting Bob get at the plaintext using a hidden passage. Furthermore, it must be hard for Eve to compute the trap-door information from the rest of the system. The function must be one way in that sense as well.

The 1976 paper of Diffie and Hellman's that Merkle saw sketched out the basic idea and some ways that one might possibly be able to produce a trap-door one-way function, but neither that paper nor "New Directions in Cryptography" had a practical asymmetric-key system. In 1977 Merkle and Hellman would invent the first such system, the "knapsack cipher," but that system eventually turned out to have a flaw that made it insecure. The honor of developing the first successful asymmetric-key system would go to someone else.

7.4 RSA

At this point, developments in public-key cryptography were moving incredibly fast. Toward the end of 1976, "New Directions in Cryptography" came into the hands of **Ron Rivest**, an assistant professor of computer science at MIT. Rivest recruited two colleagues at MIT, the theoretically inclined **Leonard Adleman** and a visiting professor from Israel, **Adi Shamir**. Rivest and Shamir were immediately excited about the prospect of an asymmetric-key system; Adleman was less so. They soon settled into a pattern where Rivest and Shamir would come up with a scheme and Adleman would break it. At first, the breaks came almost immediately, but after about 32 rounds of this, Rivest and Shamir came up with something that took Adleman all night to find the flaw. From then on, they were all in it together.

By this time Rivest, Shamir, and Adleman had started to look at a one-way function different than Diffie and Hellman's exponentiation. The easy direction takes two large numbers and multiplies them together. The hard direction is the problem of factoring, that is, taking a large number and finding its factors. We factored small numbers in Section 1.3 and it probably didn't seem very hard, but for very large numbers, we will see that factoring can be extremely difficult.

What the factoring function was missing was a trap door, and Rivest, Shamir, and Adleman didn't immediately see how to build in one. On April 3, 1977, the three of them went to a Passover seder at the home of a graduate student. As is traditional, a considerable amount of wine was drunk, and Rivest and his wife arrived home rather late. While his wife got ready for bed, Ron Rivest lay down on the couch and thought about the problem. Before he went to bed he had made the crucial breakthrough in what would become known as the **RSA cryptosystem**: the one-way function of factoring would be the trap door in the one-way function of exponentiation.

Here's how Bob sets up the asymmetric keys. He starts by selecting two different very large prime numbers, usually called p and q, which will be the secret trap-door information. The product of those numbers, called n, will be the modulus in the composite version of the exponentiation cipher (Section 6.5). Current thinking is that n should be about the same size as the modulus you would use in the Diffie-Hellman system; otherwise the factoring problem is not hard enough. As I mentioned earlier, this number should be 600 digits or more for acceptable security as of 2015. The easiest way to get 600 digits for n is to pick p and q so that they are 300 digits or more.

Bob can now find $\phi(n) = \phi(pq) = (p-1)(q-1)$ using the formula from Section 6.6. He picks an encryption exponent e such that the GCD of e and $\phi(n)$ is 1 and finds the decryption exponent $d = \bar{e} \bmod \phi(n)$. The modulus n and the encryption exponent e make up Bob's public key, which he can publish. The decryption exponent d is Bob's private key, which he needs to keep secret. He also needs to keep p, q, and $\phi(n)$ secret. In fact, he no longer needs them and can destroy his records of them if he wants.

Let's do an example using much smaller numbers than one would really use. Suppose Bob picks $p = 53$ and $q = 71$ for his primes; then $n = 53 \times 71 = 3763$ and $\phi(n) = (53 - 1) \times (71 - 1) = 3640$. He could pick $e = 17$ for his encryption exponent. Using the Euclidean algorithm, Bob can now verify that the GCD of 17 and 3640 is 1 and that the inverse

Alice   Bob

picks secret *p* and *q*
uses *p* and *q* to make public encryption key (*n*, *e*)
uses *p* and *q* to make private decryption key *d*

posts encryption key (*n*, *e*)

Figure 7.12. RSA set-up.

of 17 modulo 3640 is 1713. I'll let you fill in the details if you want; it shouldn't take very long. Bob posts *e* and *n* in a public place and keeps *p*, *q*, $\phi(n)$, and *d* secret, as shown in Figure 7.12.

When Alice wants to send Bob a message, all she has to do is look up his public modulus *n* and his public encryption exponent *e* and then encrypt a message using the exponentiation cipher. For example, with our *e* = 17 and *n* = 3763, Alice can send a ciphertext as follows:

plaintext:	ju	st	th	ef	ac	to	rs	ma	am
numbers:	10, 21	19, 20	20, 8	5, 6	1, 3	20, 15	18, 19	13, 1	1, 13
together:	1021	1920	2008	506	103	2015	1819	1301	113
to the 17th power:	3397	2949	2462	3290	1386	2545	2922	2866	2634

Bob knows the decryption exponent *d* and the public modulus *n*, so he can decipher the message by raising the ciphertext to the *d*th power modulo *n*. In our example,

ciphertext:	3397	2949	2462	3290	1386	2545	2922	2866	2634
to the 1713th power:	1021	1920	2008	506	103	2015	1819	1301	113
split apart:	10, 21	19, 20	20, 8	5, 6	1, 3	20, 15	18, 19	13, 1	1, 13
plaintext:	ju	st	th	ef	ac	to	rs	ma	am

A diagram of the whole system looks like Figure 7.13.

Alice

Bob

Picks secret *p* and *q*
Uses *p* and *q* to make public encryption key (*n*, *e*)
Uses *p* and *q* to make private decryption key *d*

Posts encryption key (*n*, *e*)

Looks up Bob's encryption key (*n*, *e*)

P

$\downarrow (n, e)$

$C \equiv P^e \text{ modulo } n$

$C \rightarrow$

C

$\downarrow (n, d)$

$P \equiv C^d \text{ modulo } n$

Figure 7.13. The whole RSA system.

Once you have seen it and grasped the math involved, the idea of RSA is really very simple. And when Rivest presented a written-up version of the system to Shamir and Adleman on the morning of April 4, it still sounded good, unlike many late-night ideas. The manuscript was attributed to Adleman, Rivest, and Shamir, in alphabetical order, which is the usual practice in mathematics and not uncommon in computer science. Adleman objected, feeling that he couldn't take credit for the idea—all he had done was to fail to shoot it down like he had the others. Rivest insisted, and eventually they settled on listing all three but with Rivest first and Adleman last. Thus, Rivest, Shamir, Adleman is the order of names that appeared on the MIT Technical Memo dated that day, the paper that was published describing the result, and the patent that was granted. And, thus, the common abbreviation for the system became RSA.

But before the academic paper was published or the patent granted, RSA managed to go public in a big way. Rivest had sent a copy of the Technical Report to Martin Gardner, who wrote a column called "Mathematical Recreations" for *Scientific American* magazine. The column was well known among both professional and amateur mathematicians. During its run from 1956 to 1981, it featured mathematical games, toys, puzzles, and pictures, including flexagons, polyominoes, tangrams, Penrose tilings, the artwork of M. C. Escher, fractals, and mathematical magic tricks. Gardner was immediately intrigued by the RSA system and set about writing a column explaining it, with Rivest's help. The column appeared in the August 1977 issue and proclaimed that public-key cryptography "is so revolutionary that all previous ciphers, together with the techniques for cracking them, may soon fade into oblivion." It included brief descriptions of the one-time pad, the contents of "New Directions in Cryptography," and RSA. It then gave a challenge to the readers: a message encrypted using a 129-digit RSA modulus with a $100 prize from Rivest, Shamir, and Adleman to the first person to break it. Rivest is quoted as estimating that it would take 40 quadrillion years to break the cipher using a 1977-era computer costing a million dollars. It also included directions for how to get a copy of the technical report by sending a self-addressed, stamped envelope to Rivest at MIT. According to Rivest, more than 3000 such requests were eventually received.

Interest in RSA remained largely confined to mathematicians, some computer scientists, and cryptography hobbyists until the invention of the World Wide Web and the explosive growth in Internet commerce in the 1990s. At that point, people realized that sending someone your credit card over the Internet was a perfect example of when you might want to send a secure communication to someone you had never met in person. Today, if you log into any secure web server, there is a very good chance that your computer has looked up the Web server's RSA public key and used it to encrypt your connection to the server.

But the Web page itself and the credit card number you want to send back are usually not directly encrypted using RSA. This is because asymmetric-key cryptography is almost always slower than symmetric-key cryptography. Instead, your computer uses the server's public key

Alice	Bob

picks secret p and q
uses p and q to make public encryption key (n, e)
uses p and q to make private decryption key d

posts encryption key (n, e)

picks a secret AES key k

looks up Bob's encryption key (n, e)

k
$\downarrow (n, e)$
k^e modulo n

k^e modulo $n \rightarrow$

k^e modulo n
$\downarrow (n, d)$
$k \equiv (k^e)^d$ modulo n

P
$\downarrow k$
C

$C \rightarrow$

C
$\downarrow k$
P

Figure 7.14. A hybrid RSA-AES system.

to encrypt some secret information that both computers can use to generate a key for a symmetric cipher such as AES. This is called a **hybrid cryptographic system**, and in practice it's very similar to a key agreement system. A simple version of a hybrid system might look like Figure 7.14. Alice is playing the part of your computer and Bob is playing the part of the server.

7.5 PRIMING THE PUMP: PRIMALITY TESTING

Before we look at how Eve can try to break an RSA-encrypted message, I want to talk a little bit about how long it takes Bob to set up the key. Notice first of all that for Bob to set up the key requires that he find two prime numbers. How do you find a prime number? The most obvious way is to pick a number and try to see if it has any factors, but we said that factoring was a hard problem. In fact, if Eve can factor n, then she will know p and q, which are Bob's secret trap-door information. So, she will be able to find d and read all the messages that were sent to Bob using that key. If it takes Bob as long to set up a new key as it does for Eve to recover it, that's bad. In that case all Eve needs to get ahead is a faster computer than Bob has.

Luckily, there are ways to find prime numbers that don't require trying to factor. Such tests have been known at least since the seventeenth century, but as a general rule they seem to have been considered impractical, either because they were too slow, sometimes slower than just trying to factor; because they worked only in special cases; or because they sometimes gave a wrong answer or no answer at all. Gauss is often cited as having separated the problem of primality testing from the problem of factoring. His quote on the subject, which is now rather famous among mathematicians, is somewhat ambiguous, however:

> The problem of distinguishing prime numbers from composite numbers and of resolving the latter into their prime factors is known to be one of the most important and useful in arithmetic. It has engaged the industry and wisdom of ancient and modern geometers to such an extent that it would be superfluous to discuss the problem at length. Nevertheless we must confess that all methods that have been proposed thus far are either restricted to very special cases or are so laborious that even for numbers that do not exceed the limits of tables constructed by estimable men . . . they try the patience of even the practiced calculator. . . . Further, the dignity of the science itself seems to require that every possible means be explored for the solution of a problem so elegant and so celebrated. For these reasons we do not doubt that the two following methods, whose efficacy and brevity we can confirm from long experience, will prove rewarding to the lovers of arithmetic.

Is Gauss referring to one problem or two? If you look at the two methods that Gauss describes, the first one factors composite numbers at the same time that it discovers that they are composite. So does the first variation of the second method. Of the very last variation that Gauss describes, he says, "...the second is superior in that it permits faster calculation, but unless it is repeated over and over again it does not produce the factors of composite numbers. It does however distinguish them from prime numbers." That is a mixed recommendation at best.

The necessary breakthrough that made RSA practical was the realization that fast primality tests could be useful even if they didn't always give the right answer. This seems to have been first pointed out by **Robert Solovay** and **Volker Strassen** in 1974, right around the time Merkle, Diffie, and Hellman began thinking about public-key cryptography. Their idea was to make a **probabilistic** primality test, that is, one that makes a random choice somewhere in the procedure. This random choice can allow a test to run very quickly, but there is a chance that it will output the wrong answer.

I'm going to show you a probabilistic test based on Fermat's little theorem. The test is similar to Solovay and Strassen's, but theirs is more complicated and more accurate. The first crucial point is that Fermat's little theorem can be used as a "compositeness test"—it can tell us for certain if a number is *not* prime. For example, suppose for the moment that you didn't know whether 15 is prime or composite. If 15 were prime, then Fermat's little theorem would tell us that for any number k between 1 and 14, $k^{14} \equiv 1$ modulo 15. So we can just start trying numbers. If $k = 2$, then $2^{14} \equiv 4$ modulo 15. But if 15 were prime, then this shouldn't happen. So we have shown that 15 is composite, and we say 2 is a **witness** for the compositeness of 15.

Not every number we try will work out so nicely. For example, if $k = 4$, then $4^{14} \equiv 1$ modulo 15, even though we now know 15 isn't prime. We say 4 is a **liar** for the Fermat test, because it implies that 15 is prime when it isn't. So, if we are testing a number n and $k^{n-1} \equiv 1$ modulo n, we can't be sure whether n is prime or whether k is a liar. This is where the random choice comes in. We will pick a bunch of different k's between 1 and $n - 1$. As soon as one is a witness, we know n is composite. If none of them is, we will say n is probably prime. The more k's we check, the more likely it is that n is prime. But unless we check a very large number of k's, we can never be certain that n is prime. And if we check too many k's, then the test no longer runs quickly enough. That may seem unsatisfactory, but for cryptographers it's good enough. After all, the humans and/or computers doing the calculations aren't perfect anyway. There's always a chance that a cosmic ray will come along at just the wrong time and hit your computer in just the wrong place. As long as the chance that your test is wrong is smaller than that, it doesn't really matter.

Let's try the test on a couple of numbers that you probably can't tell immediately whether they are prime. For each n we will pick 10 random k's unless we find a witness sooner.

Is $n = 6601$ prime?

k	k^{6600} modulo 6601
1590	1
3469	1
1044	1
3520	1
4009	1
2395	1
4740	1
4914	3773

Because $k = 4914$ is a witness, $n = 6601$ is definitely not prime.

Is $n = 7919$ prime?

k	k^{7918} modulo 7919
1205	1
313	1
1196	1
1620	1
5146	1
2651	1
3678	1
2526	1
7567	1
3123	1

We have not found any witnesses, so $n = 7919$ is probably prime. If we wanted to reduce the chance of a mistake, we would just need to test more k's.

As I mentioned, the Solovay-Strassen test is more complicated than the Fermat test, but there are more witnesses, so it is more likely to catch a composite number in the same amount of time. The test most commonly used today, however, is more accurate than either of these but almost as simple as the Fermat test. It was invented by **Michael Rabin** in 1980, based on an idea from **Gary Miller**. And just to complete the story, in 2002 the first *non*randomized primality test that was significantly faster than factoring and could be proved to always be correct was finally invented. The inventors were **Manindra Agrawal**, a professor at the Indian Institute of Technology, Kanpur, and two of his first-year graduate students, **Neeraj Kayal** and **Nitin Saxena**. This came as a great surprise to many mathematicians, who did not expect such a development for many years, if at all. However, the Rabin-Miller test is still more commonly used in cryptography, since it is considered accurate enough and it is faster in practice. A 300-digit number such as might be used in RSA can easily be tested in a few seconds with an error rate of less than 10^{-30}.

To wrap this up, Bob should have no trouble finding two prime numbers quickly. Multiplying them together is certainly fast, and the rest is just picking any old e, making sure it has a GCD of 1 with n, and finding the inverse. That's just applying the Euclidean algorithm. I mentioned in Section 1.3 that the Euclidean algorithm runs quickly. In fact, **Gabriel Lamé** proved in 1844 that the number of division steps required does not exceed 5 times the number of digits in the smaller of the 2 numbers. When you think about the 600-digit numbers used for RSA, that means less than 3000 divisions, which takes only a split second on a modern computer, or even a good handheld calculator. If Bob is unlucky, the first e he tries might be a bad key, but he'd have to be *really* unlucky to need more than 2 or 3 tries. The whole process of creating a secure RSA key usually takes less than 15 seconds on an average personal computer.

7.6 WHY IS RSA A (GOOD) PUBLIC-KEY SYSTEM?

Where does all of this leave Eve? She can look up Bob's public information just the same as Alice, so she knows n and e, as shown in Figure 7.15. And she sees C, which she knows is P^e modulo n for some P. Can she invert the function and obtain P? This is called the **RSA problem**. Like the Diffie-Hellman problem, we think it is hard, although nobody knows for sure.

Alice	Eve	Bob

picks secret p and q
uses p and q to make public (n, e)
uses p and q to make private d

posts (n, e)

looks up (n, e)

P
$\downarrow (n, e)$
$C \equiv P^e$ modulo n

$C \rightarrow$

"This must be P^e modulo n."

looks up (n, e)

"I don't know d or $\phi(n)$."
"How do I invert the function?"

C
$\downarrow (n, d)$
$P \equiv C^d$ modulo n

Figure 7.15. What Eve sees.

The most obvious way for Eve to attack the RSA problem is to factor n. Then she would know p and q, so she could calculate $\phi(n) = (p-1)(q-1)$ and find d, just the same as Bob can. We have said several times that factoring is a hard problem, but like the discrete logarithm problem, no one knows for sure. On the other hand, people have been working on factoring for even longer than on the discrete logarithm problem. Fermat, Euler, Gauss, and others worked on it before modern computers. Over 35 years' worth of mathematicians with computers have worked on it, too. They have figured out how to do much better than the obvious method of trying to divide by each prime in turn, but Eve still can't factor Bob's n nearly as fast as Bob can create it.

In August of 1993, an international team of volunteers coordinated by a few students and one professional mathematician decided to see if they could harness the power of the Internet to factor the 129-digit modulus from Martin Gardner's column. They had faster computers and better methods than in 1977, but maybe more importantly, they had *more* computers. By the time the project successfully ended on April 26, 1994, the job had been divided up between more than 600 people all over the world using more than 1600 computers, ranging from Cray supercomputers to fax machines. The computers were programmed to work on the problem only when they weren't being used for something else. After 8 months of work, the coordinators announced that the challenge had been solved. Ron Rivest awarded them the $100, which they gave to the Free Software Foundation, and announced the solution of the message:

the magic words are squeamish ossifrage

As I write this, the current factoring record is a 232-digit (768-bit) number, whose factorization was finished on December 12, 2009. This time a group of 16 researchers used specifically dedicated computing time at 8 different institutions rather than opening up the project to the Internet. The entire project took 3 months at one institution in the summer of 2005, a similar amount of time at a second institution in the spring of 2007, and about 16 months of intensive computation between August 2007 and December 2009. The researchers concluded that 1024-bit (roughly 300-digit) RSA moduli might be factorable in the next 5 years and their use should be phased out before then. As of this writing, no one has announced such a factorization, but it would not surprise me if it happened soon.

What about other things Eve could try to do besides factoring n? She could try to find $\phi(n)$ by some other method. Then she could compute d without ever knowing the secret p and q. Eve knows that $\phi(n)$ counts the number of positive whole numbers less than or equal to n that have a GCD of 1 with n, but trying the Euclidean algorithm on each one of those numbers is going to take even longer than factoring n by brute force. Plus, if Eve can find $\phi(n)$, then she can factor n automatically. How does that work? She knows

$$\phi(n) = (p-1)(q-1) = pq - p - q + 1 = n - (p+q) + 1.$$

If she knows $\phi(n)$ and n, then this equation lets her find $p + q$. But then

$$(p-q)^2 = p^2 - 2pq + q^2 = p^2 + 2pq + q^2 - 4pq = (p+q)^2 - 4n,$$

so if she knows $p + q$ and n, she can find $p - q$. Finally, if she knows $p + q$ and $p - q$, then

$$\frac{(p+q)+(p-q)}{2} = p \quad \text{and} \quad \frac{(p+q)-(p-q)}{2} = q.$$

People have tried to factor n this way and it doesn't seem to be working well, so this probably isn't a good bet for Eve.

Can Eve find d directly without finding $\phi(n)$? This too would give her a way of factoring n. If she knows d and e, she can compute $de - 1$, and since $de \equiv 1$ modulo $\phi(n)$,

$$de - 1 \equiv 0 \quad \text{modulo } \phi(n).$$

But this can happen only if $de - 1$ is a multiple of $\phi(n)$, and it turns out that there is a probabilistic algorithm that factors n even if you have only a multiple of $\phi(n)$ and not $\phi(n)$ itself.

That just leaves Eve trying to somehow solve the equation

$$C \equiv P^e \quad \text{modulo } n$$

without knowing d at all. Is that possible? It doesn't seem very likely, but after 30 years of trying no one has established either that it is or it isn't. In the sort of analogy you might see on a standardized test, the

RSA problem is to the factoring problem as the Diffie-Hellman problem is to the discrete logarithm problem: in both cases we think the two problems are equivalent, and we think both are hard, but we can't be sure about any of it.

7.7 CRYPTANALYSIS OF RSA

So if I've just said we don't know any way Eve could break RSA fast enough to be worthwhile, what is this section about? It would be more accurate to say that we don't know any way of breaking RSA *in general.* In certain cases, Eve can break the system, especially if Alice and Bob aren't careful.

The first thing to watch out for is the **small message attack**. Suppose Bob uses a modulus $n = 3763$ as in our earlier example, but to save Alice some effort in her encryption, he decides to make the encryption exponent $e = 3$ and tells her to use 0 instead of 26 for z. After all, we're not using 0 for any other letter.

Unfortunately, Alice needs to let Bob know that there are "zero zebras in Zanzibar zoos." Why is this unfortunate? Here's the encryption:

plaintext:	ze	ro	ze	br	as	in
numbers:	0, 5	18, 15	0, 5	2, 18	1, 19	9, 14
together:	005	1815	005	218	119	914
to the 3rd power:	125	2727	125	693	3098	1614

plaintext:	za	nz	ab	ar	zo	os
numbers:	0, 1	14, 0	1, 2	1, 18	0, 15	15, 19
together:	001	1400	102	118	015	1519
to the 3rd power:	1	1585	42	2364	3375	581

Now Eve knows that $e = 3$, because that's public information. And 3 is kind of small compared to 3763. So she might suspect that not all those blocks actually got wrapped around past 3763. In fact, if Eve just takes the $\frac{1}{3}$ power (or cube root) of each block using ordinary (*not* modular) arithmetic, she'll get

ciphertext:	125	2727	125	693	3098	1614
to the $\frac{1}{3}$ power:	5.00	13.97	5.00	8.85	14.58	11.73
plaintext numbers:	005	??	005	??	??	??
plaintext:	ze	??	ze	??	??	??

ciphertext:	1	1585	42	2364	3375	581
to the $\frac{1}{3}$ power:	1.00	11.66	3.48	13.32	15.00	8.34
plaintext numbers:	001	??	??	??	015	??
plaintext:	za	??	??	??	zo	??

Eve certainly can't read the whole message, but if she has a suspicion that Zanzabari zebras are under discussion, this could still be very bad for Alice and Bob. The moral is to make sure your message blocks are large enough, your encryption exponent is large enough, or both.

There is a similar **chosen-ciphertext attack** that can work if the decryption exponent is too small. Suppose Eve knows that Bob is using a modulus of $n = 4089$ and an encryption exponent of $e = 2258$, and she wants to know Bob's decryption exponent d. Instead of sending him a real message, she can send him a "ciphertext" with a mix of correctly encrypted blocks and random small numbers and hope she can find out what they decrypt to. For example,

"ciphertext":	2221	2736	1011	3	5	1474	1110	2859
to the dth power:	1612	501	1905	243	3125	2008	114	1119
"plaintext":	pl	ea	se	b?	?y	th	an	ks

The very worst thing Bob can do now is send Eve back a message, maybe encrypted using her public key, saying: "I don't understand the two blocks in the middle of your message. How do I translate 243 3125 into letters?" Even if Bob doesn't send Eve such a message, he could be in trouble if she can get hold of the decrypted numbers in some other fashion.

Now Eve knows that $243 \equiv 3^d$ modulo 4089 and $3125 \equiv 5^d$ modulo 4089. So she tries taking the base 3 logarithm of 243 using ordinary arithmetic and gets

$$\log_3(243) = 5,$$

which looks like it might be d. For some extra confirmation she tries the base 5 logarithm of 3125 and gets

$$\log_5(3125) = 5.$$

So, she has found d. Two morals here: first, if you decipher a ciphertext from someone and the plaintext seems garbled, never ever tell the person what plaintext you got. This applies to almost every cipher ever, since there are many other chosen-ciphertext attacks out there. And second, don't choose a d that's too small. In fact, there are other **low decryption exponent attacks** that make this insecure even if Eve can't use a chosen-plaintext attack.

Another possible attack is the **common modulus attack**. Suppose Bob and Dave trust each other, but they don't want to get their messages mixed up. They might decide to use the same modulus n but different values of e. This is bad.

For example, suppose $n = 3763$, Bob uses $e = 3$, Dave uses $e = 17$, and Alice sends them both the same message:

plaintext:	hi	gu	ys
numbers:	8, 9	7, 21	25, 19
together:	809	721	2519
to the 3rd power:	2214	3035	964
to the 17th power:	2019	1939	2029

Eve starts by using the Euclidean algorithm on the two values of e. If they are relatively prime, then she will be able to write 1 with a "3 times something" part and a "17 times something" part, as in Section 1.3:

$$17 = 3 \times 5 + 2, \qquad 2 = 17 - 3 \times 5,$$

$$3 = 2 \times 1 + 1, \qquad 1 = 3 - 2 \times 1$$

$$= 3 \times 6 - 17 \times 1,$$

so

$$1 = (3 \times 6) + (17 \times -1).$$

Eve knows that for the first plaintext block,

$$2214 \equiv P^3 \quad \text{modulo } 3763 \qquad \text{and} \qquad 2669 \equiv P^{17} \quad \text{modulo } 3763.$$

If she takes $2214^6 \times 2019^{-1}$ modulo 3763, then she will have

$$2214^6 \times 2019^{-1} \equiv (P^3)^6 (P^{17})^{-1} \equiv P^{(3\times6)+(17\times-1)} \equiv P^1 \equiv P \quad \text{modulo 3763.}$$

And sure enough:

ciphertext 1:	2214	3035	964
ciphertext 2:	2019	1939	2029
ciphertext 1 to the 6th power:	229	1946	897
ciphertext 2 to the −1st power:	2682	1178	523
multiplied:	809	721	2519
split apart:	8, 9	7, 21	25, 19
plaintext:	hi	gu	ys

The moral of this is not to share the same modulus, even if you trust each other.

There is also a **related message** attack if Alice sends Bob two messages that are similar but not identical, using the same n and e. This attack starts getting very difficult as soon as e is larger than 3, which is one reason to choose $e = 17$, or $e = 2^{16} + 1 = 65537$.

Speaking of related messages, there is an attack known as the **broadcast attack** that Eve can use if e people all use the same exponent e with different moduli and Alice sends them each the same message or even similar messages. Since using small e's such as 3 or 17 has enough of a speed advantage to make them common (see Section 7.4), it's best not to send similar messages to more than one person regardless. One way to make messages less similar is to carefully add random padding bits to the message before encryption and then ignore them after decryption.

Adding randomness also helps defeat a **forward-search attack**. This is a type of probable word attack that can be a problem for asymmetric-key systems in general. Suppose Eve has a guess as to the plaintext that goes with a certain ciphertext from Alice to Bob. If there is no randomness in the encryption, she can always see whether her guess is correct. That's because Eve can encrypt using Bob's public key just like anyone else! If there is no randomness in the encryption and she starts with the same plaintext as Alice, she will get the same ciphertext. If there is random padding and good diffusion, on the other hand, two different encryptions of the same message should look nothing alike. Encryption that depends on random choices is called **probabilistic encryption**; we will see another example in Section 7.8.

That's a pretty good summary of some of the types of attacks on RSA and the lessons you should draw from them. Most of the lessons fall under the category of "don't be lazy" and are fairly easy to remember given that. If you'd like to see more details, look at the references in the endnotes.

7.8 LOOKING FORWARD

Merkle's puzzles were always what you might call a proof of concept— even Merkle knew that they wouldn't work in practice. Nevertheless, they had a direct impact on the development of Diffie-Hellman key agreement. In fact, Martin Hellman has said that it really ought to be called the Diffie-Hellman-Merkle system, and the patent for the system is in all three names.

Whatever you call it, Diffie-Hellman is still very much in use as part of various security systems on the Internet. Remember, though, that since it is a key-agreement system and not an encryption system, it can't be used by itself. People did eventually come up with asymmetric-key encryption systems based on the discrete logarithm problem, and we will see some of them in Chapter 8.

RSA is also very much in current use on the Internet, probably even more than Diffie-Hellman. There are challenges to these two systems, however. One drawback they share is that they require very large keys. In Section 8.3 we will see an idea known as elliptic curve cryptography, which attempts to get the same benefits and security as Diffie-Hellman and RSA with smaller keys and possibly faster computations. There is some movement toward new systems based on elliptic curve cryptography, but for the moment the older public-key systems are still much more common.

The Snowden documents released in 2013 created another concern about security of Diffie-Hellman. Several internal NSA documents indicated that the NSA was breaking VPN traffic encrypted with the Diffie-Hellman-based security I mentioned in Section 7.2. In 2015 a team of researchers in France and the United States announced a plausible way that this could work. The attack, known as **Logjam**, has two parts.

The first part is the realization that something I said in Section 7.2 is not completely true. I said that it was fine to look up p in a table because it does not have to be kept secret. The catch is that much of the work in breaking the discrete logarithm problem can be done knowing only p, not g, A, or B. That means that if Eve knows that many people are looking up the same few primes p from the same table, she can have computations ready for those primes before any messages are even sent. When the messages are sent, she can break them much more quickly than if she were starting from scratch. When the researchers analyzed this precomputation attack, they realized that Diffie-Hellman with up to 225-digit primes was probably vulnerable to academic teams, and with up to 300-digit primes it was plausibly vulnerable to the NSA and likely other governments as well. They also found that approximately two-thirds of VPNs that they could scan preferred to use a commonly known prime of 300 digits or fewer.

The second part of the attack applies only to secure Web browsing. I mentioned that RSA is the most common way of encrypting Web connections, but Diffie-Hellman is used as well. The researchers discovered that if a Web server is using Diffie-Hellman, Eve can alter messages in such a way as to trick the system into using a smaller prime number p than Alice and Bob wanted. This is an example of a **downgrade attack**; combined with the precomputation attack, it would make a Web server vulnerable even if it used a large prime by preference. About 25% of Web sites in the study could be downgraded to one of the 10 most popular 300-digit primes, and about 8% could be downgraded to a 150-digit prime.

Incidentally, a downgrade attack on Web servers that use RSA was also discovered in 2015. This is called the FREAK attack. (FREAK stands for Factoring RSA Export Keys.) Unlike Logjam, FREAK works only against browsers and servers with certain software bugs. In general, it has become clear that Diffie-Hellman and RSA keys with less than 300 digits should not be used under any circumstances. In addition to patching bugs, most software producers are now moving to disallow these keys entirely and to encourage the use of keys at least 600 digits long.

Yet another challenge to Diffie-Hellman and RSA is related to quantum computation, which we explore in Chapter 9. We will see that if quantum computers became common, they would make both

Diffie-Hellman and RSA very insecure. We will also see two broad classes of replacement systems, with somewhat confusing names: post-quantum cryptography involves attempts to design systems for any sort of computer that could withstand quantum attacks, whereas quantum cryptography tries to take advantage of quantum physics itself to design new sorts of cryptosystems.

APPENDIX A   THE SECRET HISTORY OF PUBLIC-KEY CRYPTOGRAPHY

Perhaps fittingly, public-key cryptography turns out to have both a public history and a secret one. In 1997 the world learned that Merkle, Diffie, and Hellman hadn't been the only ones thinking about this strange idea in the early 1970s. In fact, in 1969, before any of those three had started their trip to fame, **James Ellis** also showed that public-key cryptography was possible. His discovery, unlike theirs, would stay shrouded in secrecy for almost 30 years.

And for a very specific reason: James Ellis worked for the Government Communication Headquarters (GCHQ), more or less the British equivalent of the NSA. In particular, he worked for the Communications Electronics Security Group (CESG), which was (and is) responsible for advising the British government on the security of electronic communications and data. Like Merkle and Diffie, Ellis started by thinking about the question of whether it was really necessary for two people to exchange secret messages without a secretly arranged key. Unlike Diffie, Ellis wasn't worried about trusting a third party with the key distribution. After all, he worked for an organization that specialized in that sort of thing. What he *was* worried about was the logistical problems of key distribution. If thousands of people working for a large organization needed to communicate and if it was the case that every pair needed to keep their conversation secret from every other pair, then millions of different keys would have to be dealt with.

Ellis, like everyone else, initially assumed that this situation was inevitable. But he was doing some background reading and found an anonymous paper describing a Bell Telephone voice-scrambler project from the 1940s. This was a system for analog phone lines, and the idea was that if Alice wanted to send Bob a message on a secure line, then Bob, not Alice, would be responsible for adding random noise to the line during the transmission. If Bob kept track of what the noise was, he could process the combined signal at his end to remove the noise and recover the message. Eve would be unable to understand the noisy signal, and Alice never needed to know exactly what Bob had done. Ellis understood that this analog system wasn't practical itself and couldn't be adapted exactly for digital use, but he got an important idea. Alice could send Bob an encrypted message without knowing a decryption key *if* Bob participated actively in the system.

Oddly enough, Ellis' breakthrough came to him when he was lying down, just as Ron Rivest's did. Ellis laid down in bed one night and started wondering whether it was possible to construct an asymmetric-key system for digital communications similar to the voice scrambler project. If so, only one private key for each participant would be necessary—something much easier to deal with than a symmetric-key system. Once he had properly framed the question, he had the answer within a few minutes. It was possible, and he had an idea how to do it.

Like Merkle's puzzles, Ellis' initial idea was "simple, but inefficient"; Ellis said, "It shows only that such a system is theoretically possible, and not that a practical form exists." Ellis started by supposing he had three huge tables of numbers. Ellis thought of them as machines, for reasons that will become clear in a bit, and labeled them M_1, M_2, and M_3. I like to think of them as huge books, or sets of books. In fact, think of M_2 as a whole big room full of codebooks. We haven't talked much about codes, but a codebook is really just a dictionary where words or phrases are listed in alphabetical order. Instead of a definition, each entry gives you a codegroup, say a 5-digit number, corresponding to the word or phrase. Each huge codebook in room M_2 is completely different from every other, and each one has a volume number. This room is going to be the encryption room. Room M_3, the decryption room, is very similar, except that the codebooks are in order of codegroups instead of alphabetically. Each encryption codebook in M_2 has a corresponding decryption codebook in M_3, and vice versa. For reasons that will become clear in a moment, the volume numbering system is completely different in M_2 than in M_3. Luckily for Alice and Bob, whatever insane librarian has numbered these books has also prepared a huge index volume, M_1. Volume M_1 lets you look up a decryption volume number and

Alice

Bob

"I have a message for you."

picks private decryption key d

uses d and M_1 to look up
public encryption key e

$\leftarrow e$

P

$\downarrow (M_2, e)$

C

$C \rightarrow$

C

$\downarrow (M_3, d)$

P

Figure 7.16. Ellis' public-key system.

find out the corresponding encryption volume number. *But*, and this is important, there's no inverse index going the other way.

So now when Alice sends Bob a message, she starts by asking Bob for an encryption key. Bob picks a decryption volume number d at random, looks it up in volume M_1, and sends the appropriate encryption key e to Alice, while keeping d private. Alice goes to Room M_2, finds encryption volume e, and uses it to encrypt her message, which she sends to Bob. Bob goes to room M_3, finds decryption volume d, and uses it to decrypt the message. Figure 7.16 shows the process, which should remind you a lot of things we've seen before.

What about Eve? If she has been listening in to the conversation, she knows e and the ciphertext. She has three options, none of which are good. She can go to room M_2, find volume e, and then search through it for each codegroup in the ciphertext. Since the codegroups are in no particular order, she is probably going to have to search most, if not all, of the codebook. Or, she can go to room M_3 and try to decrypt the ciphertext with every possible book until she finds one that gives a sensible plaintext. Or, lastly, she could get the index volume M_1 and search through it until she finds the decryption volume number d that corresponds to encryption volume number e. Because there's no inverse index to M_1, she is again probably going to have to search most of the book, unless she gets lucky. If both the number of volumes and the size of each volume are large enough, all these options are pretty bad.

This isn't anything like a practical system, even if it were computerized. It would be easier to store the large codebooks if they were on a computer, but Eve could search them faster, so that doesn't help. When Ellis referred to M_1, M_2, and M_3 as "machines," he was hoping that some "process" could be found that would act the same way as the codebooks or the tables without actually having to store all of the information. Despite the use of the word *machine*, this process would probably be mathematical rather than mechanical. However, Ellis was an engineer by training and didn't really feel up to the mathematical subtlety he suspected would be necessary. "Because of the weakness of my number theory," he later said, "practical implementations were left to others."

The project was not assigned much priority at CESG or GCHQ over the next few years. A few mathematicians tried to find a flaw in the reasoning, without success. A few people tried to find a practical mathematical system to implement it, also without success. This is how things stood in late 1973, when **Clifford Cocks** was hired at CESG. Unlike Ellis, Cocks was a trained mathematician, with an undergraduate degree from Cambridge and a year of graduate school at Oxford. Cocks was assigned a mentor, who one day described Ellis' idea during a tea break.

Cocks had a few things going for him as he tried to attack the problem. First, he not only had mathematical training, but he had done research in exactly the kind of mathematics that has become the foundation of public-key cryptography. Second, he hadn't seen Ellis' paper or any of the other work that had been previously done on the problem, so he could get a fresh start. Third, the problem had been posed as a puzzle rather than an assignment, so there was no pressure. And finally, as he later said, "I suppose it was actually also helpful that I wasn't doing anything that evening." That evening after work, Cocks went back to his rented room and worked out a system that was the same in all essential ways as the system that later became known as RSA. Bob's private p and q take the place of the decryption volume number in our description of Ellis' system, and his public n takes the place of the encryption volume number. In Cocks' version of RSA, the encryption exponent e is equal to n, so the encryption key only has one part. The "machines" M_2 and M_3 are modular exponentiation, and M_1 is multiplying p and q together to get n.

The security rules that went with Cocks' job forbade him from writing down anything job related while he was at home. Luckily, the system was simple enough that he still remembered it in the morning, and he wrote up a short paper at work the next day. Cocks' mentor was excited, and Ellis, when he heard, was happy but cautious. A third person who was interested was **Malcom Williamson**, Cocks' friend since childhood, who also worked at CESG. Williamson hadn't heard about Ellis' idea before and was particularly skeptical. He was so skeptical that after Cocks told him about the idea, Williamson went home and tried to prove that it *couldn't* work. That failed, of course, but late that night, after 8 or 12 hours, he realized he had an entirely different way of implementing an idea similar to Ellis'. He had discovered what we now call the three-pass protocol, which is a public-key system closely related to the Pohlig-Hellman cipher. (See Section 8.1 for the three-pass protocol.) Again, he couldn't write it down until the next day at work, and it would not be written up as a complete paper for a few months, until January 1974. In the meantime he had discussed the three-pass protocol with Ellis, who had evidently started to become less cautious and refined the ideas. After some more conversation, Williamson had yet another idea for a "cheaper and faster" method of public-key encryption, which turned out to be exactly the idea that Diffie and Hellman had.

The search for a cheaper and faster method turned out to be potentially important. The general attitude at GCHQ had by now shifted from regarding public-key encryption, or "non-secret encryption," as Ellis had dubbed it, as impossible to regarding it as impractical. Williamson, on the other hand, was having second thoughts about the entire thing. By the time he wrote up his second paper, on the key-agreement system, he wrote, "I have come to doubt the whole theory of non-secret encryption." The problem that bothered him was the inability to either prove or disprove the difficulty of the discrete logarithm problem and the factoring problem. For this reason, he said, he had delayed for 2 years in writing up this second paper. In the end, no one inside GCHQ did anything about implementing a real public-key system.

In retrospect, that was not too surprising. A government security agency was probably the wrong place for a public-key system. While it would have certainly helped with the key-distribution problem, the real advantage in a public-key system is for two people to communicate without meeting first. Two people who work for the same government agency probably won't have that issue. And such agencies are even more cautious than average when it comes to new and untried cryptographic systems. If someone in 1977 had discovered a fast way to factor numbers or solve the discrete logarithm problem, there would have been some very sorry people around MIT and Stanford. But if GCHQ or the NSA had converted their systems to public-key and a year later it had been broken, it would have been a potential national security disaster.

So nothing happened. In 1977, when Rivest, Shamir, and Adleman applied for their patent, Williamson tried to get it blocked, but his superiors decided not to do anything. In 1987 Ellis decided that "no further benefit can be obtained from continued secrecy" and wrote up his version of events in a paper. His superiors disagreed, and the paper was declared classified for the next 10 years. Finally, on December 23, 1997, GCHQ posted 5 papers on its Web site: Ellis' original paper, Cocks' paper, Williamson's two papers, and Ellis' "History of Non-Secret Encryption." Unfortunately, it was too late for Ellis, who had died on November 25, barely a month before the world was to find out what he had done.

8

Other Public-Key Systems

Now we know two ways that Alice can send Bob a secret message securely without a secure meeting first. They can use a key-agreement system to choose a secret key for a symmetric-key cipher, or they can use an asymmetric-key system, where Alice knows Bob's public encryption key but only Bob knows the private decryption key. There's a third way that uses symmetric-key cryptography to allow Alice to send Bob a message without them exchanging or agreeing on any keys at all, public or private. It's called the **three-pass protocol**; it's too inefficient for general use, but it's interesting and occasionally handy.

If we can think of asymmetric-key cryptography as a locked door with a mail slot in it, then an analogy for symmetric-key cryptography might be a suitcase with a padlock and two identical keys, as shown in Figure 8.1. If Alice wants to send a message to Bob, she puts it in the suitcase and locks it with the padlock. When Bob gets the suitcase, he unlocks the padlock with the other key and takes the message out and reads it.

Now suppose the latch of the suitcase has room to hold either or both of two padlocks independently, and Alice and Bob each have a padlock with a different key. Alice puts the message in the suitcase, puts her padlock on it, and sends it to Bob, as in Figure 8.2. This is pass 1 of the three-pass protocol.

Bob can't open Alice's padlock, because he doesn't have the key. Instead, he puts his own padlock on it and sends it back to Alice. This is pass 2, as shown in Figure 8.3.

Now Alice *unlocks* her padlock and sends the suitcase back to Bob, as shown in Figure 8.4. This is pass 3. Note that the suitcase is still locked

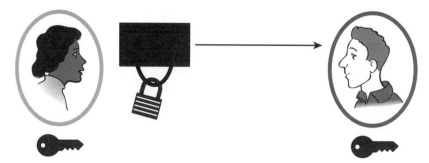

Figure 8.1. Symmetric-key cryptography.

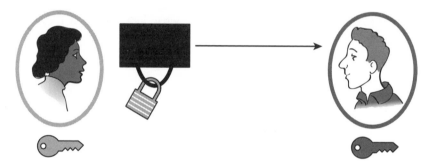

Figure 8.2. Pass 1 of the three-pass protocol.

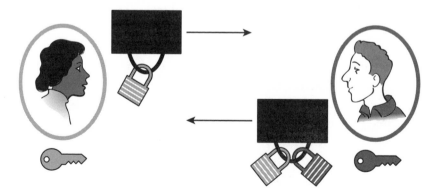

Figure 8.3. Pass 2 of the three-pass protocol.

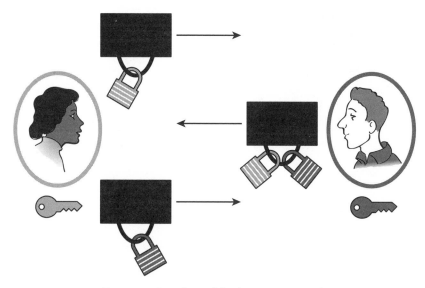

Figure 8.4. Pass three of the three-pass protocol.

with Bob's padlock, so Eve can't open it. Now Bob can unlock his own padlock and read the message. Along the way, the suitcase was never sent without a lock, and Alice and Bob never had to share or exchange any sort of key.

In order for this to work, you need a symmetric-key cipher with two specific properties. The first is that Bob's encryption and Alice's encryption can't get in each other's way—that would be as if Bob looped his padlock around Alice's, so she couldn't unlock it. In technical terms, Alice's encryption and Bob's encryption have to commute, as we discussed in Section 3.4. Doing Alice's encryption first and then doing Bob's has to be the same as doing them the other way around. Only a few of the ciphers we have studied have this property, including additive ciphers, multiplicative ciphers, and polyalphabetic and stream ciphers based on these. Affine ciphers, Hill ciphers, and transposition ciphers sometimes work, but only if Alice and Bob restrict themselves to a fairly limited set of keys. None of the symmetric-key ciphers we have seen intended for use with modern computers have this property except for the Pohlig-Hellman exponentiation cipher.

To see the other property needed, consider what happens if Alice and Bob use an additive cipher. Let a be Alice's key and b be Bob's, and

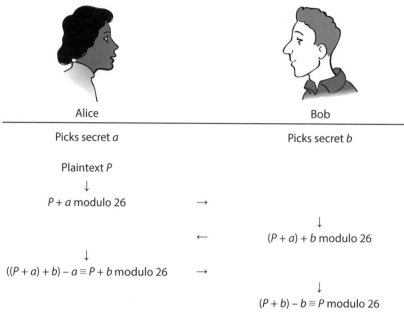

Alice	Bob
Picks secret a	Picks secret b
Plaintext P	
$\downarrow$	
$P + a$ modulo 26 $\qquad\rightarrow$	
	$\downarrow$
$\leftarrow \qquad (P + a) + b$ modulo 26	
$\downarrow$	
$((P + a) + b) - a \equiv P + b$ modulo 26 $\qquad\rightarrow$	
	$\downarrow$
	$(P + b) - b \equiv P$ modulo 26

Figure 8.5. The three-pass protocol with additive encryption.

let P be the first letter of plaintext. Then the three-pass protocol looks like Figure 8.5.

Here's the problem: after passes 1 and 2, Eve has $P + a$ and $(P + a) + b$ modulo 26. So she can mount a known-plaintext attack and recover $b \equiv ((P + a) + b) - (P + a)$ modulo 26. Then she can use b to decrypt the message from the third pass and get $P \equiv (P + b) - b$ modulo 26, as shown in Figure 8.6. So, in addition to being commutative, Alice's and Bob's encryption needs to be resistant to known-plaintext attacks. That leaves only one option from the ciphers we know, namely, the Pohlig-Hellman cipher.

Figure 8.7 shows the three-pass protocol done properly, using the Pohlig-Hellman cipher. Alice and Bob have to agree on the same large prime p, but that's all. In some ways this can be thought of as a combination of the Pohlig-Hellman exponentiation cipher and Diffie-Hellman key agreement.

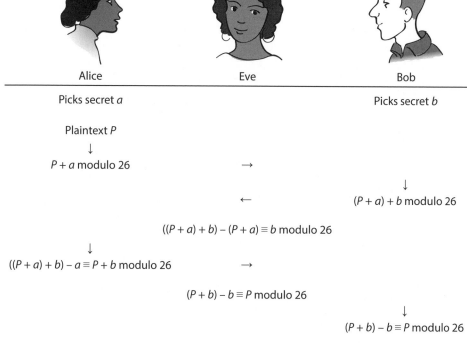

Alice	Eve	Bob
Picks secret a		Picks secret b

Plaintext P
$\downarrow$
$P + a$ modulo 26 $\qquad \rightarrow$

$\qquad\qquad\qquad\qquad\qquad\qquad \downarrow$
$\qquad\qquad\qquad\qquad\qquad\qquad (P + a) + b$ modulo 26
$\qquad\qquad\qquad\qquad \leftarrow$

$((P + a) + b) - (P + a) \equiv b$ modulo 26

$\downarrow$
$((P + a) + b) - a \equiv P + b$ modulo 26 $\qquad \rightarrow$

$(P + b) - b \equiv P$ modulo 26

$\qquad\qquad\qquad\qquad\qquad\qquad \downarrow$
$\qquad\qquad\qquad\qquad\qquad\qquad (P + b) - b \equiv P$ modulo 26

Figure 8.6. The three-pass protocol with additive encryption is insecure.

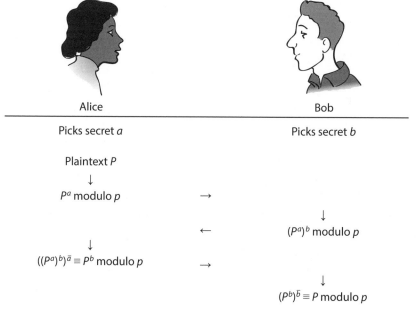

Alice	Bob
Picks secret a	Picks secret b

Plaintext P
$\downarrow$
P^a modulo p $\qquad \rightarrow$

$\qquad\qquad\qquad\qquad\qquad \downarrow$
$\qquad\qquad\qquad\qquad\qquad (P^a)^b$ modulo p
$\qquad\qquad\qquad \leftarrow$

$\downarrow$
$((P^a)^b)^{\bar{a}} \equiv P^b$ modulo p $\qquad \rightarrow$

$\qquad\qquad\qquad\qquad\qquad \downarrow$
$\qquad\qquad\qquad\qquad\qquad (P^b)^{\bar{b}} \equiv P$ modulo p

Figure 8.7. The three-pass protocol with Pohlig-Hellman encryption.

Suppose Alice wants to send Bob a message using this system. They agree to use a block size of 2 and convert letters to numbers the same way they did in Section 6.1, and they also agree to use the same prime modulus $p = 2819$ that they did in that section. Alice picks $a = 113$ for her secret key and verifies that a has an inverse modulo 2818, namely, $\bar{a} = 2419$. Bob picks $b = 87$ for his secret key, and verifies that b likewise has an inverse modulo 2818, which is $\bar{b} = 745$. Then the protocol proceeds like this:

plaintext:	te	ll	me	th	re	et	im	es
numbers:	20, 5	12, 12	13, 5	20, 8	18, 5	5, 20	9, 13	5, 19
together:	2005	1212	1305	2008	1805	520	913	519

Alice sends Bob:

to the 113th power:	1749	1614	212	774	2367	2082	2156	1473

Bob sends Alice:

to the 87th power:	301	567	48	1242	1191	1908	2486	986

Alice sends Bob:

to the 2419th power:	1808	2765	289	692	2307	2212	1561	2162

Bob decrypts:

to the 745th power:	2005	1212	1305	2008	1805	520	913	519
split apart:	20, 5	12, 12	13, 5	20, 8	18, 5	5, 20	9, 13	5, 19
plaintext:	te	ll	me	th	re	et	im	es

Now that we are using an exponential cipher, how hard is it for Eve to read the message? She could try to mount the same known-plaintext attack as before, but that would require solving the discrete logarithm problem. As in the Diffie-Hellman problem, Eve has extra information here. So as in the Diffie-Hellman problem, it might be possible to break the three-pass protocol without solving the discrete logarithm problem. No one knows of any way to do this, and it doesn't seem very likely.

In fact, it turns out that in a certain precise sense, solving the Diffie-Hellman problem and breaking the three-pass protocol are about equally difficult: if Eve can do one quickly, then she can do the other quickly, and vice versa.

The encryption system I'm calling the three-pass protocol is known by several names, including Shamir's three-pass protocol, the Massey-Omura system, and no-key cryptography. Adi Shamir invented this system in the context of a way of playing "mental poker"—that is, Alice and Bob want to play a game of poker over the phone without exchanging physical cards and without either player being able to cheat. It was published in a technical report in 1979 and then in a collection of articles dedicated to Martin Gardner in 1981. Sometime shortly after that, **James Omura**, who was then a professor of electrical engineering at UCLA, heard about the basic idea of Shamir's system but not about the use of the Pohlig-Hellman cipher and independently worked out the rest of the details. He then worked with **James Massey**, a former colleague at UCLA who had moved to the Swiss Federal Institute of Technology in Zurich, to adapt the protocol to the finite fields modulo 2 version of the Pohlig-Hellman cipher and also improve the speed of computer calculations in these fields. Massey gave a talk on these combined ideas at a major European cryptography conference in 1983, but no proceedings for the conference were published. The first time that Massey and Omura's version of the system appeared in print seems to have been their patent application, filed in 1982 and granted in 1986.

The three-pass protocol requires a lot of modular exponentiation, which makes it considerably slower than using Diffie-Hellman to agree on a key for a cipher like AES and then exchanging AES messages. It also requires sending more information back and forth, so all in all it's not very practical except in a few specialized situations, like mental poker. It's a very cool idea, though.

8.2 ELGAMAL

As we have seen, although the first practical public-key cryptography system (Diffie-Hellman key agreement) used the difficulty of the discrete logarithm problem to justify its security, the first successful asymmetric-key cryptography system used the difficulty of the factoring problem instead. It wasn't until 1984 that **Tahir Elgamal**, an Egyptian graduate student working with Martin Hellman at Stanford, came up with an asymmetric-key system related to discrete logarithms. We will see the delay is not too surprising, since **ElGamal encryption** requires a few ideas that were absent from earlier public-key systems.

Since this is an asymmetric-key system, Bob starts by setting up the keys. As in Diffie-Hellman, he picks a very large prime p and a generator g modulo p. Then he picks a private key b between 1 and $p - 1$ and computes $B \equiv g^b$ modulo p. The numbers p, g, and B become Bob's public key, which he publishes. Also as in Diffie-Hellman, p and g don't have to be secret, and there's no harm in Bob using one that someone else is already using.

Since Bob is feeling lazy in our example, he's going to keep using the values $p = 2819$ and $g = 2$ that he and Alice used for Diffie-Hellman in Section 7.2. He decides to pick the private key $b = 2798$ and calculates $B \equiv 2^{2798} \equiv 1195$ modulo 2819. He posts p, g, and B in a public place and keeps b secret.

If Alice wants to send Bob a message block with plaintext P, she looks up his public key. Then she picks a random number r between 1 and $p - 1$. This number is called a **nonce**, meaning something that is made to be only used once. She uses r to compute two more numbers, $R \equiv g^r$ modulo p and $C \equiv PB^r$ modulo p. These two numbers, R and C, together form the ciphertext block that she sends to Bob. Alice keeps r secret; in fact she is done with it and can destroy the records of it if she wants.

Why does Bob need two numbers in order to decrypt the ciphertext? The idea is that B^r is a **blind**, or **mask**, which disguises the plaintext P. In order to separate the blind from the ciphertext, Bob needs R, which is a **hint**. This idea of a blind and a hint is one of the new ideas that was necessary before a system like ElGamal could be invented.

So Alice might proceed as follows:

nonces r:	1324	2015	5	2347	2147
hints g^r:	2321	724	32	1717	2197
blinds B^r:	93	859	1175	229	1575
plaintext:	al	lq	ui	et	fo
numbers:	1, 12	12, 17	21, 9	5, 20	6, 15
together:	112	1217	2109	520	615
times blind:	1959	2373	174	682	1708
ciphertext:	2321, 1959	724, 2373	32, 174	1717, 682	2197, 1708
nonces r:	1573	2244	2064	2791	1764
hints g^r:	1050	941	1336	1573	188
blinds B^r:	2395	798	1192	1215	1786
plaintext:	rt	he	no	nc	ex
numbers:	18, 20	8, 5	14, 15	14, 3	5, 24
together:	1820	805	1415	1403	524
times blind:	726	2477	918	1969	2775
ciphertext:	1050, 726	941, 2477	1336, 918	1573, 1969	188, 2775

Notice that the ciphertext that Alice gets depends on the random nonces that she picks. That makes ElGamal a probabilistic encryption method, like we saw in Section 7.7. We will see that the nonces in El-Gamal encryption are necessary to prevent a specific attack, but they also protect against the general forward search attack mentioned in that section.

To decrypt, Bob calculates $C\overline{R^b}$ modulo p. Since $R \equiv g^r$,

$$R^b \equiv (g^r)^b \equiv (g^b)^r \equiv B^r \quad \text{modulo } p,$$

so

$$C\overline{R^b} \equiv (PB^r)\overline{B^r} \equiv P \quad \text{modulo } p,$$

and Bob gets the plaintext back.

In our example, Bob's decryption looks like this.

ciphertext:	2321, 1959	724, 2373	32, 174	1717, 682	2197, 1708
hints R:	2321	724	32	1717	2197
blinds R^b:	93	859	1175	229	1575
$C\overline{R^b}$:	112	1217	2109	520	615
split apart:	1, 12	12, 17	21, 9	5, 20	6, 15
plaintext:	al	lq	ui	et	fo
ciphertext:	1050, 726	941, 2477	1336, 918	1573, 1969	188, 2775
hints R:	1050	941	1336	1573	188
blinds R^b:	2395	798	1192	1215	1786
$C\overline{R^b}$:	1820	805	1415	1403	524
split apart:	18, 20	8, 5	14, 15	14, 3	5, 24
plaintext:	rt	he	no	nc	ex

Alice

Bob

Picks p and g

Picks secret b

Uses b to make public $B \equiv g^b$ modulo p

Posts public encryption key (p, g, B)

Looks up Bob's encryption key (p, g, B)

Picks random secret r

r

$\downarrow (p, g)$

$R \equiv g^r$ modulo p

Plaintext P

$\downarrow (p, B, r)$

$C \equiv PB^r$ modulo p

$(R, C) \rightarrow$

(R, C)

$\downarrow (p, b)$

$P \equiv C\overline{R^b}$ modulo p

Figure 8.8. The ElGamal encryption system.

Note that Bob never finds out the nonces that Alice used, which isn't generally important one way or the other. A diagram of the whole system looks like Figure 8.8.

Another way of looking at ElGamal encryption is to think of Bob's public key as the first half of a Diffie-Hellman key agreement. Alice's random nonces and hints form the second part of the key agreement, and the keys that are created are used as a one-time keystream in a multiplicative cipher modulo p. Bob uses the hints to generate the same keystream on his end and then decrypts the multiplicative cipher. If Eve has a reason to think that Alice is using the same nonce or sequence of nonces more than once, then she knows that the keystream will also have repetitions. This is essentially the same as reusing a one-time pad, and Eve can make the same sort of attack as the one we saw in Section 5.2. Assuming Alice doesn't reuse a nonce, for Eve to get P from the public p, g, and B and the ciphertext R and C is equivalent to finding $B^r \equiv g^{rb}$ modulo p from p, g, $B \equiv g^b$ modulo p, and $R \equiv g^r$ modulo p. In other words, it's exactly the same as the Diffie-Hellman problem.

Since ElGamal was never patented, unlike Diffie-Hellman and RSA, it has been a common option in free and open-source encryption programs such as Pretty Good Privacy (PGP) and GNU Privacy Guard (GPG). Now that the Diffie-Hellman and RSA patents have expired, that's no longer such a big deal, and these programs now offer both RSA and ElGamal among their encryption options. The ElGamal digital signature scheme, which is related to ElGamal encryption and was developed at the same time, has been very influential and has led to several popular variations. See Section 8.4 for more on these.

8.3 ELLIPTIC CURVE CRYPTOGRAPHY

Around 1985, two mathematicians, **Neal Koblitz** and **Victor Miller**, independently realized that many of the public-key systems that we have seen could be adapted for use with certain mathematical objects known as **elliptic curves**. The first thing you need to know about elliptic curves is that despite having a related name and despite ellipses being curves, elliptic curves are not ellipses. Ellipses look like squashed circles, have two lines of symmetry, and are all one piece, as you can see in Figure 8.9. On the other hand, elliptic curves have only one line of symmetry, have two open ends, and have either one or two pieces, as shown in Figure 8.10.

Elliptic curves are given by equations of the form

$$y^2 = x^3 + ax^2 + bx + c,$$

which first came up in the seventeenth century when mathematicians started studying the arc length of an ellipse. There are a lot of things mathematicians find interesting about elliptic curves, but the thing that will make them useful for us is that they have an "addition law"—you can "add" two points on the curve and get a third point. The way this

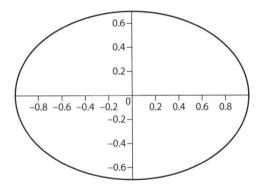

Figure 8.9. An ellipse (not an elliptic curve!).

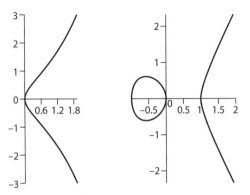

Figure 8.10. Two elliptic curves: $y^2 = x^3 + x$ and $y^2 = x^3 - x$.

works actually has very little to do with adding numbers, but it has the properties that we expect addition to have.

It's easiest to start with a graphical example: suppose we have the elliptic curve

$$y^2 = x^3 + 17.$$

Since

$$3^2 = (-2)^3 + 17$$

and

$$5^2 = 2^3 + 17,$$

the two points $P = (-2, 3)$ and $Q = (2, 5)$ are on the curve (see Figure 8.11). We need a rule to tell us how to get the point $P + Q$.

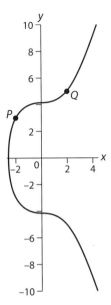

Figure 8.11. $y^2 = x^3 + 17$ with points $P = (-2, 3)$ and $Q = (2, 5)$.

We start by drawing a line through P and Q. This line always* inter-sects the curve in a third point, which we call R (see Figure 8.12). Now here's where the line of symmetry comes in: you can get a new point from R by reflecting it across the x-axis. The reflection of R, as shown in Figure 8.13, is the point we call $P + Q$.

Now for the equations: in our example, $P = (-2, 3)$ and $Q = (2, 5)$, and you can use high school geometry to calculate the equation of the line through them in point-slope form:

$$y - 3 = \frac{5 - 3}{2 - (-2)} (x - (-2)),$$

or

$$y = \tfrac{1}{2}x + 4.$$

Then we can find where this intersects $y^2 = x^3 + 17$:

$$y^2 = x^3 + 17 \qquad \text{and} \qquad y = \tfrac{1}{2}x + 4,$$

---

*Okay, almost always. Maybe you already see some exceptions. At any rate, we will get to them in a minute.

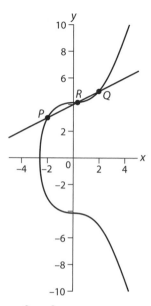

Figure 8.12. $y^2 = x^3 + 17$ with points P, Q, and R.

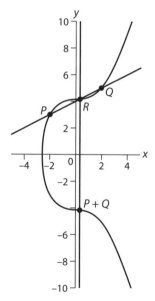

Figure 8.13. $y^2 = x^3 + 17$ with points P, Q, R, and $P + Q$.

so

$$\left(\tfrac{1}{2}x + 4\right)^2 = x^3 + 17,$$

or

$$x^3 - \tfrac{1}{4}x^2 - 4x + 1 = 0$$

which has solutions

$$x = 2, \qquad x = -2, \qquad \text{and } x = \tfrac{1}{4}.$$

We already know about the points with x-coordinate 2 and -2, so R must be the point with x-coordinate $\tfrac{1}{4}$. Then

$$y = \tfrac{1}{2}x + 4 = \tfrac{1}{2} \times \tfrac{1}{4} + 4 = \tfrac{33}{8},$$

so $R = \left(\tfrac{1}{4}, \tfrac{33}{8}\right)$. Reflecting across the x-axis is the same as multiplying the y-coordinate by -1, so the final result is $P + Q = \left(\tfrac{1}{4}, -\tfrac{33}{8}\right)$.

Why do we bother with the last reflection? For that matter, why is this procedure interesting at all? The reasons mathematicians are interested in this "addition" is because it behaves in many of the same ways as addition of numbers. For example, if P and Q are any two points on an elliptic curve, then it doesn't matter which order you draw a line through the points, so

$$P + Q = Q + P.$$

In other words, addition on elliptic curves is commutative, like addition and multiplication of numbers, but unlike the permutation products we saw in Section 3.4. It is somewhat more difficult, but also possible, to show that for any three points P, Q, and S,

$$(P + Q) + S = P + (Q + S),$$

so this addition is **associative**.

Now it's time to deal with the exceptions. The first case is easiest to deal with: what if you want to add a point, say the point Q from earlier, to itself? Here we need just a little bit of calculus. Remember that when we have two points in calculus and we let the two points approach each other until they coincide, then the line through the two points becomes a tangent line. So instead of drawing a line through two points, we draw the tangent line through Q, as in Figure 8.14, and then proceed as before. There will be one other point of intersection

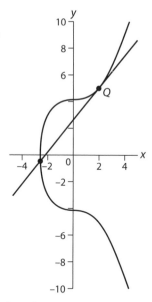

Figure 8.14. $y^2 = x^3 + 17$ with the tangent line through Q.

on the curve, and its reflection becomes $Q + Q$. The point $Q + Q$ is also called $2Q$, just like in high school algebra, and this is shown in Figure 8.15. The same logic holds if we draw a line through two distinct points and discover that instead of meeting the curve in a third point, the line is tangent to one of the two.

Aside from a little bit of calculus to find the slope of the tangent line, the equations work the same way in this case. We have $Q = (2, 5)$, and the slope of the tangent line can be found by implicit differentiation:

$$y^2 = x^3 + 17,$$

so

$$2yy' = 3x^2$$

or

$$y' = \frac{3x^2}{2y}.$$

Thus the tangent line through $(2, 5)$ is

$$y - 5 = \frac{3 \times 5^2}{2 \times 2}(x - 2),$$

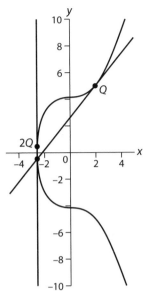

Figure 8.15. $y^2 = x^3 + 17$ with points Q and $2Q$.

or

$$y = \tfrac{6}{5}x + \tfrac{13}{5}.$$

As before, we can find where this intersects $y^2 = x^3 + 17$:

$$y^2 = x^3 + 17 \qquad \text{and} \qquad y = \tfrac{1}{2}x + 4$$

has solutions

$$x = 2, \qquad x = 2, \qquad \text{and } x = -\tfrac{64}{25}.$$

So the intersection point has x-coordinate $-\tfrac{64}{25}$ and y-coordinate

$$\tfrac{6}{5}x + \tfrac{13}{5} = -\tfrac{59}{125}$$

and the final result is $2Q = \left(-\tfrac{64}{25}, \tfrac{59}{125}\right)$.

The next exception might be a little more difficult to get your mind around. If you try to add two points P and Q that lie on a vertical line, then there is no third point of intersection. However, if you again imagine a point P' moving toward P and draw the line through P' and Q, you will see that the third point of intersection, R', has a y-coordinate

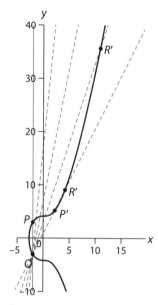

Figure 8.16. $y^2 = x^3 + 17$ with points P and Q lying on a vertical line.

that gets larger and larger or more and more negative, depending. This is shown in Figure 8.16. When P' coincides with P, we say that the point of intersection is at infinity and we write $P + Q = \infty$. The point at infinity is considered to be its own reflection, so we don't have to worry about that part.

Furthermore, any time we have a vertical line, we consider ∞ to be one of the points of intersection of the line with the curve. By symmetry, the other two points must be reflections of each other across the x-axis. So we can also consider Figure 8.16 as showing us that $P + \infty$ is the reflection of Q across the x-axis, which is P itself. This illustrates two more ways in which elliptic curve addition is like adding numbers. First, it has an **identity**, a point that acts like zero for addition, in that adding it to another point doesn't change the original point. In other words, $P + \infty = P$ for any P that you choose. Secondly, every point has an additive inverse that cancels it out, because if Q is the reflection of P, then $P + Q$ is the identity. To emphasize the similarity with negatives of numbers, we use $-P$ to denote the reflection of P. Then we can write $P + (-P) = \infty$, or $P - P = \infty$.

Now that you have seen the addition law for elliptic curves, you might have started to get an idea about how they could be useful in cryptography. But in order to be really useful, we need to introduce one more thing, which is the "wraparound" idea from Chapter 1. To do this, we will pick a prime number p and treat two points as the same if their coordinates are the same modulo p.

For example, take the curve

$$y^2 = x^3 + 17$$

again and the prime $p = 7$. We saw that the point $P = (-2, 3)$ is on the curve, but modulo 7 this is the same as the point $(5, 3)$. It's not true that

$$5^2 = 3^3 + 17,$$

but it is true that

$$5^2 \equiv 3^3 + 17 \quad \text{modulo 7,}$$

so we say that $(5, 3)$ is on the curve modulo 7. Likewise, the point $Q = (2, 3)$ is on the curve modulo 7. What about $P + Q = \left(\frac{1}{4}, -\frac{33}{8}\right)$? Well, the equivalent of $\frac{1}{4}$ modulo 7 is $\overline{4}$, which is 2, and the equivalent of $-\frac{33}{8}$ is

$$-33 \times \overline{8} \equiv 2 \times \overline{1} \equiv 2 \times 1 \equiv 2 \quad \text{modulo 7,}$$

so $P + Q \equiv (2, 2)$ modulo 7, and you can check that

$$2^2 \equiv 2^3 + 17 \quad \text{modulo 7.}$$

This confirms that $P + Q$ is on the elliptic curve modulo 7. If we encounter a point where we need to find the inverse of something and we can't, then we consider that point to be ∞. The geometric picture of elliptic curves with which we started no longer makes very much sense when we are working modulo p, but all the formulas that are necessary to do addition still work and all of the properties that we have discussed still hold. So we can talk about addition on an elliptic curve modulo p without any problems.

Although we have been calling our method of combining points on elliptic curves addition, there is one important way in which it is less like addition of numbers and more like multiplication. That's because there is a discrete logarithm problem for elliptic curve addition. Remember that the discrete logarithm problem for numbers is where Eve is given numbers C and P and a prime number p and has to find a whole number e such that

$$C \equiv P^e \quad \text{modulo } p.$$

For the rest of this section I am going to call that the modular exponentiation discrete logarithm problem to distinguish it from what I am about to introduce.

Remember that $2P$ means $P+P$ using the addition law for the elliptic curve. In general, eP means P added to itself e times using the addition law, and $0P$ means ∞, because it's the identity for the addition law. Then the **elliptic curve discrete logarithm problem** is where Eve is given the equation of an elliptic curve and points C and P and a prime number p, and she has to find a whole number e such that

$$C \equiv eP \quad \text{modulo } p.$$

Like the modular exponentiation discrete logarithm problem, we think the elliptic curve discrete logarithm problem is hard, but we don't know for sure. In fact, the elliptic curve version seems to be even harder, since some of the methods we know to solve the modular exponentiation discrete logarithm problem don't seem to work on elliptic curves.

Now we pretty much have all the ingredients for elliptic curve cryptography. As Neal Koblitz tells the story, in 1984 he was involved in research on elliptic curves as professor at the University of Washington, when he got a letter from another mathematician about a method of factoring large integers using elliptic curves. Since Koblitz was well aware that factoring integers was important for the security of RSA, this got him thinking about elliptic curves and factoring. Before he had any results, however, Koblitz left for a previously scheduled trip to the Soviet Union, where he spent several months. While he was there he hit on the idea of using the elliptic curve discrete logarithm to construct a cryptographic system, but of course no one in the Soviet Union could talk to an American about cryptography. Koblitz wrote a letter to another mathematician in the United States describing his idea and a month later got a response. Not only was Koblitz' idea a good one, but it was so good that Victor Miller, who was working at IBM, had independently had the same idea. In the end, both Koblitz and Miller published papers written in 1985 on the topic.

Miller's paper explained how Alice and Bob can do **elliptic curve Diffie-Hellman key agreement**. They need to pick some public information, namely, a specific elliptic curve and a very large prime number p—but not as large as in the modular exponentiation Diffie-Hellman system, since we think the elliptic curve version is harder to break. Experts think that a security level equivalent to the 600-digit prime I mentioned for modular exponentiation Diffie-Hellman in Section 7.2 would be given by a prime "only" about 70 digits long for the elliptic curve system.

Then Alice and Bob need to find a point G. Unlike in the case of numbers modulo p, it may not be possible to find a G that generates all of the points on an elliptic curve, but it should generate a large number of them. As with the generators modulo p that are needed for modular exponentiation Diffie-Hellman (see Section 7.2), these items could be looked up in a table if Alice and Bob don't want to bother computing them.

For the secret information, Alice picks a number a and Bob picks a number b. Then Alice computes the point on the elliptic curve $A \equiv aG$ modulo p and sends it to Bob, and Bob computes $B \equiv bG$ modulo p and sends it to Alice. Finally, Alice computes aB modulo p, which is the same as abG modulo p, and Bob computes bA modulo p, which is the same as $baG \equiv abG$ modulo p, so once again Alice and Bob have a shared piece of secret information that they can use as a secret key. A diagram of the system looks like Figure 8.17.

In order to get the shared secret, Eve would have to solve the **elliptic curve Diffie-Hellman problem**, which is figuring out abG from aG and bG. As in the case of numbers modulo p, we think that this is probably as hard as the elliptic curve discrete logarithm problem, which we think is hard, but we aren't sure about any of it. These problems haven't been considered for as long as the other hard problems we have discussed, but it's still been more than 25 years, and Eve hasn't had a lot of luck. The current record for finding a discrete logarithm on an elliptic curve modulo p is for a curve modulo the 34-digit, or 112-bit, prime

$$p = \frac{2^{128} - 3}{11 \times 6949}.$$

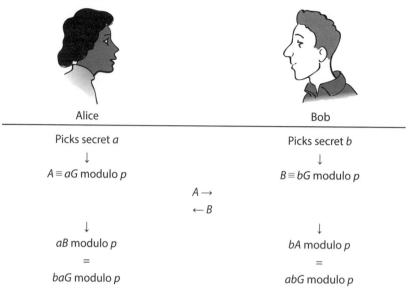

Figure 8.17. Elliptic curve Diffie-Hellman key agreement.

The computation ran (with interruptions) for about 6 months on a cluster of more than 200 PlayStation 3 game consoles.

The Diffie-Hellman system isn't the only public-key cryptographic system that has an analog over elliptic curves. RSA doesn't, because while there are prime numbers and prime polynomials, there doesn't seem to be any good way of defining prime points on an elliptic curve. Therefore, the factoring problem doesn't seem to have any good equivalent. The three-pass protocol and ElGamal encryption, on the other hand, are based on discrete logarithms and therefore have elliptic curve analogs, as Koblitz explained in his paper. **Elliptic curve ElGamal encryption** is a straightforward adaptation of the modular exponentiation version and is shown in Figure 8.18.

The **elliptic curve three-pass protocol**, on the other hand, has a slight catch to it. In addition to the things we have discussed, it is now necessary for Alice and Bob to know the number of points on the elliptic curve modulo p. This is because they need the following equivalent of the Euler-Fermat theorem.

Alice   Bob

Picks an elliptic curve, p, and G

Picks secret b

Uses b to make private decryption key
$B \equiv bG$ modulo p

Posts public encryption key
(curve, p, G, B)

Looks up Bob's encryption key
(curve, p, G, B)

Picks random secret r

r

$\downarrow$ (curve, p, G)

$R \equiv rG$ modulo p

Plaintext P
(represented as a point on the elliptic curve)

$\downarrow$ (curve, p, B, r)

$C \equiv P + rB$ modulo p

$(R, C) \rightarrow$

(R, C)

$\downarrow$ (curve, p, b)

$P \equiv C - bR$ modulo p

Figure 8.18. The elliptic curve ElGamal encryption system.

Theorem (The Elliptic Curve Euler-Fermat Theorem) *For any elliptic curve and any prime p, let f be the number of points on the curve (including ∞) which are distinct modulo p. Then for any point P on the elliptic curve,*

$$fP \equiv \infty \quad \text{modulo } p.$$

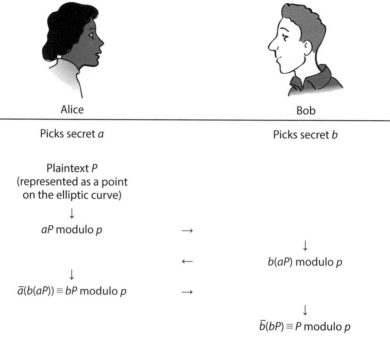

Alice | Bob

Picks secret *a* | Picks secret *b*

Plaintext *P*
(represented as a point
on the elliptic curve)
↓
aP modulo *p*   →

                    ↓
          ←   *b*(*aP*) modulo *p*

↓
$\bar{a}(b(aP)) \equiv bP$ modulo *p*   →

                    ↓
          $\bar{b}(bP) \equiv P$ modulo *p*

Figure 8.19. The elliptic curve three-pass protocol.

Once again, if Alice and Bob don't feel like calculating *f*, they can find a curve and a *p* for which they can look it up, and if they do want to compute *f*, there are fairly fast techniques for it.

The elliptic curve Euler-Fermat theorem is useful to us because since

$$fP \equiv \infty \equiv 0P \quad \text{modulo } p,$$

computations with *a* in an expression *aP* on an elliptic curve modulo *p* actually work modulo *f*, in the same way that computations with *a* in the expression k^a modulo *n* actually work modulo $\phi(n)$. So *a* and *b* in Figure 8.19 need to have inverses modulo *f*, and $\bar{a}$ and $\bar{b}$ in the figure need to be taken modulo *f*. Given that, everything proceeds as you would expect.

Elliptic curve cryptographic systems took a while to catch on, but they have been attracting more attention in recent years. I'll say more about why when we look forward at the end of this chapter.

8.4 DIGITAL SIGNATURES

In Section 7.2, I quoted Whitfield Diffie as saying that his and Martin Hellman's invention of public-key cryptography was the result of "two problems and a misunderstanding." We haven't talked about the second problem yet, which is the problem of authentication: how can the recipient of a digital message be sure of who the sender was? Symmetric-key cryptography solves this problem, but only in a limited way. If Alice and Bob have a shared secret key that no one else knows and Bob gets a message encrypted using that key, then he knows only Alice could have sent it. However, there are situations in which this isn't good enough. If Alice and Bob have not been able to exchange a secret key, then they can't use it to prove who they are any more than they can use it to keep a secret. Furthermore, suppose Alice and Bob do have a secret key, which Bob can use to verify that Alice sent a particular message. Now what if Bob wants to prove to a third party that Alice sent the message? At the very least he would have to reveal a secret encryption key, which is often undesirable. Then Bob would have to prove that Alice really had the key and hadn't given it to anyone else, which might be difficult without Alice's cooperation. And even if Bob did achieve all that, he still can't prove that he didn't write the message himself, since he knew the key the same as Alice.

What we need is a **digital signature**, which acts like a handwritten signature on a document: it should be difficult to forge, and it should be difficult to remove from one document and attach it to some other document. It's not good enough to scan your handwritten signature and attach it to the bottom of an email or a file, because it is usually easy for someone to copy that part of the file and attach it to another part. Or, someone could simply get any piece of paper with your signature and scan it and attach it themselves.

In Diffie and Hellman's first paper, before they knew how to make an asymmetric-key encryption system, they already understood how such a system could be used to provide "a time and message dependent digital signature which cannot be forged even when past signatures have been seen." We need a couple of assumptions: first, that it is possible to treat a plaintext message as if it were a ciphertext and, second, that even if encryption and decryption are done out of order, you still get the same message back. These aren't always true, but sometimes they are.

If Alice wants to send Bob a signed message, she takes the message and applies her *decryption* key to it, as if it were a ciphertext. Since Alice's decryption key is private, she is the only one who can do this. When Bob receives the message, he can apply Alice's encryption key, which cancels out the decryption. Alice's encryption key is public, so Bob doesn't have to share a secret with Alice in order to verify the signature. If he gets back a recognizable message, then he knows it must have come from Alice. In some cases it might be a good idea for Alice to send the unsigned message as well as the signed message so Bob can compare them. Remember that we are not trying to keep the message secret, just authenticate it. In addition, Bob can demonstrate to a third party, such as Carol, that Alice signed the message. Since anyone can get Alice's public encryption key, Carol can assure herself that Bob isn't giving her a fake key. And since Bob doesn't have Alice's private decryption key, Carol knows Bob couldn't have signed the message himself and claimed Alice did it.

The RSA system was the first one that could actually be used for a digital signature in this way, so let's use it for our example. Alice is going to send Bob a signed message. Her private primes are $p = 59$ and $q = 67$, making her public modulus $n = 3953$. Her public encryption key in this case might be more descriptively called a **verification key**, and we will represent it by v. Alice would like this key to be small for speed reasons, so she chooses $v = 5$. She calculates $\phi(n) = (p-1) \times (q-1) = 3828$, which means her private decryption key, or **signing key**, is $\bar{5}$ modulo $3828 \equiv 2297$. We will call this σ. (That's the Greek letter sigma, for "signing".) As usual, she posts n and v in a public place and keeps the rest secret. Then to sign the message M, she sends Bob the signature $S \equiv M^\sigma$ modulo n:

message:	ev	er	yw	he	re	as	ig	nx
numbers:	5, 22	5, 18	25, 23	8, 5	18, 5	1, 19	9, 7	14, 24
together:	522	518	2523	805	1805	119	907	1424
to the 2297th power:	2037	2969	369	3418	3746	1594	1551	1999

Alice	Bob

Picks secret p and q
Uses p and q to make public verification key (n, v)
Uses p and q to make private signing key σ

Posts verification key (n, v)

$$M$$
$$\downarrow (n, \sigma)$$
$$S \equiv M^\sigma \text{ modulo } n$$

$$S \rightarrow$$

Looks up Alice's verification key (n, v)

$$S$$
$$\downarrow (n, v)$$
$$M \equiv S^v \text{ modulo } n$$

Figure 8.20. The RSA digital signature system.

Bob can recover the message and check the signature by calculating $M \equiv S^v$ modulo n:

signature:	2037	2969	369	3418	3746	1594	1551	1999
to the 5th power:	522	518	2523	805	1805	119	907	1424
split apart:	5, 22	5, 18	25, 23	8, 5	18, 5	1, 19	9, 7	14, 24
message:	ev	er	yw	he	re	as	ig	nx

Since the message makes sense, Bob concludes that it is a genuine message from Alice. The entire RSA digital signature scheme looks like Figure 8.20.

There are some other desirable things that can be added to this scheme. Since anyone can verify the signature and recover the message, there is no secrecy provided by a digital signature. However, if

Alice wants her message to Bob to be signed *and* encrypted, that's no problem. Alice has a private p, q, and σ and a public n and v. Bob has a private p, q, and d and a public n and e. Note that Alice's p, q, and n will be different than Bob's. After Alice applies her private signing key σ, she can encrypt the whole thing with Bob's public encryption key e. When Bob gets the message, he first decrypts it with his private decryption key d and then verifies the signature with Alice's public verification key v.

Another common way that digital signatures are combined with public-key encryption is in a **certificate**. One issue we have not really discussed is the question of how Bob knows that Alice's public key, either for encryption or verification, really belongs to her. He needs to make sure it wasn't posted somewhere by Eve in an attempt to fool people into sending messages that Eve can read. Before Alice posts her public key, she can have it signed by Trent, a trusted authority. Trent gives Alice a certificate, which is essentially a statement of what Alice's public key is. The certificate is signed with Trent's private signature key. If Bob has Trent's public verification key, he can verify the signature and get some assurance that Alice's public key is correct. If Bob doesn't have Trent's key already, he can get a certificate for Trent signed by someone else, and so on. This is called a **certificate chain**. Web browsers use certificates like these to verify that secure Web sites really belong to the organizations to which they claim to belong. The certificate chain stops when it reaches one of a set of public keys that are built into the browser and (hopefully) were confirmed when the browser software was written.

Incidentally, certificates based on RSA digital signatures are by far the most common on the web. This is probably because Netscape, the first Web browser to use certificates, had only one built-in certificate, which was signed by RSA Data Security, Inc. The certificate services division of RSA Data Security was later spun off into a company called VeriSign, which is now owned by Symantec. Symantec also owns several other companies that issue Web certificates, and it was still the leading source for certificates on the Internet, at least as of 2013. Popular browsers such as Internet Explorer, Firefox, Chrome, and Safari also support certificates based on another system, the Digital Signature Algorithm. I'll say a little bit more about the Digital Signature Algorithm shortly.

We have seen how a digital signature protects Alice and Bob from Frank the forger's obvious attack, which is to make up a message and try to get Bob to think it came from Alice. There are a couple of attacks that are more subtle, and defending against them requires a couple of other additions to the scheme. The first attack is called a **replay attack**. Frank listens in while Alice sends Bob a signed message and records it. Then later he replays the message, sending it to Bob as if it had come from Alice. Bob verifies the signature and concludes that it came from Alice, since she did originally sign it. If the message is a simple one, such as meet me at eight o'clock or send me file X, then Bob may not see anything wrong with getting the same message twice at different times, and it could potentially cause a lot of trouble. Or, Frank might have managed to intercept the message the first time so Bob only gets it once, but at the wrong time. The standard solution for this is simply to include a **timestamp** as part of the message, so that a message can't be repeated or delayed. This timestamp needs to be added to the message before the signature so that Frank can't change it without invalidating the signature. Then Alice and Bob need to make sure they have synchronized clocks, which leads to another whole set of issues.

Another type of attack is called an **existential forgery**. We saw earlier that the fact that Eve can encrypt any plaintext leads to a forward search attack. An existential forgery attack is made possible by the corresponding fact that Frank can verify any signature. In this attack Frank takes a random string of numbers or bits and applies Alice's verification key to it as if it were a signature. He then sends Bob the "message" that was produced by the verification key and the "signature" that he started with. The message will be a bunch of random numbers or bits, not anything like English words (or any other language). But the signature will correctly verify as Alice's. In some circumstances, such as if Bob is expecting a certificate containing nothing but a signed public key, this could cause Bob a great deal of trouble and perhaps even lead to a security breach. The defense against a forward search attack is to add randomness to the encryption process. Conversely, the defense against an existential forgery is to add structure, thus reducing the randomness. If, for instance, Bob knows that the certificate should contain not only the public key but Alice's name, a timestamp, or both, then it's very unlikely that Frank will be able to try enough random signatures to produce a message that Bob will believe.

The RSA digital signature scheme is an example of a **reversible digital signature**, sometimes also called a **digital signature with message recovery**, because the verification process reverses the signature processes and returns the original message. There are also **nonreversible digital signature** schemes, which produce a signature that cannot be used to recover the original message. In this case, Alice always needs to send both the message and the signature to Bob. Sometimes a scheme like this is called a **digital signature with appendix**, because the signature is often sent as an appendix to the message.

Not being able to reverse the signature sounds inconvenient, but it does have some advantages. One is that the signature can be much shorter than the message, which makes the calculations faster. Also, Alice can give Bob the signature of a message at one point in time to prove that she knows a piece of information and only later reveal the message containing the information.

An example of a nonreversible digital signature is the **ElGamal signature scheme**, which is closely related to ElGamal encryption and was developed at the same time. This system is illustrated in Figure 8.21. The ElGamal signature scheme has been very influential and has led to several popular variations, including the **Digital Signature Algorithm (DSA)**. The DSA was the first digital signature system endorsed by NIST, in 1994, and is still a federal standard. It was somewhat controversial when first proposed but now seems to be generally accepted. There is also an **elliptic curve ElGamal digital signature scheme** and the **Elliptic Curve Digital Signature Algorithm (ECDSA)**. Like the DSA, the ECDSA is a federal standard, as of 2000.

One company's (mis)use of the ECDSA made a fairly big splash, at least among people interested in cryptography, at the end of 2010. Sony used ECDSA in its PlayStation 3 video game console, which was released in 2006. The digital signature was used to identify code that had been approved by Sony to run on the console and prevent unauthorized code from being run. Unfortunately, it seems that Sony had neglected to observe an important fact about the ECDSA. Like ElGamal encryption and ElGamal digital signatures, the DSA and ECDSA use a random nonce. And as we noted in Section 8.2, if a nonce is reused, the system

Alice

Bob

Picks p and g
Picks secret a
Uses a to make public $A \equiv g^a$ modulo p

Posts public verification key (p, g, A)

Picks random secret r

r
$\downarrow (p, g)$
$R \equiv g^r$ modulo p

message M
$\downarrow (p, a, r, R)$
$S \equiv \bar{r}(M - aR)$ modulo $p - 1$

$(R, S, M) \rightarrow$

Looks up Alice's verification key (p, g, A)
(R, S, M)
$\downarrow (p, g, A)$
Is $A^R R^S \equiv g^M$ modulo p?
If it is, the signature is valid.

Figure 8.21. The ElGamal digital signature scheme.

is insecure. In late 2010 a group of hackers revealed that Sony was using the same nonce for every signature. This made it possible for them to recover Sony's private signing key and create their own software for the PlayStation 3. Soon another hacker had also recovered the key and published it on his Web site. Sony filed a lawsuit against all these hackers, which was settled out of court in April 2011.

8.5 LOOKING FORWARD

The three-pass protocol, as I said, is currently much too slow to use except in certain very specialized situations, which probably rarely occur in practice. If someone were to come up with a symmetric-key cipher that was commutative and resistant to known-plaintext attacks but competitive in speed with modern block ciphers, it would suddenly make the three-pass protocol very attractive. Currently that doesn't seem particularly likely.

ElGamal encryption, in both the original and elliptic curve versions, turns out to be subject to an **adaptive chosen-ciphertext attack**, where Eve is trying to read a ciphertext that Alice has sent Bob. If Eve can trick Bob into deciphering a related ciphertext (this is the adaptive part) and revealing what it decrypts to, then Eve can recover the original message. A number of variations on ElGamal encryption have been proposed to remedy this. One of the simpler ones, the **Diffie-Hellman integrated encryption scheme** (DHIES), uses the same blind and hint as ElGamal but combines the blind and the message using symmetric encryption rather than modular multiplication. Its elliptic curve version, the **elliptic curve integrated encryption scheme** (ECIES), has attracted attention due to the fact that shorter elliptic curve keys seem to provide the same level of security as longer RSA and modular exponentiation DLP-based keys, making elliptic curve systems potentially faster and more convenient for the same level of security. ECIES has been endorsed by a committee of the Japanese government and a number of industry committees, although not by the US Government.

I mentioned that an advantage of using elliptic curves is the shorter key sizes. This is convenient in many situations, especially where there is very little memory, such as smart cards or radio frequency identification tags. It would be even more convenient if the keys could be broken into smaller pieces that didn't have to be operated on all at once. This can be done with a more general type of curve, known as a hyperelliptic curve. Hyperelliptic curves are given by equations of the form

$$y^2 = x^n + a_{n-1}x^{n-1} + a_{n-2}x^{n-2} + \cdots + a_2 x^2 + a_1 x + a_0,$$

where n is larger than 4. Compared to elliptic curves, these curves have a rather more complicated addition law, which operates on sets of points rather than one point at a time. The size of the whole set of points making up a key is similar to the size of the key of an elliptic curve, but some of the calculations needed for **hyperelliptic curve cryptography** can be done one point at a time.

Another advantage of elliptic curves is that they sometimes have extra useful structure beyond the addition law. For instance, some elliptic curves have a "pairing" function, which has the property (among others) that for some point G on the elliptic curve and any two integers a and b,

$$f(aG, bG) = f(G, G)^{ab}.$$

This can be used in a **tripartite Diffie-Hellman key agreement** to let three people agree on a piece of secret information: if Alice chooses a secret a and a public $A = aG$, Bob chooses a secret b and a public $B = bG$, and Carol chooses a secret c and a public $C = cG$, then

$$f(B, C)^a = f(A, C)^b = f(A, B)^c = f(G, G)^{abc},$$

and all three of them can calculate the secret. Another possible use of pairing functions is in **identity-based encryption**. The idea here is that when Alice wants to send a message to Bob, instead of looking up his public key she can just generate it herself from his e-mail address or some other piece of public information. Not only is this convenient, but Alice doesn't have to worry so much about Eve tampering with whatever source Alice got the key from. Then Alice carries out a set of computations similar to ElGamal encryption, but involving the pairing of Bob's public key with the public key of Trent, the trusted authority. Bob deciphers the message using the pairing and a secret key unique to him but generated by Trent using Trent's secret key. For the details, see the references in the endnotes.

In 2005, the NSA announced the "Suite B" set of approved cryptographic algorithms for communicating classified and other sensitive data both within and to the US Government. These algorithms originally included AES for symmetric-key cryptography, elliptic curve Diffie-Hellman and one other elliptic curve algorithm for key agreement, the Elliptic Curve Digital Signature Algorithm, and an algorithm for helping create short, nonreversible digital signatures. AES and the short-signature algorithm are already pretty much commercial as well as government standards, and the NSA was clearly hoping that the elliptic curve algorithms would follow. The NSA particularly mentioned the speed and security advantages of elliptic curve cryptography that come from having a smaller key size.

Despite these advantages and the push from the NSA, elliptic curve cryptography has been slow to catch on commercially. One reason is that cryptographers are inherently conservative and tend to stick with systems as long as they do not appear to be broken—the longer people have been unsuccessfully trying to break a system, they feel, the less likely it is that there will be a nasty surprise tomorrow.

Two recent developments have cast more doubt on the adoption of elliptic curve algorithms. The first is related to a system for generating random numbers that could be used for generating secret keys, secret information for public key systems, or random choices for probabilistic encryption. The system, known as the **Dual Elliptic Curve Deterministic Random Bit Generator** (Dual EC DRBG), was first published in 2004 and was adopted as a NIST recommended standard in 2006, along with three other random-number-generation systems. As the name suggests, Dual EC DRBG uses two elliptic curves. For that and other reasons, the system was much slower than the other three, which seemed odd. Also, researchers discovered early on that the random numbers had a small bias, which ought to have disqualified it as a standard unless it had some other superiority that was not obvious. Finally, the default setup for the standard included some arbitrary choices that were never explained. Ever since the DES S-boxes, unexplained choices in a system have made cryptographers very suspicious that something might have been done to weaken it.

Then in 2007, two researchers from Microsoft showed how someone who knew a certain relationship between two of these choices could use it to predict the supposedly random numbers produced by the system after watching the output for just a short time. That kind of back door would allow anyone who knew the relationship to break any cryptographic system that had used the system to generate secret information.

At this point it was already suspected that the NSA had rigged the standard so that they could break it. The suspicion remained in the background, however, until the Snowden release in 2013. Documents in that release suggest that the NSA originally came up with the Dual EC DRBG and successfully pushed for it to be included in the standard and also an international standard. Eventually NIST removed the system from its recommended standards, but not before at least one well-known cryptographer advocated staying away from elliptic-curve systems entirely, saying they "have constants that the NSA influences when they can."

In 2015, a second development weakened the argument for adopting elliptic curve algorithms. In August of that year, the NSA announced preliminary plans to replace Suite B with a new set of algorithms that would be resistant to the sort of quantum computer we will talk about in the next chapter. Most elliptic curve algorithms, unfortunately, would be vulnerable to this sort of computer if it were made practical. What algorithms might be under consideration to replace them has not been revealed, but we will take a look at some of the candidates in Section 9.2. In the meantime, for those who have not yet adopted elliptic curve algorithms, the NSA recommended "not making a significant expenditure to do so at this point." Instead, Diffie-Hellman and RSA were added to the list of acceptable algorithms for classified and sensitive data during the transition period. NIST followed suit with a report on the state of quantum-resistant cryptography, finalized in April 2016. According to the report, new standards for quantum-resistant algorithms will be developed by a process similar to the AES competition but with the likelihood of NIST endorsing multiple candidates in various categories. The submission deadline is planned for late in 2017, followed by 3 to 5 years of public scrutiny before final standards are announced.

......9.......

The Future of Cryptography

9.1 QUANTUM COMPUTING

As we have said several times, the security of the public-key ciphers in current use relies on the apparent difficulty of solving some well-known mathematical problems, such as the discrete logarithm problem and factoring. No one has been able to find an easy way to solve these problems, but no one has been able to prove that there isn't one. So, it's always possible that someone will come along tomorrow and announce that they've discovered a new mathematical technique that will break all these codes.

Even if a new mathematical technique doesn't appear, it's also possible that a new kind of computer could make these codes insecure. The most likely candidate is a computer based on quantum physics. Even though no one has publicly demonstrated a quantum computer capable of solving problems of nontrivial size, researchers in the last couple of decades have started figuring out how to write programs for such computers. This new field is known as **quantum computing**, and it might have important repercussions for cryptography.

The most famous thing that makes quantum physics different from classical physics is probably **superposition**, the idea that a quantum particle, like Schrödinger's hypothetical cat, can be in more than one state at the same time. In 1935, the physicist **Erwin Schrödinger** asked what would happen if a cat was placed in a sealed box where it couldn't be seen or heard. A small amount of radioactive material was also placed in the box, such that in the course of an hour there was a 50% chance that an atom would decay and a 50% chance that nothing would happen. If a Geiger counter in the box detected a decay, it would activate an automatic cat-food dispenser; otherwise nothing would happen. At the

Figure 9.1. Is Schrödinger's cat hungry or fed?

end of an hour, is the cat hungry or fed? (See Figure 9.1.) According to quantum physics, until we open the box to find out, the atom has both decayed and not decayed, so the cat is both hungry and fed. Just like the cat can be in two different states at the same time, a **quantum bit**, or **qubit** (pronounced "cue-bit"), can be both 0 and 1 at the same time instead of having to be either one or the other.

A lot of descriptions of quantum computing make this sound like an instant solution to all our problems. Suppose we want to factor a number using a quantum computer: say, 4. We start by setting a bunch of qubits to the binary representation of the number: 100. Now we take another bunch of qubits and set them to a possible factor. But instead of trying each possible factor one by one, we'll set the factor qubits to

$$\left\{ \begin{array}{c} 0 \\ \text{or} \\ 1 \end{array} \right\} \left\{ \begin{array}{c} 0 \\ \text{or} \\ 1 \end{array} \right\},$$

so each qubit is simultaneously 0 and 1, and together they make the "qunumber"

$$\left\{ \begin{array}{c} 00 \\ \text{or} \\ 01 \\ \text{or} \\ 10 \\ \text{or} \\ 11 \end{array} \right\},$$

which is 0, 1, 2, or 3. We are looking for factors less than 4, so we are good so far. Then we can divide 4 by the qunumber, and we keep the quotient if it's a whole number less than 4 or output 0 if not. We get the qubit representation

$$\left\{ \begin{array}{c} 00 \\ \text{or} \\ 00 \\ \text{or} \\ 10 \\ \text{or} \\ 00 \end{array} \right\}.$$

Now we have to remember the other part of Schrödinger's thought experiment: until we open the box, the cat is both hungry and fed, but once we open it and look inside, the cat instantaneously "collapses" into one state or the other. Applying this to our quantum factoring says that when we examine the output of the quantum computer, we will sometimes get the answer 00, which is not a nontrivial divisor and isn't helpful. And sometimes we will get the answer 10, which is helpful because 2 is a divisor. But we can do that with a probabilistic algorithm, so we haven't gained anything just by using superposition.

However, there's another aspect of quantum physics we can use. Consider the setup in Figure 9.2. A single subatomic particle, such as an electron or photon, is sent in the direction of a beam splitter. In the case of a photon, for instance, the beam splitter might be a half-silvered mirror. If we do this repeatedly, half the time the particle passes through the beam splitter in the same direction and half the time it bounces it off in the other direction, which is exactly what we observe at the detectors (Figure 9.3). So far the particle has behaved entirely according to probabilistic principles.

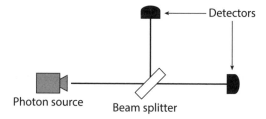

Figure 9.2. An experiment with one beam splitter.

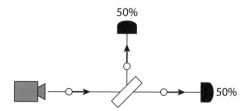

Figure 9.3. Each detector registers half the time.

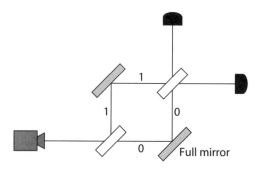

Figure 9.4. An experiment with two beam splitters.

Now consider the set up in Figure 9.4, which has two beam splitters and two fully reflective barriers, such as fully silvered mirrors, that the particle always bounces off of. If each beam splitter passes the particle through half the time, there are two possible paths to each detector, and they are all equally likely. So we would still expect the particle to end up at each detector half the time. But that's *not* necessarily what happens. Instead, depending on the exact details of the setup, we might see the particle at the same detector all of the time. (See Figure 9.5.)

The explanation is that somehow each particle is *both* going through *and* bouncing off each splitter, causing it to **interfere** with itself when

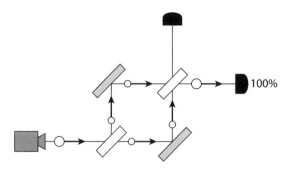

Figure 9.5. One detector never registers and the other one always does!

the paths rejoin. The mirrors can be arranged so that in one case the interference causes the particle to cancel itself out and in the other case the particle reinforces itself, so that one detector never registers the particle and the other one always does. We need to take advantage of interference as well as superposition in order to fully take advantage of quantum computing.

I'm not going to get into the details, but suffice it to say that in 1985, the physicist **David Deutsch** described the first **quantum algorithm** that could solve a computational problem using a quantum computer faster than it could be solved using a conventional computer. The particular problem wasn't very interesting, but the techniques were extremely important. They led directly to the first "useful" quantum algorithm in 1994, when **Peter Shor** discovered an algorithm to (probabilistically) factor numbers using a quantum computer faster than any known algorithm for conventional computers. In fact, this algorithm can factor numbers roughly as fast as any algorithm, quantum or conventional, can find large prime numbers. So widespread use of this algorithm would make RSA completely insecure. Shor's paper also showed how to solve the discrete logarithm problem quickly, so Diffie-Hellman and all the other systems based on that and variations on it, including the elliptic curve discrete logarithm problem, would also be insecure.

Progress toward large quantum computers has been slow, but it seems to have picked up lately. The limiting factor at the moment is the number of qubits that we can construct and keep stable. In 2001, a team of IBM scientists and Stanford graduate students announced that they had used a quantum computer with 7 qubits to factor the number 15, which is the smallest number for which Shor's algorithm works. In 2012 a group in the United Kingdom found a way to factor 21 using fewer qubits, and another group in China factored 143 using only 4 qubits and a different algorithm. In 2014 it was announced that the identical 4-qubit computation that factored 143 can be used to factor numbers as large as 56,153, but only if they have a special form.

If quantum computers could make all common public-key cryptography insecure, what about symmetric-key systems? The situation there isn't quite as dramatic, but quantum computers would have an effect there, too. In 1996, **Lov Grover**, an Indian-American computer scientist at AT&T Bell Labs, invented a quantum algorithm for (probabilistically) searching a database much faster than you can do it with a classical computer. In particular, if you have N things to search through, such as N keys for a symmetric system, then Grover's algorithm can do it in only $\sqrt{N}$ steps. The smallest-size AES key, to be very explicit, has 128 bits, so a brute-force attack with a conventional computer needs to search through 2^{128} keys. Grover's algorithm would need to do only the equivalent of searching through $\sqrt{2^{128}} = 2^{64}$ keys. The use of quantum computers immediately cuts the effective key size of a symmetric-key cipher in half, at least as far as brute-force search is concerned. The NSA is now recommending 256-bit AES keys as part of its transition to a new set of quantum-resistant algorithms, as mentioned in Section 8.5.

9.2 POSTQUANTUM CRYPTOGRAPHY

What would cryptographers do if quantum computers became common? For symmetric-key systems, raising key sizes seems to be sufficient at the moment. For public-key systems, research is being done into what is often called **postquantum cryptography**, although a more descriptive name might be **quantum-resistant cryptography**. These are systems based on problems that are not known to be easily solvable using a quantum computer. Instead of relying on factoring or discrete logarithms, they rely on problems such as solving systems of multivariable polynomials, finding the shortest distance from a point to an n-dimensional skewed grid of other points, or finding the closest bit

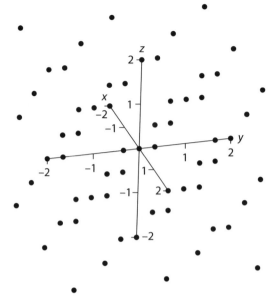

Figure 9.6. A three-dimensional lattice.

string to a set of other bit strings. These methods typically have not been used in the past because they are less efficient. They are getting better, however, and as we saw in Section 8.5, at least the NSA and NIST think that it is time to move toward adopting them.

As an example, let's take a look at **lattice-based cryptography**. A **lattice** is an evenly spaced grid of points in an n-dimensional space equipped with coordinate axes. A three-dimensional lattice, for example, is shown in Figure 9.6. There are two standard lattice problems that are thought to be hard, even for quantum computers. In each case, the lattice is specified by n points that generate it, as shown in Figure 9.7. Generating the lattice means starting at the origin of the coordinate axes and extending the given points into a regularly spaced grid. The **shortest vector problem** is the problem of finding a point in a lattice as close as possible to the origin of the axes, given the generators of the lattice. For two dimensions, this is shown in Figure 9.8. In the **closest vector problem**, we are given generators for the lattice and another point not in the lattice. The goal is to find a point in the lattice as close as possible to the given point. This is shown in two dimensions in Figure 9.9.

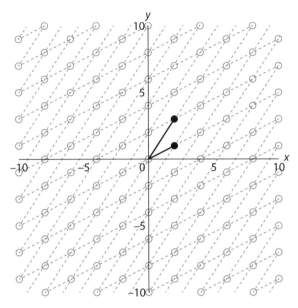

Figure 9.7. Two points (solid circles) and the lattice generated by them (open circles).

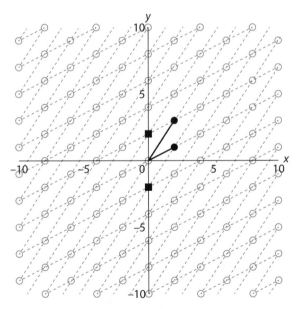

Figure 9.8. The shortest vector problem: Two generators (solid circles), the lattice generated by them (open circles), and the closest points in the lattice to the origin (solid boxes).

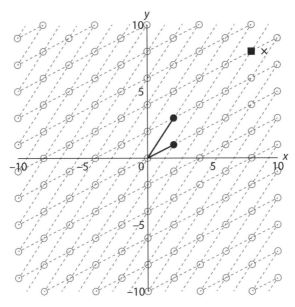

Figure 9.9. The closest vector problem: Two generators (solid circles), the lattice generated by them (open circles), a point not on the lattice (cross), and the closest point in the lattice to it (solid box).

These lattice problems probably don't look especially hard, and in fact the examples shown aren't. There are two things necessary to make lattice problems harder. One is to increase the number of dimensions. A practical cryptographic system would need to use lattices in 500 dimensions or more. Even then finding the right point isn't very hard if the angles in the grid are close to 90°. So the second thing necessary is to make the angles far from right angles, as in Figure 9.10. In two dimensions you can probably tell by visual inspection that there is a different set of generators for the same lattice, giving a grid with much more convenient angles. But if you can imagine a lattice in 500 dimensions with angles like the ones in the figure, you might start to see the issues involved in solving cryptographic lattice problems.

Let's focus on the closest vector problem, since we will use it in our example cryptographic system. In 1984, **László Babai** pointed out that it was easy to approximately solve the problem by exploiting the connection between lattice generation and the same sorts of equations we saw dealing with the Hill cipher in Section 1.6. Keeping with

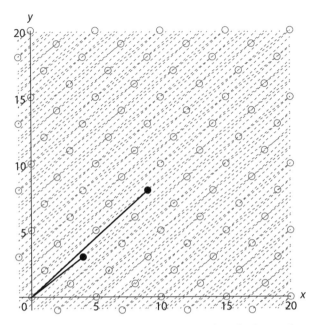

Figure 9.10. A lattice where the angles in the grid are far from right angles.

two-dimensional examples, suppose a lattice is generated by the points (k_1, k_3) and (k_2, k_4). Then any point in the lattice can be represented by taking two integers s and t and finding the point

$$s(k_1, k_3) + t(k_2, k_4) = (sk_1 + tk_2, sk_3 + tk_4).$$

On the other hand, if you have a point (x, y) in the lattice and want to know how it was generated, you can find out by setting

$$(x, y) = (sk_1 + tk_2, sk_3 + tk_4)$$

and solving the equations

$$x = sk_1 + tk_2,$$
$$y = sk_3 + tk_4.$$

This is basically the same system of two equations in two unknowns that we saw in Section 1.6, and the methods used to solve the system there work here also. You get back the integers s and t used to represent

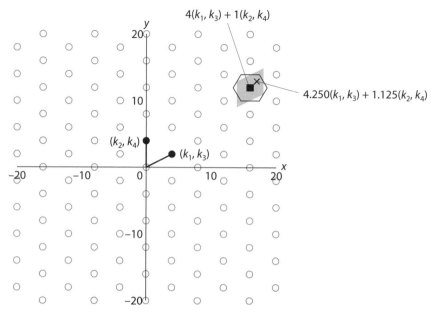

Figure 9.11. Solving the CVP using Babai's method: Babai's method rounds the given point not in the lattice (cross) to a point on the lattice (solid box). This is the correct answer in the case of the given point. In fact, the method rounds everything in the gray parallelogram to the same solid box. The points within the outlined hexagon are the points *actually* closest to the solid box. The large overlap shows that Babai's method is usually correct for this lattice.

the lattice point. If you have n dimensions, then you get n equations in n unknowns to solve, and it works the same way.

What happens if you try this with a point not in the lattice? You can still get s and t, but they will not be integers. If you round s and t each to the closest integer, you get a likely candidate for the closest lattice point to the point not on the lattice. For example, in Figure 9.11, the point designated by a cross can be written as $4.250(k_1, k_3) + 1.125(k_2, k_4)$. It is rounded to $4(k_1, k_3) + 1(k_2, k_4)$, which is designated by the solid box.

Now, if the angles in the lattice grid are close to right angles, then this rounded point is probably the closest lattice point to the given one. If the angles in the grid are far from right angles, as shown in Figure 9.12, then Babai's method is likely to find a point on the lattice that is close—but *not* the closest—to the given point. In the figure, the point designated by a cross can be written as $2.4(k_1, k_3) - 1.4(k_2, k_4)$. It is rounded to $2(k_1, k_3) - 1(k_2, k_4)$, which is designated by the open box. However, the

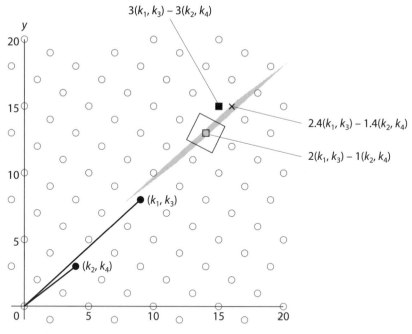

Figure 9.12. Babai's method on a bad lattice: The solid box is the point in the lattice closest to the given point (cross), but Babai's method rounds the given point to the open box. This is not the correct answer in the case of the given point. In fact, the method rounds everything in the gray parallelogram to the same open box. The points within the outlined square are the points *actually* closest to the open box. The small overlap shows that Babai's method is usually *not* correct for this lattice.

actual closest point to the cross is the point designated by a solid box, which is $3(k_1, k_3) - 3(k_2, k_4)$. The same ideas hold in n dimensions, and the more dimensions involved, the harder it is to find the correct point.

How do we make that into a asymmetric-key cryptographic system? Suppose that Bob knows both a "good" set of generators and a "bad" set of generators for the same lattice, as in Figure 9.13. The good set makes a grid with angles close to right angles. The bad set makes a grid on the same points, but with angles far from right angles. The bad generators are going to be Bob's public key, and the good generators are his private key. For a two-dimensional example, the public key might be $(50, 40)$ and $(58, 46)$, which have an angle of $0.24°$ between them. The private key might be $(2, 4)$ and $(4, -2)$, which have an angle of $90°$ between them. Remember that in real life, we would be using many more dimensions.

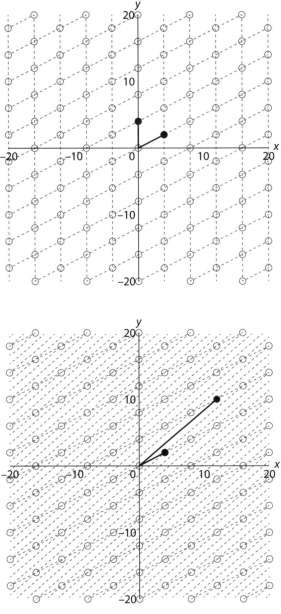

Figure 9.13. A good set of generators (left) and a bad set of generators (right) for the same lattice.

When Alice wants to send Bob a message, she turns the message into numbers and uses the numbers and the bad generators to find a point in the lattice. One difference between this cipher and some of our previous ones is that this cipher is actually more secure if each "block" consists of a very small amount of information. So, in our example we will split each letter into two decimal digits, which we will treat as two different numbers. Each pair of numbers will give us a lattice point.

plaintext:	l	a	t	t	i
numbers:	12	1	2 0	2 0	9
split apart:	1, 2	0, 1	2, 0	2, 0	0, 9
lattice points:	166, 132	58, 46	100, 80	100, 80	522, 414

plaintext:	c	e	n	o	w
numbers:	3	5	14	15	23
split apart:	0, 3	0, 5	1, 4	1, 5	2, 3
lattice points:	174, 138	290, 230	282, 224	340, 270	274, 218

Then Alice adds a small random nonce to each point, similar to the use of the blind in the ElGamal system from Section 8.2. The result is a point that is near a lattice point but no longer in the lattice itself. This is the ciphertext that Alice sends to Bob. For example,

plaintext:	l	a	t	t	i
numbers:	12	1	2 0	2 0	9
split apart:	1, 2	0, 1	2, 0	2, 0	0, 9
lattice points:	166, 132	58, 46	100, 80	100, 80	522, 414
nonce:	1, 1	1, 1	$-1, 1$	$1, -1$	1, 1
ciphertext:	167, 133	59, 47	99, 81	101, 79	523, 415

plaintext:	c	e	n	o	w
numbers:	3	5	14	15	23
split apart:	0, 3	0, 5	1, 4	1, 5	2, 3
lattice points:	174, 138	290, 230	282, 224	340, 270	274, 218
nonce:	1, 1	1, 1	1, 1	$-1, 1$	$-1, -1$
ciphertext:	175, 139	291, 231	283, 225	339, 271	273, 217

To decrypt the point, Bob uses Babai's method and the good generators in the private key to find a lattice point, which is almost certain to be the one that Alice used. Then he can work backward to find the original plaintext.

ciphertext:	167, 133	59, 47	99, 81	101, 79	523, 415
Babai's s and t:	43.3, 20.1	15.3, 7.10	26.1, 11.7	25.9, 12.3	135.3, 63.1
rounded:	43, 20	15, 7	26, 12	26, 12	135, 63
lattice points:	166, 132	58, 46	100, 80	100, 80	522, 414
numbers:	1, 2	0, 1	2, 0	2, 0	0, 9
together:	12	1	20	20	9
plaintext:	l	a	t	t	i

ciphertext:	175, 139	291, 231	283, 225	339, 271	273, 217
Babai's s and t:	45.3, 21.1	75.3, 35.1	73.3, 34.1	88.1, 40.7	70.7, 32.9
rounded:	45, 21	75, 35	73, 34	88, 41	71, 33
lattice points:	174, 138	290, 230	282, 224	340, 270	274, 218
numbers:	0, 3	0, 5	1, 4	1, 5	2, 3
together:	3	5	14	15	23
plaintext:	c	e	n	o	w

Eve can also try to find the correct lattice point, but she has only the bad generators. So she can try to use Babai's method, but she will most likely come up with the wrong lattice point:

ciphertext:	167, 133	59, 47	99, 81	101, 79	523, 415
Eve's s and t:	1.6, 1.5	.6, .5	7.2, −4.5	−3.2, 4.5	.6, 8.5
rounded:	2, 2	1, 1	7, −5	−3, 5	1, 9
lattice points:	216, 172	108, 86	60, 50	140, 110	572, 454
numbers:	2, 2	1, 1	7, −5	−3, 5	1, 9
together:	22	11	??	??	19
plaintext?:	v	k	??	??	s

ciphertext:	175, 139	291, 231	283, 225	339, 271	273, 217
Eve's s and t:	.6, 2.5	.6, 4.5	1.6, 3.5	6.2, .5	1.4, 3.5
rounded:	1, 3	1, 5	2, 4	6, 1	1, 4
lattice points:	224, 178	340, 270	332, 264	358, 286	282, 224
numbers:	1, 3	1, 5	2, 4	6, 1	1, 4
together:	13	15	24	61	14
plaintext?:	m	o	x	??	n

Alice	Bob

Bob:
- Picks a dimension n
- Picks a secret set of generating points $b_1, \ldots, b_n$
- Uses $b_1, \ldots, b_n$ to make public set of generating points $B_1, \ldots, B_n$ for the same lattice
- Posts public encryption key $B_1, \ldots, B_n$

Alice:
- Looks up Bob's encryption key $B_1, \ldots, B_n$
- Picks randon small secret point r
- Start with plaintext numbers $P_1, \ldots, P_n$

↓

- Calculate ciphertext point
  $$C = P_1 B_1 + P_2 B_2 + \cdots + P_n B_n + r$$

$C \rightarrow$

C

↓

- Get rounded C using Babai's method and $(b_1, \ldots, b_n)$

↓

- Solve
  $$(\text{rounded } C) = P_1 B_1 + P_2 B_2 + \cdots + P_n B_n$$
  to get $P_1, \ldots, P_n$

Figure 9.14. The GGH encryption system.

In order to find the correct point in the lattice, Eve has to solve the closest vector problem. If the generators are bad enough and the number of dimensions is high enough, we believe that this will be hard, even if Eve has a quantum computer.

This system is known as the Goldreich-Goldwasser-Halevi, or **GGH, cryptosystem**, after **Oded Goldreich**, **Shafrira Goldwasser**, and **Shai Halevi**, the three Israeli computer scientists who invented it in 1997. The whole system looks like Figure 9.14. Unfortunately, in 1997 it was discovered that the system is insecure in practice. Alice's blind has to be small compared to the size of the lattice or else the closest point that Bob finds won't be the one Alice started with. But it turns out that Eve can use that information to make the problem much easier to solve than the standard closest vector problem.

There are other lattice-based cryptographic systems that have not yet been broken, and many of them use elements similar to GGH. The most promising lattice-based system is known as **NTRU**. It was invented in 1996 by three researchers at Brown University: **Jeffrey Hoffstein**, **Jill Pipher**, and **Joseph Silverman**. NTRU was originally described using other sorts of mathematics but was later shown to be equivalent to a system using lattices. It's never been definitively revealed what NTRU stands for, but rumors suggest that it might be Number Theorists aRe Us or Number Theorists aRe Useful. When asked about it, Jeff Hoffstein once replied, "It stands for whatever you want."

Both GGH and NTRU also have digital signature systems associated with them; see the endnotes for more information.

9.3 QUANTUM CRYPTOGRAPHY

Another possibility is that the same quantum physics that could allow us to build quantum computers could also allow us to protect against quantum-computational attacks. **Quantum cryptography** is the study of how to use a combination of quantum physical laws and cryptographic cleverness to create cryptographic systems. The first examples of this were originally proposed by **Stephen Wiesner** when he was a graduate student in physics at Columbia University in the late 1960s. Wiesner proposed two ideas: The first was a way to send two messages at the same time such that the recipient could choose to receive either one, but not both. The second idea was a way to make currency with a serial number that could not be copied and, therefore, could not be forged. Like Ralph Merkle, Wiesner encountered almost universal incomprehension and disbelief from his professors and colleagues, and his paper was repeatedly rejected by scientific journals. The paper would not be published until 1983.

Figure 9.15. Polarized photons.

One person who did appreciate Wiesner's paper was **Charles Bennett**. Bennett knew Wiesner from when they were undergraduates at Brandeis together and had studied chemistry, physics, and mathematics before settling on computer science. Thus, he was ideally suited to understand quantum cryptography. Somewhere along that career path, Wiesner showed Bennett a copy of his manuscript. As Wiesner had hoped, Bennett was fascinated. He thought about it on and off for the next 10 years or so, but he didn't really know what to do with the idea until he ran into **Gilles Brassard** while swimming at a hotel beach during a conference in 1979. Bennett knew that Brassard was giving a talk on cryptography and immediately started to explain Wiesner's ideas. Brassard had read an account of Bennett's work in one of Martin Gardner's columns but had no way of connecting the name to the man who was swimming on the beach! Eventually they got proper introductions made and started working together on quantum cryptography—leading, among other things, to the **BB84 protocol**.

The BB84 protocol, which comes from the names of Bennett and Brassard and the year it was first published, is a key agreement system, which, like Wiesner's systems, uses polarized photons to convey information. One can think of the polarization of a photon as the direction in which it vibrates; if the photon is traveling toward you and parallel to the ground, it could be vibrating left and right as you look at it, up and down, or somewhere in between. (See Figure 9.15.) In order to detect the polarization of a photon, we need a polarized filter. Such a filter is designed to let photons through only if they are vibrating in the same direction as the photon. So, in Figure 9.16, the filter lets through photons vibrating in the vertical direction and not those vibrating in the horizontal direction.

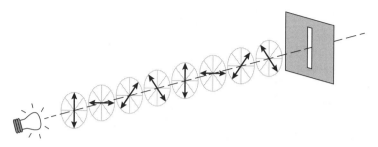

Figure 9.16. Polarized photons approaching a filter.

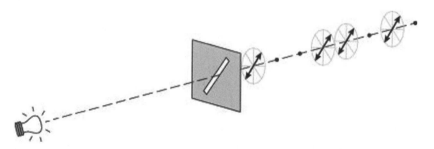

Figure 9.17. Polarized photons after passing through a filter.

The interesting part is what happens when a photon is vibrating diagonally, say at a 45° angle. According to quantum physics, we can think of diagonal vibration as a superposition of a vertically vibrating state and a horizontally vibrating state. We can think of the filter as causing the photon to collapse into one state or the other at random. So, if a bunch of photons are vibrating diagonally, half of them will get through and half will not, which is not too surprising. But you might expect the ones that get through to be still tilted, or at "half-strength," or something similar, and this is *not* true. Once a photon passes through a vertically polarized filter, it looks just like any other vertically polarized photon. There's no way to tell whether it was originally vertically polarized or diagonally polarized. Likewise, if the photon doesn't go through, there is no way to tell whether it was horizontally polarized or whether it was diagonally polarized and just unlucky. Figure 9.17 shows the same photons as Figure 9.16, but after they have attempted to pass through the filter.

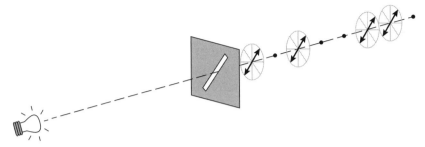

Figure 9.18. Polarized photons after passing through a diagonal filter.

One last note and then we are ready to go. Just as a diagonally polarized state is a superposition of vertically and horizontally polarized states, we can equally well think of a vertically or horizontally polarized photon as being in a superposition of two diagonally polarized states (upper left to lower right and lower left to upper right). Thus, if a vertically or horizontally polarized photon attempts to pass through a diagonally polarized filter, as in Figure 9.18, half the time it will get through and half the time it will be blocked. And if it does get through, it will be indistinguishable from any other diagonally polarized photon.

Now we are ready for the BB84 protocol. Alice and Bob need to have a communications link over which Alice can send Bob single polarized photons and an ordinary (not necessarily single-photon-grade) two-way means of communication. Eve might be able to overhear one or both links. Alice starts by choosing two sets of random bits. The first set controls whether Alice is going to use a vertical and horizontal scheme, which we will denote ⊞, or a diagonal scheme, which we will denote ⊠. In the ⊞ scheme, a vertically polarized photon (↕) will stand for the bit 1 and a horizontally polarized photon (↔) will stand for the bit 0. In the ⊠ scheme, a lower-left-to-upper-right polarized photon (↗) will stand for 1 and an upper-left-to-lower-right polarized photon (↘) will stand for 0. The second set of random bits controls the bits that get sent using the chosen schemes. In the following example, I'm not going to list the first set of bits because only the schemes chosen are important.

Alice's schemes:	⊠	⊠	⊠	⊠	⊞	⊞	⊞	⊠	⊞	⊠
Alice's bits:	0	1	0	0	0	0	0	1	0	0
Alice's photons:	↘	↗	↘	↘	↔	↔	↔	↗	↔	↘

Now Bob also chooses a random set of bits. He uses this set to select schemes in which to detect the photons by using a polarized filter. If Bob's scheme matches Alice's for some photon, then he will detect the photon correctly and convert it back into the correct value of the bit. If not, then the photon will collapse into a state at random and Bob will get a random value of the bit. This value may or may not match Alice's.

Alice's schemes:	⊠	⊠	⊠	⊠	⊞	⊞	⊞	⊠	⊞	⊠
Alice's bits:	0	1	0	0	0	0	0	1	0	0
Alice's photons:	↘	↗	↘	↘	↔	↔	↔	↗	↔	↘
Bob's schemes:	⊠	⊠	⊞	⊞	⊞	⊞	⊞	⊞	⊠	⊠
Bob's photons:	↘	↗	↕	↔	↔	↔	↔	↕	↘	↘
Bob's bits:	0	1	1	0	0	0	0	1	0	0

Remember, at this stage neither Alice nor Bob knows which bits have been correctly received.

Now Alice and Bob open their two-way line of communication. For each of the bits, Alice tells Bob what scheme she used but *not* what bit she sent. Bob tells her if he used the same scheme, and when the schemes match, they keep the bit. Otherwise they throw it away.

Alice's schemes:	⊠	⊠	⊠	⊠	⊞	⊞	⊞	⊠	⊞	⊠
Alice's bits:	0	1	∅	∅	0	0	0	1	∅	0
Alice's photons:	↘	↗	↘	↘	↔	↔	↔	↗	↔	↘
Bob's schemes:	⊠	⊠	⊞	⊞	⊞	⊞	⊞	⊞	⊠	⊠
Bob's photons:	↘	↗	↕	↔	↔	↔	↔	↕	↘	↘
Bob's bits:	0	1	1	∅	0	0	0	1	∅	0

As you can see from the example, occasionally they have to throw out bits which by random chance did match. There's no way around that. Nevertheless, on the average about half the schemes will match, so Alice and Bob can keep about half the bits. They can then use these bits as the secret key to a secure, nonquantum, symmetric-key cipher, just like in any key-agreement system. If by chance they had to throw away so many bits that they don't have enough for their chosen symmetric-key cipher, they can go back to the protocol and gather more bits the same way.

What if Eve is listening in on the two communications links? She can also choose a set of random schemes and try to detect the photons just the same as Bob can:

Alice's schemes:	⊠	⊠	⊠	⊠	⊞	⊞	⊞	⊠	⊞	⊠
Alice's bits:	[0]	[1]	0̸	0̸	[0]	[0]	[0]	1̸	0̸	[0]
Alice's photons:	↘	↗	↘	↘	↔	↔	↔	↗	↔	↘
Bob's schemes:	⊠	⊠	⊞	⊞	⊞	⊞	⊞	⊞	⊠	⊠
Bob's photons:	↘	↗	↕	↔	↔	↔	↔	↕	↘	↘
Bob's bits:	[0]	[1]	1̸	0̸	[0]	[0]	[0]	1̸	0̸	[0]
Eve's schemes:	⊞	⊞	⊞	⊠	⊠	⊞	⊞	⊠	⊞	⊠
Eve's photons:	↔	↔	↕	↘	↗	↔	↔	↗	↔	↘
Eve's bits:	0	0	1	0	1	0	0	1	0	0

She can also listen in on Alice and Bob's conversation and find out which of her schemes matched Alice's and/or Bob's. Unfortunately for her, the only bits that do her any good are the ones where she matched *both* Alice and Bob's schemes.

Alice's schemes:	⊠	⊠	⊠	⊠	⊞	⊞	⊞	⊠	⊞	⊠
Alice's bits:	[0]	[1]	0̸	0̸	[0]	[0]	[0]	1̸	0̸	[0]
Alice's photons:	↘	↗	↘	↘	↔	↔	↔	↗	↔	↘
Bob's schemes:	⊠	⊠	⊞	⊞	⊞	⊞	⊞	⊞	⊠	⊠
Bob's photons:	↘	↗	↕	↔	↔	↔	↔	↕	↘	↘
Bob's bits:	[0]	[1]	1̸	0̸	[0]	[0]	[0]	1̸	0̸	[0]
Eve's schemes:	⊞	⊞	⊞	⊠	⊠	⊞	⊞	⊠	⊞	⊠
Eve's photons:	↔	↔	↕	↘	↗	↔	↔	↗	↔	↘
Eve's bits:	0	0	1̸	0̸	1	[0]	[0]	1̸	0̸	[0]

If Eve matches Alice but Bob doesn't, then the bit gets thrown out. And if Bob matches Alice but Eve doesn't, then Eve has no idea whether her bit was correct or not. On average, Eve will be guaranteed to correctly intercept about half the bits that Alice and Bob use in the end. And by chance, about half the rest will be correct, but Eve doesn't know which ones, so there's not much she can do with that. Eve has managed to reduce Alice and Bob's effective key size by half, but as long as Alice and Bob have taken that into account, they're in good shape.

In fact, things are even worse for Eve than they appear. I left out the fact that when Eve intercepts the photon using the wrong detector, she will collapse it into a different state, which will affect what Bob receives. So what will actually happen will be something like this:

Alice's schemes:	⊠	⊠	⊠	⊠	⊞	⊞	⊞	⊠	⊞	⊠
Alice's bits:	[0]	[1]	0̸	0̸	[0]	[0]	[0]	1̸	0̸	[0]
Alice's photons:	↘	↗	↘	↘	↔	↔	↔	↗	↔	↘

Eve's schemes:	⊞	⊞	⊞	⊠	⊠	⊞	⊞	⊠	⊞	⊠
Eve's photons:	↔	↔	↕	↘	↗	↔	↔	↗	↔	↘
Eve's bits:	0	0	1̸	0̸	1	[0]	[0]	1̸	0̸	[0]

Bob's schemes:	⊠	⊠	⊞	⊞	⊞	⊞	⊞	⊞	⊠	⊠
Bob's photons:	↘	↘	↕	↔	↕	↔	↔	↕	↘	↘
Bob's bits:	[0]	[0]	1̸	0̸	[1]	[0]	[0]	1̸	0̸	[0]

When Eve has guessed wrong and collapsed the photon, then half the time Bob will receive it incorrectly. If Alice and Bob have reason to think that Eve might be listening, all they have to do is choose a random sampling of bits that *ought* to agree and reveal them over the public channel. If they do agree, then they throw those bits out and use the rest for their key. Either Eve is not listening or she has gotten very lucky. If they don't agree, then Eve is listening and they need to start over or find another way to communicate.

For the next 5 years or so after BB84 was published, nothing much happened in the field of quantum cryptography. Eventually, Bennett and Brassard decided that they needed to build a working prototype of the system in order to get people to take their idea seriously. With the help of three students, Bennett and Brassard performed the first-ever key agreement by quantum cryptography in late October 1989, on the tenth anniversary of Bennett and Brassard's initial meeting. The quantum transmissions took place over a distance of 32.5 centimeters and therefore had little practical value, but they proved it could be done.

Bennett and Brassard achieved their goal of getting researchers interested in their idea, and people soon began to build systems on a practical scale. By 2014, a team from the University of Geneva and Corning Incorporated was able to implement a quantum key-distribution protocol over a fiber-optic cable 307 kilometers long, which is long enough to be practical in almost all the fiber-optic networks used today.

Secret key bits were generated at 12,700 bits per second, which could potentially be enough even for a one-time pad system. In 2006, on the other hand, a team from various institutions in Europe and Asia implemented BB84 using laser transmissions through the open air between two of the Canary Islands, over a distance of 144 km. The researchers suggest that this is reasonably comparable to a transmission between the ground and a low-orbiting satellite; although the distance to the satellite would be longer, the amount of atmospheric interference ought to be less.

Even before these experiments, a demonstration of the commercial possibilities of quantum cryptography—although of a nature perhaps more dramatic than truly useful—took place on April 21, 2004, when the first bank transfer protected by quantum cryptography was transmitted from the City Hall of Vienna, Austria, to the headquarters of Bank Austria Creditanstalt elsewhere in the city. The necessary fiber-optic cables (about 1.5 km long) were specially laid through the Vienna sewer system. Several companies now have quantum-cryptographic equipment for sale or in development, and various multibuilding computer networks protected by quantum cryptography have been set up by researchers in the United States, Austria, Switzerland, Japan, and China, among others. The Japanese network, for example, has 6 links ranging from 1 kilometer to 90 kilometers long. In 2010, secret key bits generated at 304,000 bits per second were used to encrypt live video using a one-time pad on one of the 45-kilometer links. At this point the expensive equipment required is probably not justified by most organizations' security needs, but in 2013 a nonprofit research and development contractor in Ohio installed what they called the first commercial quantum key-distribution system in the United States. "I don't know that everyone will [adopt QKD]," says a researcher there, "but I do think that companies and organizations that have very high-value data will."

Of course, the adoption of quantum cryptography will not mean the end of cryptanalysis. Most of the cryptanalytic attacks discussed in this book fall into what is sometimes called pure cryptanalysis. This loosely defined term refers to techniques that require little or no information besides the plaintexts and/or ciphertexts under consideration and perhaps the language the plaintexts are in. For starters, this excludes probable-word attacks, which generally require Eve to know something about the context of the messages in addition to the messages themselves. These techniques also assume that Eve doesn't have any way of getting information about the inner workings of Alice and Bob's cryptographic techniques or machines, only the input and output. Finally, pure cryptanalysis assumes that Alice and any machines she uses have done their job of encryption exactly like they are supposed to. Cryptanalytic attacks that use knowledge about the inner workings of the cryptographic processes, including possible mistakes that Alice might make or Eve might force, are called **implementation attacks**.

It's generally agreed that if Eve doesn't have any access to the inner workings of Alice and Bob's equipment and everything works exactly as it's supposed to, BB84 is secure from anything Eve can try to do. One thing that doesn't always work like it's supposed to is building a transmitter that is certain to produce exactly one photon at a time. Many systems use a very weak laser. When it's fired, the laser usually produces no photons at all, sometimes produces one photon, and occasionally produces more than one photon. If a pulse has no photons, then Bob won't receive anything and he and Alice will agree to throw out that bit as if Bob had chosen the wrong detection scheme. If a pulse had more than one photon, then they will all be polarized the same way, and it doesn't matter how many or which ones Bob detects.

However, Eve can take advantage of this variation in the number of photons using the **photon number–splitting attack**. This attack is based on the fact that while Eve can't determine the polarization of a photon without disturbing it, she can determine the *number* of photons in a pulse without changing their polarizations. So if Alice's laser sends more than one photon, Eve can very carefully split off one of the photons and send the rest on to Bob. Since in the real world some photons are usually lost in transmission anyway, Alice or Bob may very well not realize what Eve is doing. Eve can then keep her captured photons in some sort of quantum storage device until she has a chance to listen to Alice and Bob exchange detection schemes, after which she can use the correct detection scheme on the correct photons.

Using only the rare multiphoton pulses won't get Eve very many bits of the key, but it gets worse. Eve can also simply block some or all of the single-photon pulses—again, Alice and Bob won't know that this was done deliberately, as opposed to being accidental losses. If Eve blocks the right number of single-photon pulses and intercepts the right number of multiphoton pulses, she can obtain much or all of Alice and Bob's final key without them realizing it.

There are several ways for Alice and Bob to defend against this attack, including developing better photon generators and making modifications to BB84. Another promising defense is the use of **decoy pulses**, which Alice deliberately makes have more or fewer photons than usual. While sending her photons, Alice randomly intersperses the regular pulses, which will be used to compute the key as usual, with these decoy pulses. During the two-way, nonquantum, part of Alice and Bob's communication, in addition to revealing the polarization schemes, Alice also reveals which pulses were decoy pulses. If Eve has been using photon number splitting, then the rate at which the decoy pulses are "lost" in transmission will be different from the rate of loss of regular pulses. If the difference is large enough, Alice and Bob can conclude that Eve is listening in and take appropriate action.

Photon number splitting is essentially a **passive attack**, which Eve can carry out with minimal interference to Alice and Bob's communications. Several other attacks on quantum cryptography also make use of peculiarities of the equipment used by Alice and Bob but require Eve to interfere more actively with Alice and/or Bob's equipment or lines of communications. A number of these **active attacks** have been shown to be successful against commercially sold systems. In the **bright illumination attack**, for example, Eve attacks Bob's detectors with specially tailored pulses of bright laser light. Certain detectors can be blinded and even fooled into thinking they are picking up Alice's photons this way.

9.4 LOOKING FORWARD

Edgar Allan Poe famously wrote that, "it may be roundly asserted that, human ingenuity cannot concoct a cipher which human ingenuity cannot resolve." In theory he has been proven wrong: when executed properly under the proper conditions, techniques such as the one-time pad and the BB84 protocol can be proved to be secure against any possible attack by Eve. In real-life situations, however, Poe was undoubtedly right. Every time an "unbreakable" system has been put into actual use, some sort of unexpected mischance eventually has given Eve an opportunity to break it. The race between the cryptographers and the cryptanalysts goes on, as it surely will as long as people try to send secret messages. And as long as people remain interested in things like power, money, and relationships, I'm pretty sure the secret messages are going to keep coming.

■ ■ ■ LIST OF SYMBOLS ■ ■ ■

Note: Some symbols, like C, P, and k are used so frequently
that I have only listed the first or first few instances.

■ ■ ■ NOTES ■ ■ ■

(Page xi) **"each equation ... would halve the sales"**: Stephen W. Hawking, *A Brief History of Time: From the Big Bang to Black Holes* (Toronto; New York: Bantam, 1988), p. vi.

(Page xi) **"mathematics and muddle"**: J.W.S. "Ian" Cassels (1922–2015), former head of pure mathematics at Cambridge University, quoted in Bruce Schneier, *Applied Cryptography*, 2d ed. (New York: Wiley, 1996), p. 381.

Chapter 1 Introduction to Ciphers and Substitution

(Page 1) **"A code consists of ...."**: David Kahn, *The Codebreakers*, rev. ed. (New York: Scribner, 1996), p. xvi.

(Page 2) **probably wasn't the original inventor**: Edgar C. Reinke, "Classical cryptography," *The Classical Journal* 58:3 (1962).

(Page 2) **"There are also letters ...."**: Suetonius, *De Vita Caesarum, Divus Iulius* (*The Lives of the Caesars, The Deified Julius*; c. 110 CE), paragraph LVI.

(Page 2) **x goes to A**: Actually, Caesar's Roman alphabet didn't have either a *w* or a *z*, but the idea was the same.

(Page 2) **"And you too, Brutus"**: "Et tu, Brute," in Latin. William Shakespeare, *Julius Caesar* (1599), act 3, scene 1, line 77.

(Page 3) **Gauss codified wraparound**: In Carl Friedrich Gauss, *Disquisitiones arithmeticae* (New Haven and London: Yale University Press, 1966), Section I.

(Page 3) **changing letters into numbers**: Note that it was decades after Gauss before anyone thought of applying modular arithmetic to cryptography, as far as we know. There is evidence that Charles Babbage (who will appear several times in Chapter 2) did so starting in the 1830s. (Ole Immanuel Franksen, "Babbage and cryptography. Or, the mystery of Admiral Beaufort's cipher," *Mathematics and Computers in Simulation* 35:4 (1993), p. 338–39) The first person to publish work involving modular equations and cryptography seems to have been the **Marquis Gaëtan Henri Léon de Viaris**, in 1888. De Viaris is also known for inventing some of the first printing cipher machines. (Kahn, *The Codebreakers*, p. 240.)

(Page 2–3) **encipher and decipher: Decode** and **decrypt** are defined analogously. Note that some older books use decrypt when a modern cryptologist would say cryptanalyze—this older usage is also standard in some other languages, so you may see it in books that are translated into English.

(Page 4) **Caesar's point of view**: There is also some evidence that Caesar may have at times used shifts other than 3 and other more complicated ciphers. (Reinke, "Classical cryptography.")

(Page 4) **"The system . . . "**: Auguste Kerckhoffs, "La cryptographie militaire, I," *Journal des sciences militaires* IX (1883).

(Page 4) **mostly used by militaries and governments**: As you might guess from the title of Kerckhoffs' essay, *La Cryptographie Militaire*!

(Page 4) **advantages to not keeping your system secret**: There's one other advantage to *not* keeping your system secret, which has become widely appreciated more recently. The more people who have tried out your system, the more likely it is that any deficiencies will be discovered. This same basic idea is an important part of the open source software movement.

(Page 5) **Augustus' system**: Suetonius, *The Divine Augustus*, paragraph LXXXVIII.

(Page 5) **shift cipher or additive cipher**: Many ciphers have more than one name, especially if you can describe them both with and without modular arithmetic. I will generally use the term involving modular arithmetic unless I am trying to make a point.

(Page 7) **multiplicative cipher**: Multiplicative cipher is really just another name for the decimation method.

(Page 10) **shift left k letters**: Or, you could shift right $26 - k$ letters, since $26 - k$ is the same as $-k$ modulo 26.

(Page 11) **$\overline{3}$**: There's not really a single standard notation for this number. Both $\overline{3}$ and 3^{-1} are common. Gauss merely called it "$\frac{1}{3}$ (mod 26)". (Gauss, *Disquisitiones arithmeticae*, Article 31.)

(Page 17) **"atbash" cipher**: In the Hebrew alphabet, the first letter is *aleph*, which is encrypted into the last letter, *tav*, and the second letter, *bet*, is encrypted into the second-last letter, *shin*. In Hebrew, those four letters would spell *atbash*.

(Page 17) **atbash in Jeremiah**: Kahn, *The Codebreakers*, pp. 77–78. The atbash cipher also got some play in the book *The Da Vinci Code*. (Dan Brown, *The Da Vinci Code*, 1st ed. (New York: Doubleday, 2003), Chapters 72–77.)

(Page 18) **al-Kindi**: Ibrahim A. Al-Kadi, "Origins of cryptology: The Arab contributions," *Cryptologia* 16 (1992).

(Page 20) **Hill cipher**: Lester S. Hill, "Cryptography in an algebraic alphabet," *The American Mathematical Monthly* 36:6 (1929)

(Page 24) **affine Hill cipher**: Since the addition step works independently and differently on each letter, it could be considered an example of a polyalphabetic cipher like the ones we will see in Chapter 2.

(Page 24) **the most common digraph**: Parker Hitt, *Manual for the Solution of Military Ciphers* (Fort Leavenworth, KS: Press of the Army Service Schools, 1916), Table IV.

(Page 24) **the most common trigraph**: Hitt, *Manual*, Table V.

(Page 24) **Hill's machine**: Louis Weisner and Lester Hill, "Message protector," United States Patent: 1845947, 1932. http://www.google.com/patents?vid=1845947

(Page 24) **polyalphabetic substitution ciphers via mechanical devices**: The Enigma machines used by the German military in World War II were examples of these; we will see them in more detail in Section 2.8.

(Page 24) **regained substantial importance**: See for example, Section 4.5.

(Page 26) **almost as easy to break**: The affine Hill cipher has 6 key numbers, so Eve needs 6 equations, which means 3 blocks of plaintext. In general, she needs 1 more block of plaintext than the block size to break an affine Hill cipher, which is not much of an improvement.

CHAPTER 2 POLYALPHABETIC SUBSTITUTION CIPHERS

(Page 29) **Arab homophony**: Al-Kadi, "Origins of cryptology."

(Page 29) **Mantuan homophony**: Kahn, *The Codebreakers*, 107.

(Page 31) **homophones for consonants**: Kahn, *The Codebreakers*, 108.

(Page 32) **expected frequency table for English text**: Henry Beker and Fred Piper, *Cipher Systems* (New York: Wiley, 1982), Table S1.

(Page 32) **William Friedman**: Ronald William Clark, *The Man Who Broke Purple: The Life of Colonel William F. Friedman, Who Deciphered the Japanese Code in World War II* (Boston: Little Brown, 1977).

(Page 33) **Elizebeth Friedman**: Elizebeth Smith Friedman's first name was spelled with an *e* in the third syllable because her mother disliked the idea of her being nicknamed Eliza. Clark, *Man Who Broke Purple*, p. 37.

(Page 34) **William Friedman and the index of coincidence**: Although there is no doubt that Friedman came up with the idea of the index of coincidence, it should be pointed out that the version given here was formulated by his assistant, Solomon Kullback.

(Page 35) **alphabet of 26 letters**: Dealing with a different number of ciphertext letters would change the exact numbers but not the basic idea.

(Page 35) **don't pick exactly same A again**: It doesn't really seem fair to allow us to pick exactly the same letter both times, since of course they will match! In the cases we considered before, we had a very large amount of text, so the chance of picking exactly the same letter twice was so small we didn't worry about it.

(Page 35) **phi test**: William Friedman, *Military Cryptanalysis. Part III, Simpler Varieties of Aperiodic Substitution Systems* (Laguna Hills, CA: Aegean Park Press, 1992), p. 94. Friedman and Kullback used the Greek letter phi to refer to the actual number of coincidences; that would be the numerator of our index of coincidence, or the index times 322×321.

(Page 35) **simple substitution cipher**: The plaintext is from Mark Twain, *The Adventures of Tom Sawyer* (1876), Chapter 2.

(Page 36) **homophonic cipher**: The plaintext is from Twain, *Adventures of Tom Sawyer*, Chapter 5.

(Page 36) **frequency analysis in Europe**: The technique was probably known earlier but not published. Kahn, *The Codebreakers*, p. 127.

(Page 37) **52 cells**: Alberti had 24 letters in his Latin alphabet, and he also had some cells with numbers that he used for codenumbers. I am not going to worry about those in order to focus on the polyalphabetic part of the machine.

(Page 37) **"Not in regular order . . . ."**: Kahn, *The Codebreakers*, p. 128.

(Page 37) **ciphertext alphabet**: This is a multiplication cipher we saw earlier.

(Page 39) **weakness in Alberti's cipher**: Kahn, *The Codebreakers*, p. 136.

(Page 39) **addition is done before the multiplication**: I leave it as an exercise to the interested reader to show that this really *is* a $kP + m$ cipher, just not the same as the one you get if you multiply first and then add. We see more of this sort of thing in Section 3.3.

(Page 39) **first printed book on architecture**: *De Re Aedificatoria*, published in 1485.

(Page 40) **Trithemius' stranger writings**: See, for example, Thomas Ernst, "The numerical-astrological ciphers in the third book of Trithemius's Steganographia," *Cryptologia* 22:4 (1998); Jim Reeds, "Solved: The ciphers in book III of Trithemius's Steganographia," *Cryptologia* 22:4 (1998).

(Page 40) **back to the starting position**: In fact, Trithemius left out the last line, but we will need it later.

(Page 41) **other tables in Trithemius**: C. J. Mendelsohn, "Blaise de Vigenère and the 'Chiffre Carré,' " *Proceedings of the American Philosophical Society* 82:2 (1940), p. 118.

(Page 41) **Bellaso's life**: Augusto Buonafalce, "Bellaso's reciprocal ciphers," *Cryptologia* 30:1 (2006).

(Page 41) **Bellaso's key letters**: Most modern authors start by labeling the plaintext alphabet with the key A and leave off the last line. It should shortly become clear why I did it this way. At any rate, Bellaso was well aware that it didn't matter how you arranged the key letters.

(Page 42) **"tre teste di leone"**: The Bellaso family coat of arms was "Azzurro a tre teste di leone d'oro poste di profilo e linguate di rosso" (On a blue field three red-tongued gold lion heads in side view). Augusto Buonafalce, "Bellaso's reciprocal ciphers."

(Page 43) **key numbers plus the plaintext numbers**: In other words, the encryption equation is $C \equiv P + k$ modulo 26. For the tabula aversa, it would be $C \equiv k - P$ modulo $26 \equiv 25P + k$ modulo 26.

(Page 43) **"sporting his clothes . . ."**: Buonafalce, "Bellaso's reciprocal ciphers."

(Page 43) **combination of tabula recta and repeating-key**: It's not actually clear who did first think of putting the two together. Quite possibly Bellaso thought of it and immediately rejected it in favor of his more complicated system.

(Page 45) **ciphertext is effectively random**: We see further implications of this in Section 5.2.

(Page 45) **Babbage**: Franksen, "Babbage and cryptography."

(Page 45) **Kasiski**: Kahn, *The Codebreakers*, p. 207.

(Page 46) **factor**: Factor means the same thing as divisor, but for some reason it's a more commonly used term when discussing the Kasiski test.

(Page 48) **kappa test**: Friedman actually developed the kappa test to solve a slightly different cipher, which we shall see in Section 5.1.

(Page 48) **"Here is Edward Bear . . . ."**: A. A. Milne, *Winnie-the-Pooh*, reissue ed (New York: Puffin, 1992), Chapter 1.

(Page 48) **"The Piglet lived . . ."**: Milne, *Winnie-the-Pooh*, Chapter 3.

(Page 49) **no particular reason for other than random**: It's not quite true that two ciphertexts with different keys will have coincidences totally at random, but it's close enough for this test.

(Page 50) **percentage of coincidences**: Note that we have been very fortunate in our choice of plaintexts. The difference between 3.8% of 50, which is "about 2," and 6.6% of 50, which is "about 3," is really not enough to reliably distinguish the cases within the usual margin of error. One ought to use at least 100 letters of text, and 2 or 3 times that would be better.

(Page 50) **slide the plaintext 4 steps**: I've also "wrapped around" the upper version of the text when it ended, which won't affect our argument either way but gives us a little longer text to work with.

(Page 50) **other common meanings**: Slide usually refers to a particular device used in various sorts of substitution ciphers, and shift is generally reserved for additive ciphers.

(Page 53) **adding up the frequencies**: William Friedman, *Military Cryptanalysis. Part II, Simpler Varieties of Polyalphabetic Substitution Systems* (Laguna Hills, CA: Aegean Park Press, 1984), pp. 21, 40. In fact, this is a special case of the chi test, which we see (and justify) in Section 5.1.

(Page 55) **brute-force search**: It doesn't really in particular matter if the ciphers are additive, as long as Eve has a relatively limited number of options.

(Page 55) **polyalphabetic cipher**: The plaintext is from Lewis Carroll, *Alice's Adventures in Wonderland* (1865), Chapter 1.

(Page 57) **Babbage**: Franksen, "Babbage and cryptography," p. 337. One more modern technique for breaking multiple products of repeating-key encryption involves a superimposition using the length of one of the keys and then looking at the "differences" of the rows. One of the keys will cancel out, leaving a difference of plaintexts encrypted with a single key. Techniques related to those of Section 5.1 can be used to analyze the "differenced" plaintext and extract the second key. For a full description, see Alan G. Konheim, *Cryptography, A Primer* (New York: Wiley, 1981), Sections 4.11–15.

(Page 58) **Hagelin**: Kahn, *The Codebreakers*, pp. 425–26

(Page 58) **use of M-209**: Robert Morris, "The Hagelin cipher machine (M-209): Reconstruction of the internal settings," *Cryptologia* 2:3 (1978), says "This machine was in wide use in the U.S. Army for tactical purposes until the early 1950's." Kahn, *The Codebreakers*, photo facing p. 846 (described on p. 1151) shows a picture of an American soldier using an M-209 at Hypochong, Korea, in October 1951.

(Page 60) **inactive positions of lugs**: Actually, it is not entirely clear from the photos I have seen of the C-362 (Jerry Proc, "Hagelin C-362," http://www.jproc.ca /crypto/c362.html.) how many, if any, inactive positions there are. In fact, there seem to have been several different versions of the C-36, which may have had different numbers of lugs and/or positions. The M-209 definitely had two inactive positions.

(Page 61) **C-36 repeating-key substitutions**: Technically the first substitution is by a tabula aversa, and the rest are by tabula recta. More importantly the product cipher is still a repeating-key cipher, and in fact it is a reciprocal tabula aversa cipher.

(Page 61) **pin and lug settings**: These are the actual lug settings for the version of the C-36 with fixed lugs, according to Frédéric André, "Hagelin C-36," http://fredandre .fr/c36.php?lang=en.

(Page 62) **"Bork, bork, bork!"**: ABC, "The Muppet Show: Sex and Violence," Television, 1975.

(Page 62) **key settings on the C-36**: Strictly speaking, the wheel starting positions could remain the same and the pins could all be changed to compensate. However, changing the starting positions is much easier. Therefore, it was a common way to add extra variation to the key.

(Page 63) **distinguishing the active pins statistically**: One way is to use the chi test, which we see in Section 5.1.

(Page 63) **ciphertext-only attack on Hagelin machines**: Wayne G. Barker, *Cryptanalysis of the Hagelin Cryptograph* (Laguna Hills, CA: Aegean Park Press, 1981), especially Chapter 5; Beker and Piper, *Cipher Systems*, Section 2.3.7 has a slightly different way to determine the lug settings.

(Page 63) **known-plaintext attack on Hagelin machines**: Barker, *Cryptanalysis of the Hagelin Cryptograph*, especially Chapter 6; and Beker and Piper, *Cipher Systems*, Section 2.3.5–2.3.6; the latter closely follows Morris, "The Hagelin cipher machine." Barker, *Cryptanalysis of the Hagelin Cryptograph*, also has several other attacks using various types of information.

(Page 64) **recent research**: Karl de Leeuw, "The Dutch invention of the rotor machine, 1915–1923," *Cryptologia* 27:1 (2003).

(Page 64) **four others**: See Friedrich L. Bauer, "An error in the history of rotor encryption devices," *Cryptologia* 23:3 (1999), for the time line of these four, but note that this was written before van Hengel and Spengler's work was brought to light.

(Page 64) **evidence that Koch had access**: de Leeuw, "Dutch invention." It is not clear to me if Scherbius could have seen the Dutch patent application before he filed his own.

(Page 64) **independent inventions**: Damm's rotor, in particular, does not work quite the same way as the others. See, for example, Friedrich Bauer, *Decrypted Secrets*, 3rd, rev., updated ed. (Berlin [u.a.]: Springer, 2002), Section 7.3.

(Page 64) **multiplicative ciphers in rotors**: There's no particular reason to use multiplicative ciphers in rotors, and in fact there's some reason not to—for starters, there aren't really enough of them. However, it will make it easy to write formulas and the general principle isn't really very different.

(Page 65) **following a rotor through**: If you feel like simplifying the formula, you will find that it's actually an affine cipher, but that isn't really important for our discussion.

(Page 66) **once every 26 letters**: Note that in most versions of the famous German Enigma rotor machine, the motion is more complicated that this. For details on some of the kinds of Enigma machines and their differences, including rotor motions, see David H. Hamer, Geoff Sullivan, and Frode Weierud, "Enigma variations: An extended family of machines," *Cryptologia* 22:3 (1998).

(Page 68) **the equations are nested**: The equations would be even more complicated if we had used a more complicated rotor wiring—you might be tempted to simplify the equations somewhat by multiplying through, but with a more practical system you can't even do that.

(Page 70) **key settings for the Enigma**: Later complications included as many as 8 possible rotors, out of which 3 (or, in some cases, 4) could be chosen, rotating and/or reconfigurable reflectors, and variations on the mechanism that determined how often the rotors after the first turned.

(Page 70) **Enigma**: There are many excellent descriptions of the Enigma and its early history. Those that I have consulted include Józef Garliński, *The Enigma War: The Inside Story of the German Enigma Codes and How the Allies Broke Them*, hardcover, 1st American ed. (New York: Scribners, 1980), Chapters 1–2 and

Appendix; Bauer, *Decrypted Secrets*, Section 7.3; and Konheim, *Cryptography*, Sections 5.6–5.7.

(Page 70) **determining rotor wirings**: Kahn, *The Codebreakers*, pp. 973–74; see also Garliński, *Enigma War*, Appendix; and Bauer, *Decrypted Secrets*, Section 19.6. Many other methods were devised for more or less specialized circumstances.

(Page 71) **determining key settings**: Kahn, *The Codebreakers*, pp. 975–76; see also Garliński, *Enigma War*, Appendix; and Bauer, *Decrypted Secrets*, Section 19.6. For probable word attacks, see, for example, Bauer, Section 19.7. The Poles and British developed some very important forerunners of modern computers in order to carry out the necessary brute force searches.

(Page 71) **probable words**: For a little more on this technique, see Section 5.1.

(Page 71) **modern attack on rotors**: See Konheim, *Cryptography*, Sections 5.4–5.5 and 5.8–5.9 for details.

(Page 71) **Van Hengel and Spengler**: de Leeuw, "Dutch invention."

(Page 72) **Hebern**: Kahn, *The Codebreakers*, pp. 417–20.

(Page 72) **Scherbius**: Kahn, *The Codebreakers*, pp. 421–22; David Kahn, *Seizing the Enigma*, 1st ed. Boston: Houghton Mifflin, 1991, pp. 31–42.

(Page 72) **lack of profits for the inventors**: Another well-known rotor machine, the British Typex, was explicitly based on the Enigma during World War II. (Louis Kruh and C. A. Deavours, "The Typex Cryptograph," *Cryptologia* 7:2 (1983).) Similarly, the Soviets introduced their Fialka rotor machine in 1956. (Paul Reuvers and Marc Simons, "Fialka," http://www.cryptomuseum.com/crypto/fialka/) Presumably, neither country considered compensating the original rotor machine inventors. The Japanese World War II cipher machine known as RED to the Americans also was a rotor machine, with elements similar to Damm's. The more well-known PURPLE machine, however, used a different principle. For more on the Japanese machines, see, for example, Alan G. Konheim, *Computer Security and Cryptography* (Hoboken, NJ: Wiley-Interscience, 2007), Chapter 7.

(Page 73) **Damm and Hagelin**: Kahn, *The Codebreakers*, pp. 425–27.

Chapter 3 Transposition Ciphers

(Page 75) **authenticity of the scytale**: Thomas Kelly, "The myth of the skytale," *Cryptologia* 22 (1998). Another possibility is that the scytale was authentic but worked in an entirely different way. See, for instance, Reinke, Classical cryptography.

(Page 75) **"The dispatch-scroll . . . "**: Plutarch, *Plutarch's Lives* (London; New York: Heinemann; Macmillan, 1914), Lysander, Chapter 19.

(Page 76) **"Go tell the Spartans . . . "**: Attributed by Herodotus to Simonides of Ceos. Translated by William Lisle Bowles. Quoted in Edward Strachey, "The soldier's duty," *The Contemporary Review* XVI (1871).

(Page 78) **only four possibilities**: If we used numbers that were not prime instead of 3 and 11, there would be a few more. Can you tell how many?

(Page 78) **methods of reading out of the rectangle**: Hitt, *Manual*, Chapter V, Case 1, p. 26–27.

(Page 78) **Friedman's 1941 manual**: William Friedman, *Advanced Military Cryptography* (Laguana Hills, CA: Aegean Park Press, 1976).

(Page 80) **"permits of no variation . . . "**: Hitt, *Manual*, Chapter V, Case 1-i, p. 29.

(Page 80) **"they do not depend on a key ... "**: Hitt, *Manual*, Chapter V, Case 1, p. 30.

(Page 80) **the Earl of Argyll's cipher**: David W. Gaddy, "The first U.S. Government Manual on Cryptography," *Cryptologic Quarterly* 11:4 (1992).

(Page 80) **Abraham Lincoln's cipher example**: Kahn, *The Codebreakers*, Chapter 7, p. 215. See David W. Gaddy, "Internal struggle: The Civil War," pages 88–103 of *Masked Dispatches: Cryptograms and Cryptology in American History, 1775–1900*, 3rd ed. National Security Agency Center for Cryptologic History, 2013 for more details on the origin of this cipher system.

(Page 81) **al-Kindi's transpositions**: Al-Kadi, "Origins of cryptology.".

(Page 81) **ibn ad-Duraihim's transpositions**: Al-Kadi, "Origins of cryptology."

(Page 81) **examples of ibn ad-Duraihim's transpositions**: Kahn, *The Codebreakers*, p. 96.

(Page 81) **"Drink to the rose ... "**: Al-Hasan ibn Hani al-Hakami Abu Nuwas, "Don't cry for Layla," Princeton Online Arabic Poetry Project, https://www.princeton.edu/~arabic/poetry/layla.swf.

(Page 82) **ways to notate permutations**: Warning: Some mathematicians prefer to use a notation based on where the letters *go* instead of where they *come from*. We will find our version more convenient, especially when we see cipher operations that repeat or drop some message elements, later in this section and in Chapter 4.

(Page 82) **"The battle and the sword ... "**: Abu at-Tayyib Ahmad ibn al-Husayn al Mutanabbi, "al-Mutanabbi to Sayf al-Dawla," Princeton Online Arabic Poetry Project, http://www.princeton.edu/~arabic/poetry/al_mu_to_sayf.html.

(Page 83) **inverse of a permutation**: Note that the numbers 4132 have reappeared. This is not a coincidence—can you figure out the connection?

(Page 83) **HDETS REEKO NTSEM WELLW**: The plaintext is from al Mutanabbi, "al-Mutanabbi to Sayfal-Dawla."

(Page 84) **functions**: Yes, this really is the same concept as the functions you learned about in high school, only it acts on letters and positions instead of real numbers. We will talk more about this in Section 4.3.

(Page 84) **trivial permutation**: Can you figure out how to write the trivial permutation?

(Page 85) **expansion functions**: In fact, cryptographers often call these functions expansion permutations, despite the fact that they are not permutations at all. I think expansion function is a good compromise.

(Page 86) **compression function**: Or compression permutation.

(Page 88) **permutation products are not commutative**: Just to make things a little more confusing, not all mathematicians write permutation products in the same order. Some people prefer to do the right permutation first and then the left instead of the way we did it. If you're one of those people, please don't write me any nasty letters!

(Page 88) **encrypt only with expansion functions, ... :** Unless you do something really fancy, like we will see in Section 4.3.

(Page 90) **cipher corresponding to poetry**: No, that's not a mistake—this time the keyword and the permutation give us the same numbers. Can you see why?

(Page 92) **first appearance of keyed columnar transposition**: John (J. F.) Falconer, *Rules for Explaining and Decyphering All Manner of Secret Writing, Plain and*

Demonstrative with Exact Methods for Understanding Intimations by Signs, Gestures, or Speech..., 2nd ed. (London: Printed for Dan. Brown... and Sam Manship..., 1692), p. 63.

(Page 92) **John Falconer**: Kahn, *The Codebreakers*, p. 155.

(Page 92) **ciphers based on keyed columnar transposition**: For example, see numerous references in Kahn, *The Codebreakers*.

(Page 92) **decrypting a keyed columnar transposition quickly**: Note the "shoes and socks" principle again. Alice writes the plaintext without using the key and reads the ciphertext off using the key. So Bob reverses the process by writing the ciphertext in *using* the key and reading it off *without* using the key.

(Page 96) **Nihilist transposition cipher**: Kerckhoffs, "La cryptographie militaire, I," pp. 16–17. The Nihilist transposition cipher should not be confused with the Nihilist substitution cipher, which is something else.

(Page 96) **double transposition in World War II**: Kahn, *The Codebreakers*, p. 539. To be exact, this was generally the "incompletely filled rectangle" variation explained in the sidebar on page 104. See also Leo Marks, *Between Silk and Cyanide*, 1st US ed (New York: Free Press, 1999) for much more on ciphers used by British and Allied agents during World War II.

(Page 97) **frequency of the letters will be the same**: Barring the addition of some rather unusual nulls.

(Page 97) **approximately 38.1% of them will be vowels**: I'm counting only a, e, i, o, and u as vowels and always counting them as vowels. You can argue this, but it doesn't really matter as long as you are consistent.

(Page 98) **variance**: If you are familiar with the standard deviation, the variance is the square of the standard deviation. But the variance will be a little easier to use for our situation.

(Page 99) **10-letter word with no a's, e's, i's, o's, or u's**: I haven't actually been able to find any 10-letter words like this. The only 11-letter word I've been able to find is "twyndyllyng," an obsolete term for a small twin. Maybe you know some others.

(Page 99) **hopelessly jumbled**: If this were a permutation cipher instead of a keyed columnar transposition, hopelessly jumbled would probably be an exaggeration. Most likely, we would have nonconsecutive bits of 2 or maybe 3 plaintext rows mixed together on each ciphertext row. Nevertheless, this statistical method still works quite well.

(Page 101) **only ones that are really likely to follow column I**: Of course, we should really account for the possibility that column I is the last column. In that case we could look for columns that could precede it or look for a column that would follow it after wrapping around, which would shift each letter down one in this example.

(Page 102) **digraph frequencies**: We are using the table of Hitt, *Manual*, Table IV, as in Section 1.6.

(Page 102) **adding the frequencies**: William Friedman, *Military Cryptanalysis. Part IV, Transposition and Fractionating Systems* (Laguna Hills, CA: Aegean Park Press, 1992), p. 5.

(Page 102) **it's wrong mathematically**: Friedman, *Military Cryptanalysis. Part IV*, p. 6.

(Page 102) **using logarithms**: In Friedman, *Military Cryptanalysis. Part IV*, p. 6, note 5, he describes this as, "A suggestion for which the author is indebted to Mr. A. W.

Small, junior cryptanalyst in this office. The principle makes practicable the use of tabulating machinery for the purpose of speeding up and facilitating the matching of columns in the anagramming process." For tabulating machinery, we would now read computers.

(Page 103) **we will use** log .0001: We can't use the logarithm of 0 because the logarithm of 0 is undefined.

(Page 103) **the closer a log weight is to 0**: Because 0 is the logarithm of 1.

(Page 103) **appears only in column II**: Or perhaps column V if we have wrapped around to the next line.

(Page 103) **keyed columnar transposition cipher**: The plaintext is from Howard Roger Garis, *Uncle Wiggily's Adventures* (New York: A. L. Burt, 1912), Story I.

(Page 103) **guess at the keyword**: Note that there's no way of telling exactly what the keyword used to generate a permutation was. For example, both the keyword WORD and the keyword IDEA give you the same cipher—try it and see!

(Page 105) **superimposition for transposition ciphers**: In fact, the use of the contact method for a permutation cipher looks very much like the use of superimposition we saw for repeating-key ciphers, and the technique of multiple anagramming we are about to see looks very much like the version of superimposition we see in Section 5.1.

(Page 106) **multiple anagramming**: The plaintexts are the titles of a series of books by Howard Garis.

(Page 107) **form of a rotation**: The reason we put $k + 1$ at the beginning of the row instead of k is so that the stupid key is $k = 0$, which is convenient.

(Page 107) **Madryga**: W. E. Madryga, "A High Performance Encryption Algorithm," in *Proceedings of the 2nd IFIP International Conference on Computer Security: a Global Challenge*, edited by James H. Finch and E. Graham Dougall (Amsterdam: North-Holland, 1984).

(Page 107) **RC5**: Ronald L. Rivest, "The RC5 encryption algorithm," in Bart Preneel (ed.), *Fast Software Encryption* (Springer Berlin Heidelberg, 1995).

(Page 107) **RC6**: Ronald L. Rivest et al., "The RC6™ block cipher," NIST, August 1998, series AES Proposals. RC5 and RC6, incidentally, were invented—with help, in the case of RC6—by Ron Rivest, whom we will meet again in Chapter 7. RC6 was a finalist in the Advanced Encryption Standard competition, which I will talk about in Chapter 4.

(Page 107) **Akelarre**: Gonzalo Alvarez et al., "Akelarre: A new block cipher algorithm," in Stafford Tavares and Henk Meijer (eds.), *Proceedings of the SAC '96 Workshop* (Kingston, ON: Queen's University, 1996).

(Page 107) **Madryga flawed**: Alex Biryukov and Eyal Kushilevitz, "From differential cryptanalysis to ciphertext-only attacks," in Hugo Krawczyk (ed.), *Advances in Cryptology—CRYPTO '98* (Springer Berlin Heidelberg, 1998).

(Page 107) **Some attacks on RC5**: B. S., Kaliski and Yiqun Lisa Yin, "On the security of the RC5 encryption algorithm," RSA Laboratories (September 1998).

(Page 107) **Comparison of RC6 to AES**: James Nechvatal et al., Report on the development of the Advanced Encryption Standard (AES), NIST (October 2000).

(Page 107) **Akelarre partly based on RC5**: Alvarez et al., *"Akelarre."*

(Page 108) **Attacks on Akelarre**: Niels Ferguson and Bruce Schneier, "Cryptanalysis of Akelarre," in Carlisle Adams and Mike Just (eds.), *Proceedings of the SAC '97 Workshop* (Ottawa, ON: Carleton University, 1997); Lars R., Knudsen and Vincent Rijmen, "Ciphertext-only attack on Akelarre," *Cryptologia* 24:2 (2000). The second of these papers includes the attack, which essentially bypasses everything but the rotation. An earlier version of that paper was called "Two rights sometimes make a wrong," due to the combining of elements of two strong ciphers into a weak one.

(Page 108) **process very much like anagramming**: Although it is actually somewhat easier for two reasons. First, the number of columns is known, since a cipher with a variable number of columns would be comparatively rather difficult to implement on a computer. Second, since we know the permutation is a rotation, there are many fewer possibilities to try.

CHAPTER 4 CIPHERS AND COMPUTERS

(Page 109) **Polybius**: Polybius, *The Histories* Cambridge, MA: Harvard University Press, 1922–1927, Book X, Chapters 43–47.

(Page 109) **using torches to send coded messages**: This was a practice with great longevity—the most famous example to Americans would be "one if by land, two if by sea."

(Page 109) "**It is as follows . . .** ": Polybius, *Histories*, X.45.7–12.

(Page 110) **Polybius' cipher doesn't have a key**: To be fair, it's not clear that Polybius was even interested in the secrecy of the message—he seems most concerned with just getting messages quickly and accurately across long distances.

(Page 112) **multiple tables for ternary numerals**: Or we could use 3-dimensional tables, but those are difficult to print in a book like this one.

(Page 112) **ternary table with grouped digits**: The similarity with the base 9 table is not a coincidence. It comes from the similarity between the formulas $r \cdot 9 + c$ and $r \cdot 9 + (c_1 \cdot 3 + c_2)$ for the letter in row r and column c.

(Page 113) **modern English example**: Bacon used only 24 letters in his alphabet, treating i and j as the same and u and v as the same, and he started with a as 00000 instead of 00001. He also used the symbols a and b instead of 0 and 1. In fact, it's not clear that he thought of his strings of a's and b's as numbers at all. On the other hand, he did put them in the same order that binary numerals would come in.

(Page 113) **biformed alphabet**: Francis Bacon, *Of the Advancement and Proficience of Learning* (Oxford: Printed by Leon Lichfield, Printer to the University, for Rob Young and Ed Forrest, 1640), Book VI, Chapter I, Part III.

(Page 114) **Gauss and Weber**: William V. Vansize, "A new page-printing telegraph," *Transactions of the American Institute of Electrical Engineers* 18 (1902), p. 22.

(Page 114) **Baudot**: Vansize, "New page-printing telegraph," p.22.

(Page 114) **Vernam had Baudot's code**: To be perfectly accurate, this was not Baudot's original code but a revised version.

(Page 114) **noncarrying addition**: Noncarrying addition can also be thought of as vector addition modulo 2 for those who are really into that. For those with computer programming experience, you might know it as bitwise exclusive-or, aka XOR.

(Page 115) **Vernam's method**: Gilbert Vernam, "Secret signaling system," U.S. Patent: 1310719, 1919, http://www.google.com/patents?vid=1310719.

(Page 116) **straddling checkerboard**: As usual, a real system would mix up the order of the letters and/or digits according to some key.

(Page 116) **"most interesting and practical"**: Friedman, *Military Cryptanalysis. Part IV*, p. 97.

(Page 116) **GedeFu 18**: Michael van der Meulen, "The road to German diplomatic ciphers—1919 to 1945," *Cryptologia* 22:2 (1998), p. 144.

(Page 116) **called it ADFGVX**: David Kahn, "In memoriam: Georges-Jean Painvin," *Cryptologia* 6:2 (1982), p. 122. When first introduced, the square was 5 × 5, and only the letters ADFGX were used. The ciphertext letters were apparently chosen to provide an early example of error correction since their Morse code equivalents were different enough not to be easily confused.

(Page 117) **outline of general method**: In the original version of M. Givierge, *Cours de cryptographie* (Paris: Berger-Levrault, 1925).

(Page 117) **identical beginnings or endings**: Kahn, *The Codebreakers*, p. 344. In modern terms, we would call these differential attacks, and we will see them again in Section 4.4.

(Page 117) **division into columns**: Friedman, *Military Cryptanalysis. Part IV*, pp. 123–24.

(Page 118) **diffusion**: C. E. Shannon, "Communication theory of secrecy systems," *Bell System Technical Journal* 28:4 (1949).

(Page 118) **confusion**: Shannon, "Communication theory." The common definition of confusion has mutated somewhat over the years. Schneier, *Applied Cryptography*, p. 237, for instance, defines it as "obscur[ing] the relationship between the plaintext and the ciphertext," for example, through substitution.

(Page 118) **avoid high-frequency letters clustering**: Otherwise it may be possible to distinguish which letters in the ciphertext designate rows and which designate columns. This information can be used in a similar way to vowels and consonants to find the number of columns. Then one can attempt to anagram the columns into "digraphic" combinations whose phi test index of coincidence matches a monoalphabetic cipher. A complete description can be found in Friedman, *Military Cryptanalysis. Part IV*, pp. 124–43.

(Page 120) **"Speaking loosely . . ."**: Shannon, "Communication theory," p. 712.

(Page 121) **depend on a key *k***: U and V could also depend on two different keys for more security.

(Page 122) **didn't really start thinking about Shannon's principles**: At least, as far as the public record is concerned. We still know very little about what organizations like the NSA were doing during this time period.

(Page 122) **Feistel through 1944**: Steven Levy, *Crypto*, 1st paperback ed (New York: Penguin (Non-Classics), 2002), p. 40.

(Page 122) **Feistel 1944–1967**: Kahn, *The Codebreakers*, p. 980.

(Page 122) **perhaps because of NSA pressure**: Whitfield Diffie and Susan Landau, *Privacy on the Line*, updated and expanded edition (Cambridge, MA: MIT Press, 2010), p. 57.

(Page 124) **128 bits**: Feistel was apparently thinking the same thing, although most of his contemporaries thought that 64 bits was plenty at the time. See Horst Feistel, "Cryptography and computer privacy," *Scientific American* 228:5 (1973).

(Page 126) **32 groups of 4**: Feistel, "Cryptography and computer privacy."

(Page 127) **example of an SP-network**: The example will have one small exception, which I will point out, to the SP-network structure.

(Page 127) **avalanche effect**: Feistel, "Cryptography and computer privacy," p.23.

(Page 127) **3-bit example**: Kwangjo Kim, Tsutomu Matsumoto, and Hideki Imai, "A recursive construction method of S-boxes satisfying strict avalanche criterion," in CRYPTO '90: *Proceedings of the 10th Annual International Cryptology Conference on Advances in Cryptology*, edited by Alfred Menezes and Scott A. Vanstone. (Berlin/Heidelberg, New York: Springer-Verlag, 1991).

(Page 128) **128-bit S-boxes**: Eight-bit S-boxes are common as of this writing and 16-bit S-boxes are not unheard of, but one expects that by the time we are ready to easily manufacture and/or program fast 128-bit S-boxes, we will need even larger ones!

(Page 128) **adding the round key modulo 2**: Occasionally the round key is added modulo something else or combined in some other way.

(Page 130) **Lucifer**: Levy, *Crypto*, p. 41. Apparently Lucifer was a pun on an earlier name, Demon, which was originally merely short for Demonstration. The reason for the abbreviation was simply that the computer system they were using couldn't handle 13-letter file names!

(Page 130) **IBM 2984**: Diffie and Landau, *Privacy on the Line*, p. 251.

(Page 130) **soliciting proposals**: Actually, IBM let the initial response date in 1973 pass, and it wasn't until 1974 that IBM's chief scientist offered DSD-1 as a candidate to NBS. Since no responses to NBS's first call for proposals had even remotely met the standards, NBS immediately reopened the solicitation. Levy, *Crypto*, pp. 51–52.

(Page 130) **NSA didn't want to design DES**: Diffie and Landau, *Privacy on the Line*, p. 59.

(Page 130) **NBS requested help**: Schneier, *Applied Cryptography*, p. 266.

(Page 130) **reduction from 128 bits to 64**: Levy, *Crypto*, p. 58, quotes Walt Tuchman, the head of the IBM product development group.

(Page 131) **some people at IBM suspected**: Notably Alan Konheim, who headed the mathematical team. Levy, *Crypto*, p. 59.

(Page 131) **error-checking mechanism**: Walt Tuchman again. Levy, *Crypto*, p. 58.

(Page 131) **48-bit key**: Thomas R. Johnson, *American Cryptology during the Cold War, 1945–1989; Book III: Retrenchment and Reform, 1972–1980* (Center for Cryptologic History, National Security Agency, 1995), p. 232. The relevant sentence is redacted in the version posted on the NSA Web site but can be found in the version posted at http://cryptome.org/0001/nsa-meyer.htm.

(Page 131) **differential attack**: Eli Biham and Adi Shamir, *Differential Cryptanalysis of the Data Encryption Standard* (New York: Springer, 1993), p. 7. This differential attack is somewhat similar to the attack on the ADFGVX cipher we mentioned in Section 4.2, although the high level of diffusion makes it much harder to carry out.

(Page 131) **particularly resistant**: Biham and Shamir, *Differential Cryptanalysis*, pp. 8–9.

(Page 131) **S-boxes had been redesigned**: Don Coppersmith, by personal email cited in Eli Biham, "How to make a difference: Early history of differential cryptanalysis," slides from invited talk presented at Fast Software Encryption, 13th International Workshop, 2006, http://www.cs.technion.ac.il/~biham/Reports/Slides/fse2006

-history-dc.pdf, and publicly in D. Coppersmith, "The Data Encryption Standard (DES) and its strength against attacks," *IBM Journal of Research and Development* 38:3 (1994).

(Page 131) **kept secret until rediscovered**: Coppersmith, "Data Encryption Standard." There was still some lingering suspicion that the NSA had put some kind of "back door" in the S-boxes to weaken them, but in general Coppersmith's explanation was accepted by the cryptographic community.

(Page 132) **purpose of P-boxes**: Schneier, *Applied Cryptography*, p. 271.

(Page 134) **linear cryptanalysis**: Schneier, *Applied Cryptography*, p. 293.

(Page 134) **linear cryptanalysis not known**: Or if it was, they chose for some reason not to do anything about it. Coppersmith, "Data Encryption Standard."

(Page 134) **1728 custom chips**: According to The Electronic Frontier Foundation, "Frequently Asked Questions (FAQ) about the Electronic Frontier Foundation's 'DES cracker' machine," http://w2.eff.org/Privacy/Crypto/Crypto_misc/DESCracker /HTML/19980716_eff_des_faq.html. Other sources give between 1536 and 1856 chips.

(Page 134) **time and cost**: Electronic Frontier Foundation, " 'DES cracker' machine."

(Page 135) **DES was breakable**: Susan Landau, "Standing the test of time: The Data Encryption Standard," *Notices of the AMS* 47:3 (March 2000).

(Page 135) **request for nominations**: "Announcing request for candidate algorithm nominations for the Advanced Encryption Standard (AES)," *Federal Register* 62:177 (1997).

(Page 135) **foreign national reviewers**: Susan Landau, "Communications security for the twenty-first century: The Advanced Encryption Standard," *Notices of the AMS* 47:4 (April 2000). When the AES selection process started, it was still generally illegal to export cryptographic software with keys longer than 40 bits, even DES, outside the United States. NIST, however, allowed any foreign national to obtain software implementations of the AES candidates just as long as they registered with NIST and promised not to pass on the algorithms.

(Page 135) **three public conferences**: Including one outside the United States, in Rome, Italy.

(Page 135) **at least one non-US national**: Landau, "Communication security."

(Page 135) **"Rijndael"**: As you would guess, the cipher designers combined their names to name the cipher. According to Rijmen, "If you're Dutch, Flemish, Indonesian, Surinamer or South-African, it's pronounced like you think it should be. Otherwise, you could pronounce it like 'Reign Dahl,' 'Rain Doll,' 'Rhine Dahl.' We're not picky. As long as you make it sound different from 'Region Deal.'" Vincent Rijmen, "The Rijndael page," http://www.ktana.eu/html/theRijndaelPage.htm. Also quoted in Wade Trappe and Lawrence C. Washington, *Introduction to Cryptography with Coding Theory*, 2nd ed. (Upper Saddle River, NJ: Prentice Hall, 2005), pp. 151–152. Most English-speaking people seem to say "Rhine Dahl," or just "A-E-S."

(Page 135) **AES block size**: The original Rijndael submission allowed block sizes of 192 and 256 bits as well as 128, but NIST decided not to include them in the standard. Note that increasing the block size does not necessarily increase the security of a cipher.

(Page 137) **one gigantic P-box**: They have particularly cited the high cost of implementation of large P-boxes in modern ciphers. See, for example, p. 75 and

p. 131 of Joan Daemen and Vincent Rijmen, *The Design of Rijndael*, 1st ed. (Berlin/Heidelberg, New York: Springer, 2002).

(Page 137) **dispersion**: This could be considered a form of diffusion, although it does not provide the avalanche effect which is now considered highly desirable. It should also remind you of the transposition ciphers using rectangles from Section 3.2.

(Page 138) **Hill cipher encryption**: The AES designers call the transformation a D-box, for diffusion. Daemen and Rijmen, *The Design of Rijndael*, p. 22. The last round leaves out the Hill cipher step. For technical reasons, this allows a more efficient implementation of the decryption algorithm. Daemen and Rijmen, pp. 45–50.

(Page 138) **DES S-boxes were "human-made"**: Coppersmith, "Data Encryption Standard."

(Page 139) **not so complicated in a different way**: This turns out to have been somewhat controversial. The AES S-box gives good protection against differential and linear attacks, but there have been other attacks proposed that may be able to take advantage of the higher-level simplicity of the AES S-box. See, for example, Daemen and Rijmen, *The Design of Rijndael*, p. 156.

(Page 140) **published list**: According to Joan Daemen and Vincent Rijmen, AES Proposal: Rijndael (NIST, September 1999), series AES Proposals. , p. 25, the designers used the list from R. Lidl and H. Niederreiter, *Introduction to Finite Fields* (Cambridge, UK: Cambridge University Press, 1986), p. 378.

(Page 140) **prime modulo 2**: A polynomial can have factors when you are working in modular arithmetic even though it doesn't have any in ordinary arithmetic. For example, $x^2 + 1$ is prime in ordinary arithmetic but not modulo 2, since $(x + 1) \times (x + 1) = x^2 + 2x + 1 = x^2 + 1$.

(Page 140) $x^8 + x^4 + x^3 + x + 1$: Daemen and Rijmen, *The Design of Rijndael*, p. 16. Since the flap about the DES S-boxes, it's been considered very important for cipher designers to explain any time they choose an arbitrary number, polynomial, and so on, exactly where they got it. This helps convince people that they haven't slipped in any back doors. Numbers with this sort of explanation are sometimes called "nothing up my sleeve" numbers.

(Page 141) **AES polynomial arithmetic**: The technical term for this type of polynomial arithmetic modulo a prime polynomial and a prime number is **finite field arithmetic**.

(Page 142) **some concern**: Nechvatal et al., "Report on the Development of the AES," p. 28.

(Page 142) **XSL not better than brute force**: Carlos Cid and Ralf-Philipp Weinmann, "Block ciphers: Algebraic cryptanalysis and Gröbner bases," in Massimiliano Sala, Shojiro Sakata, Teo Mora, Carlo Traverso, and Ludovic Perret (eds.), *Gröbner Bases, Coding, and Cryptography* (Springer Berlin Heidelberg, 2009), p. 313.

(Page 142) **polynomial attacks in the future**: Cid and Weinmann, "Block Ciphers," p. 325.

(Page 142) **known-key and related-key attacks**: See, for example, Niels Ferguson, et al., *Cryptography Engineering* (New York: Wiley, 2010), p. 55 and the references cited there.

(Page 142) **not used in the way they were intended**: See, for example, Schneier, *Applied Cryptography*, p. 447, for an example of a situation where a known-key attack could be applied, and Ferguson et al., *Cryptography Engineering*, pp. 323–24 for an example of a related-key attack on the WEP (Wired Equivalent Privacy) algorithm originally used to protect wireless local area computer networks.

(Page 142) **2011 attack on AES**: Andrey Bogdanov, Dmitry Khovratovich, and Christian Rechberger, "Biclique cryptanalysis of the full AES," in Dong Hoon Lee and Xiaoyun Wang (eds.), *Advances in Cryptology—ASIACRYPT 2011* (Springer Berlin Heidelberg, 2011). Some reports of this attack suggested that all 2^{88} sets of texts would need to be stored in memory at once, which would be wildly impractical. This does not, in fact, seem to be the case.

(Page 142) **unreasonably long time**: Dave Neal, "AES encryption is cracked," *The Inquirer* (August 17, 2011).

(Page 143) **reevaluate AES**: NIST, Announcing the Advanced Encryption Standard (AES), NIST, November 2001. It's not clear whether any formal reevaluations have been done.

(Page 143) **NBS document**: NBS, "Guidelines for Implementing and Using the NBS Data Encryption Standard," April 1981.

(Page 143) **draft proposal for format-preserving encryption**: Morris Dworkin, "Recommendation for block cipher modes of operation: Methods for format-preserving encryption," NIST, July 2013.

(Page 143) **report from April 2015**: Morris Dworkin and Ray Perlner, Analysis of VAES3 (FF2), 2015. The report, from two researchers at NIST, ends with "[t]he authors acknowledge the National Security Agency for notifying NIST in general terms that FF2 might not meet NIST's security requirements."

(Page 144) **homomorphic encryption in 1978**: Ronald L. Rivest, Len Adleman, and Michael L. Dertouzos, "On Data Banks and Privacy Homomorphisms," in Richard A. DeMillo, David P. Dobkin, Anita K. Jones, and Richard J. Lipton (eds.), *Foundations of Secure Computation* (New York: Academic Press, 1978).

(Page 144) **first fully homomorphic system**: Craig Gentry, "Fully homomorphic encryption using ideal lattices," in *Proceedings of the Forty-first Annual ACM Symposium on Theory of Computing*, Association for Computing Machinery Special Interest Group on Algorithms and Computation Theory (ACM, 2009). Gentry and some colleagues soon developed a simpler version of the original scheme. A description of the second scheme, with a nice extended analogy based on allowing workers to assemble jewelry without being able to steal the raw materials, can be found in Craig Gentry, "Computing arbitrary functions of encrypted data," *Communications of the ACM* 53:3 (2010). Both of these schemes, and most of the fully homomorphic schemes proposed since, are asymmetric-key cryptographic schemes of the sort we discuss in Section 7.3. These systems tend to be easier to make homomorphic since they are easier to manipulate mathematically. As Gentry points out, however, a fully homomorphic encryption scheme can be either symmetric or asymmetric. For a relatively simple (but not practical) symmetric scheme see Jeffrey Hoffstein et al., *An Introduction to Mathematical Cryptography*, 2nd ed. (New York: Springer, 2014), Example 8.11.

(Page 144) **two government agencies and at least one company**: NSA Research Directorate staff, "Securing the cloud with homomorphic encryption," *The Next Wave* 20:3 (2014).

(Page 144) **NSA document**: Spiegel Staff, "Prying eyes: Inside the NSA's war on Internet security," *Spiegel Online* (2014).

(Page 144) **full document**: NSA, "Summer mathematics, R21, and the Director's Summer Program," *The EDGE: National Information Assurance Research Laboratory (NIARL) Science, Technology, and Personnel Highlights*, 2008, http://www.spiegel.de /media/media-35550.pdf.

CHAPTER 5 STREAM CIPHERS

(Page 145) **"sweet spot"**: Mendelsohn, *Blaise de Vigenère and the "Chiffre Carré"*, for example, p. 127.

(Page 146) **keytext**: One contemporary source suggests that an 1892 work of Arthur Hermann was the first to put this into definitive form. André Lange and Émile-Arthur Soudart, *Treatise on Cryptography* (Washington, D.C.: US Government Printing Office, 1940), pp. 31, 87.

(Page 146) **"Dorothy lived in the midst . . . "**: L. Frank Baum, *The Wonderful Wizard of Oz* (Chicago: George M. Hill, 1900), Chapter 1.

(Page 146) **"A slow sort of country . . . "**: Lewis Carroll, *Through the Looking-Glass, and What Alice Found There* (1871), Chapter 2.

(Page 148) **"Mowgli was far and far through the forest . . . "**: Rudyard Kipling, *The Jungle Book* (1894), Chapter 1.

(Page 148) **too few messages**: Remember that the more ciphertext you have, the better letter frequency analysis works. This also applies to the brute force with "frequency sums" technique from Section 2.6.

(Page 148) **chi test and cross-product sum**: To be exact, Friedman and Kullback used the Greek letter chi to refer to the numerator of what I am calling the cross-product sum. The chi test and cross-product sum, like the phi test, first appear in Solomon Kullback, *Statistical Methods in Cryptanalysis* (Laguna Hills, CA: Aegean Park Press, 1976). The algebraic equivalence is shown in Friedman, *Military Cryptanalysis. Part III*, pp. 66–67.

(Page 150) **multiple ciphertexts with the same running-key**: The plaintexts are taken from the chapter titles of a famous book by Robert Louis Stevenson. Not all of them start at the beginning of the title and some are parts of two titles run together.

(Page 152) **just for variety**: It's not really just for variety. It also makes things work out slightly more easily, but it's not really important—the technique still works with tabula recta; it just takes a little more trial and error.

(Page 152) **ciphertext with keytext from a common book**: The keytext and plaintext are from Rudyard Kipling, *Just So Stories* (1902), Chapters 1 and 7.

(Page 153) **probable words**: If you happened to already know where I got my plaintext from in this example, you would want to consider the words *best* and *beloved*.

(Page 154) **Frank Miller**: Steven M. Bellovin, "Frank Miller: Inventor of the one-time pad," *Cryptologia* 35:3 (2011). Miller's system was similar to the German Foreign Office system described on page 155, except without the modular arithmetic.

(Page 154) **some disagreement**: Kahn, *The Codebreakers*, pp. 397–401, tells the story and holds that Mauborgne made the crucial decision. Steven M. Bellovin, "Vernam, Mauborgne, and Friedman: The one-time pad and the index of coincidence," Department of Computer Science, Columbia University, May 2014, lays out the case for both Vernam and Mauborgne and sides with Vernam.

(Page 154) **never be reused**: One model of teletypewriter actually had a blade that sliced the tape in half after reading it to make sure it couldn't be reused. Kahn, *The Codebreakers*, p. 433.

(Page 155) **three cryptologists**: Werner Kuze, Rudolf Schauffler, and Erich Langlotz. Kahn, *The Codebreakers*, p. 402.

(Page 155) **German diplomatic one-time pad**: Kahn, *The Codebreakers*, p. 402–3.

(Page 155) **one-time pad was unbreakable**: Bellovin, "Vernam, Mauborgne, and Friedman," credits Friedman as the first to really understand why.

(Page 155) **Shannon's proof**: Shannon, "Communication theory." This is the same famous paper in which he defined confusion and diffusion; see Sections 4.2 and 4.3. Apparently Vladimir Kotelnikov also developed the theory of perfect security in the Soviet Union in 1941, but his work is still classified. Natal'ya V. Kotel'nikova, "Vladimir Aleksandrovich Kotel'nikov: The life's journey of a scientist," *Physics-Uspekhi* 49:7 (2006); Vladimir N. Sachkov, "V. A. Kotel'nikov and encrypted communications in our country," *Physics-Uspekhi* 49:7 (2006); Sergei N. Molotkov, "Quantum cryptography and V. A. Kotel'nikov's one-time key and sampling theorems," *Physics-Uspekhi* 49:7 (2006).

(Page 156) **how to exchange random key material**: Unlike a running-key cipher, Alice and Bob can't just both pick up identical copies of the same book.

(Page 157) **fall-back system**: Kahn, *The Codebreakers*, p. 401.

(Page 157) **red phone**: Kahn, *The Codebreakers*, p. 715–16.

(Page 157) **Soviet spy one-time pads**: Kahn, *The Codebreakers*, p. 663–64.

(Page 157) **Cardano**: Cardano is better known to most mathematicians as one of the first people to discover a general formula for solving cubic equations.

(Page 159) **Treatise on Ciphers**: Blaise de Vigenère, *Traicté des Chiffres, ou Secrètes Manières d'Escrire* (*Treatise on Ciphers, or Secret Methods of Writing*) Paris: A. L'Angelier, 1586.

(Page 159) **"a worthless cracking of the brain"**: Vigenère's opinion of cryptanalysis was that it was "un inestimable rompement de cerveau." Vigenère, *Traicté des Chiffres*, p. 12r.

(Page 159) **extra step**: Vigenère also presented the possibility that the ciphertext could be altered again after adding the keystream.

(Page 160) **"waste all your oil"**: Another Vigenère comment on the practice of cryptanalysis. Vigenère, *Traicté des Chiffres*, p. 198r, quoted in Mendelsohn, *Blaise de Vigenère and the "Chiffre Carré."*

(Page 161) **16 text characters**: See Sidebar 4.1.

(Page 164) **other repeating-key ciphers**: You can see that the distinctions between progressive ciphers, repeating-key ciphers, and key autokey ciphers are somewhat fluid.

(Page 165) **addition modulo 10**: Remember that we can also think of this as noncarrying addition.

(Page 165) **Soviet World War II cipher**: Alex Dettman et al., *Russian Cryptology During World War II*. (Laguna Hills, CA: Aegean Park Press, 1999), p. 40. The initialization vector and key that I used are the dates of the beginning and end of the Battle of Stalingrad. The plaintext also refers to that battle.

(Page 165) **only 5 key digits:** Plus the initialization vector, but the way we have set up this system, Alice and Bob don't even necessarily have to keep the initialization vector secret. It should be different for every message, though. Ferguson, Schneier, and Kohno, *op. cit.*, p.69.

(Page 166) **some experts suggest not using OFB:** Ferguson et al., *Cryptography Engineering*, p.71.

(Page 167) **counter initialization vector requirements:** Ferguson et al., *Cryptography Engineering*, p.70.

(Page 167) **useful for data files:** Schneier, *Applied Cryptography*, p. 206.

(Page 168) **"multiply like rabbits":** Fibonacci's original presentation of the Fibonacci sequence was in the context of a problem about rabbit reproduction.

(Page 168) **Gromark cipher:** Gromark stands for GROnsfeld with Mixed Alphabet and Running Key. W. J. Hall, "The Gromark cipher (Part 1)," *The Cryptogram* 35:2 (1969). The Gronsfeld cipher is just a name for variants of a tabula recta cipher using a key of numbers instead of letters, such as we use here and in the key autokey cipher of the previous section. Our version doesn't actually use a mixed alphabet, and we are making a distinction between a running key and an autokey cipher. So it might be more accurate to call it a "Grotrak" cipher, or maybe a "Grolfak" cipher.

(Page 168) **VIC cipher:** David Kahn, "Two Soviet Spy Ciphers," in *Kahn on Codes* (New York: Macmillan, 1984).

(Page 169) **linear equations:** Note that the Hill cipher also uses linear equations; this will become relevant when we talk about the cryptanalysis of LFSRs.

(Page 170) **feedback:** This also happens in plaintext feedback mode, ciphertext feedback mode, and output feedback mode.

(Page 171) **LSFRs in software:** See, for example, Schneier, *Applied Cryptography*, p. 378 for more on this variation.

(Page 171) **as far back as 1952:** Maybe earlier; the AFSAY-816 voice-encryption device from the late 1940s used "shift registers," which were very likely LFSRs. Thomas R. Johnson, *American Cryptology during the Cold War, 1945–1989; Book I: The Struggle for Centralization, 1945–1960* (Center for Cryptologic History, National Security Agency, 1995), p. 220; David G. Boak, "A history of U.S. communications security" (Volume I). National Security Agency, July 1973, p. 58.

(Page 171) **KW-26:** Melville Klein, *Securing Record Communications: The TSEC/KW-26* (Center for Cryptologic History, National Security Agency, 2003).

(Page 173) **decimal equivalent:** Alice can't necessarily convert her plaintext bits back to characters using ASCII, because some of the numbers (such as 9) might not represent printable characters.

(Page 173) **modulo-2 LFSR with four cells and a period of 15:** Can you find it?

(Page 173) **LFSRs with maximum period:** See, for example, Solomon Golomb, *Shift Register Sequences*, Rrev. ed. (Laguna Hills, CA: Aegean Park Press, 1982), Section III.3.5.

(Page 173) **2j pairs of plaintext and ciphertext bits:** Note that 2j pairs is not a large number compared to $2^j - 1$, the length of the period. In practice, j is likely to be less than 100, but even $2^{30} - 1$ is already about 10 billion.

(Page 174) **finding the initialization vector:** Using these equations is not really the quickest way to find the initialization vector, but it's easy and it works.

(Page 175) **harder to analyze**: See, e.g., Schneier, *Applied Cryptography*, p. 412.

(Page 175) **options for adding nonlinearity**: See Schneier, *Applied Cryptography*, Section 16.4, for lots of examples of the last two options.

(Page 175) **A5 ciphers**: There at least three different A5 ciphers. A5/1 was intended for use in the United States and Europe. A5/2 is a weaker version intended for markets outside the Organization for Economic Co-operation and Development. Elad Barkan and Eli Biham, "Conditional estimators: An effective attack on A5/1," in *Selected Areas in Cryptography* (Berlin/Heidelberg: Springer, 2006). A5/3 is an entirely different cipher designed for 3G phones and does not use LFSRs. A5/4 seems to be the same as A5/3 with a longer key.

(Page 175) **disagreement among intelligence agencies**: Ross Anderson, "A5 (Was: HACKING DIGITAL PHONES)," Posted in uk.telecom (Usenet group), June 17, 1994, http:// groups.google.com/group/uk.telecom/msg/ba76615fef32ba32.

(Page 175) **efficiency may have played a role**: Ross Anderson, "On Fibonacci Keystream Generators," in *Fast Software Encryption* (Berlin/Heidelberg: Springer, 1995).

(Page 175) **British university**: Schneier, *op. cit.*, p. 389.

(Page 175) **almost-complete description posted**: Anderson, *"A5 (Was: HACKING DIGITAL PHONES)."*

(Page 175) **complete design reverse-engineered, posted, and confirmed**: See Alex Biryukov, Adi Shamir, and David Wagner, "Real Time Cryptanalysis of A5/1 on a PC," in *Fast Software Encryption* (Berlin/Heidelberg: Springer, 2001), Abstract and Introduction. The reverse-engineering was done by Marc Briceno, of the Smart Card Developers Association.

(Page 176) **A5/1 key setup**: In actual GSM phones, the key setup is a little more complicated, but that's not really important for our purposes. See Barkan and Biham, "Conditional estimators."

(Page 176) **each LFSR shifts 3/4 of the time**: Assuming each combination of bits is equally likely.

(Page 176) **cuts down on the period**: W. G. Chambers and S. J. Shepherd, "Mutually clock-controlled cipher keystream generators," *Electronics Letters* 33:12 (1997).

(Page 176) **careful use**: W. Chambers, "On random mappings and random permutations," in *Fast Software Encryption* (Berlin/Heidelberg: Springer, 1995).

(Page 177) **as early as 1994**: Anderson, *"A5 (Was: HACKING DIGITAL PHONES)."*

(Page 177) **1997 paper**: Jovan Dj. Golic, "Cryptanalysis of alleged A5 stream cipher," in *Advances in Cryptology—EUROCRYPT '97*, edited by Walter Fumy (Konstanz, Germany: Springer-Verlag, 1997).

(Page 177) **considerably refined**: See Barkan and Biham, "Conditional estimators," for a summary of the various papers.

(Page 177) **various logistical reasons**: Audio data or file transfers would need to be carefully synchronized; raw digital data collection requires access to the phone itself or a computer connected to it, and soon.

(Page 177) **2006 correlation-type attack**: Barkan and Biham, "Conditional estimators."

(Page 178) **2003 precomputation attack**: Elad Barkan, et al., "Instant ciphertext-only cryptanalysis of GSM encrypted communication," in *Advances in Cryptology— CRYPTO 2003* (Berlin/Heidelberg: Springer, 2003).

(Page 178) **project to create these tables**: Chris Paget and Karsten Nohl, "GSM: SRSLY?" Slides from lecture presented at 26th Chaos Communication Congress, 2009, http://events.ccc.de/congress/2009/Fahrplan/events/3654.en.html.

(Page 178) **showed some partial successes**: Frank A. Stevenson, "[A51] Cracks beginning to show in A5/1...," Email sent to the A51 mailing list, May 1, 2010, http://lists.lists.reflextor.com/pipermail/a51/2010-May/000605.html.

(Page 178) **GSM Association**: GSM Association, "GSMA statement on media reports relating to the breaking of GSM encryption," Press release, December 30, 2009, http:// gsmworld.com/newsroom/press-releases/2009/4490.htm.

(Page 178) **"process" A5/1**: NSA, "GSM Classification Guide," September 20, 2006, https://s3.amazonaws.com/s3.documentcloud.org/documents/888710/gsm -classification-guid-20-sept-2006.pdf.

(Page 178) **generally taken**: Craig Timberg and Ashkan Soltani, "By cracking cellphone code, NSA has ability to decode private conversations," *The Washington Post* (December 13, 2013).

(Page 178) **major wireless carriers**: Ashkan Soltani and Craig Timberg, "T-Mobile quietly hardens part of its U.S. cellular network against snooping," *The Washington Post* (October 22, 2014).

(Page 178) **"identify new stream ciphers..."**: The ECRYPT Network of Excellence, "Call for stream cipher primitives, version 1.3," 2005, http://www.ecrypt.eu.org /stream/call.

(Page 178) **eSTREAM**: For more on the eSTREAM project, see Matthew Robshaw and Olivier Billet (eds.), *New Stream Cipher Designs: The eSTREAM Finalists* (Berlin, New York: Springer, 2008) and the project's Web site: "eSTREAM: the eSTREAM stream cipher project." http://www.ecrypt.eu.org/stream/index.html.

(Page 179) **NIST-approved modes**: NIST Computer Security Division, "Computer Security Resource Center: Current modes." http://csrc.nist.gov/groups/ST/toolkit /BCM/current_modes.html.

(Page 179) **authentication**: Some of these authentication modes are designed for specialized situations rather than messages in general. We will talk about a different view of authentication in Section 8.4.

(Page 179) **CBC-MAC**: Computer Data Authentication, NIST, May 1985.

(Page 179) **two different keys**: If she uses the same key for CBC and CBC-MAC, then the MAC is not secure. See, for example, Ferguson et al., *Cryptography Engineering*, p. 91.

(Page 180) **Trivium**: For more on the design and specifications of Trivium, see Christophe De Cannière and Bart Preneel, "Trivium," in Matthew Robshaw and Olivier Billet (eds.), *New Stream Cipher Designs* (Berlin, New York: Springer, 2008).

(Page 180) **nonlinear operations**: This is nonlinear because the keystream bits are directly multiplied instead of being multiplied by constants and then added.

CHAPTER 6 CIPHERS INVOLVING EXPONENTIATION

(Page 182) **jamming the numbers together**: If this doesn't seem mathematical enough to you, think of the number for a plaintext block as $P = 100P_1 + P_2$. But it doesn't really matter.

(Page 184) **Pierre de Fermat**: Michael Mahoney, *The Mathematical Career of Pierre de Fermat (1601–1665)* (Princeton NJ: Princeton University Press, 1973).

(Page 184) **how you might have discovered it**: Since Fermat didn't have Gauss' idea of modular arithmetic and probably didn't know much cryptology either, he probably had something else in mind. But who knows? The first published proof was apparently by Leonhard Euler in 1741. The proof given here is more or less the one in James Ivory, "Demonstration of a theorem respecting prime numbers," *New Series of The Mathematical Respository*. Vol. I, Part II (1806).

(Page 185) **cancel $1 \times 2 \times 3 \times \cdots \times 12$**: Or, if you prefer, multiply each side by $\frac{1}{1 \times 2 \times 3 \times \cdots \times 12}$.

(Page 188) **Pohlig-Hellman exponentiation cipher**: M. E. Hellman and S. C. Pohlig, "Exponentiation cryptographic apparatus and method," United States Patent: 4424414, 1984, http://www.google.com/patents?vid=4424414.

(Page 188) **invention of the Pohlig-Hellman cipher**: Although first written in 1976, the paper describing the cipher wasn't published until 1978, by which time the ideas contained in it were well known in the cryptographic community. S. Pohlig and M. Hellman, "An improved algorithm for computing logarithms over GF(p) and its cryptographic significance (corresp.)," *IEEE Transactions on Information Theory* 24 (1978). For the story of the delay, see Martin Hellman, "Oral history interview by Jeffrey R. Yost," Number OH 375. Charles Babbage Institute, University of Minnesota, Minneapolis, 2004, pp. 43–44, http://purl.umn.edu/107353. Both Pohlig and Hellman are now better known for other ideas related to public-key cryptography. Hellman is best known for his part in the Diffie-Hellman key agreement system, which we see in Section 7.2. Pohlig is best known for his part in the Silver-Pohlig-Hellman algorithm for computing discrete logarithms (see Section 6.4). That algorithm was first published by Pohlig and Hellman in the same paper as their exponentiation cipher, although according to that paper it had been independently discovered by Roland Silver. Pohlig and Hellman, "Improved algorithm."

(Page 189) **Alice needs only 46 multiplications**: In fact, we could do even better if we convert 769 to a binary numeral, but this is good enough to get the idea.

(Page 189) **Eve will need all 768 multiplications**: Actually, with the best-known techniques, Eve can go somewhat faster than this, but still not nearly as fast as Alice and Bob can.

(Page 189) **35 years on the discrete logarithm problem**: Much more if you count precomputer investigations. Gauss, for example, made tables of discrete logarithms, which he called "indices." Gauss, *Disquisitiones arithmeticae*, Articles 57–59.

(Page 189) **no one knows for sure**: Actually, it is possible that someone knows and isn't telling. If so, the NSA would be most likely, but it could be another government or even some other organization. The same thing goes for the Diffie-Hellman problem that we see in Section 7.2, the factoring problem from Section 7.4, and the RSA Problem from Section 7.6.

(Page 190) **composite numbers**: As I implied in Section 1.3, every positive whole number can be written as the product of primes. Therefore, every positive whole number other than 1 is either prime or composite. Mathematicians consider 1 to be neither prime nor composite.

(Page 190) **"Decomposing Composers"**: Monty Python, "Decomposing composers," *Monty Python's Contractual Obligation Album.* Charisma Records, 1980.

(Page 193) **Fermat was in the seventeenth**: For the kind of mathematics in this chapter, anyway.

(Page 193) **Euler's 1763 paper**: Leonhard Euler, "Theoremata Arithmetica Nova Methodo Demonstrata," *Novi Commentarii Academiae Scientiarum Petropolitanae* 8 (1763).

(Page 193) **function we now write $\phi(n)$**: This notation seems to have been introduced later by Gauss. Gauss, *Disquisitiones arithmeticae*, Article 38.

(Page 193) **Euler phi function**: Not to be confused with Friedman's phi from Section 2.2.

(Page 194) **we have to add it back in**: This taking-out and adding-back-in procedure is often known as the **principle of inclusion-exclusion**.

(Page 196) **find the inverse**: If Alice made a mistake and picked a bad key, Bob will find that out in this step.

(Page 197) **decryption does always work properly**: Most books prove only the case of two distinct primes, because that is what is needed for RSA. (See Section 7.4.) However, the proof in S. C. Coutinho, *The Mathematics of Ciphers* (Natick, MA: AK Peters, Ltd., 1998), pp. 166–67 (Section 11.3) or Robert Edward Lewand, *Cryptological Mathematics* (The Mathematical Association of America, 2000), pp. 156–57 (Theorem 4.1) is very readable and generalizes easily to more primes. The proof in Thomas H. Barr, *Invitation to Cryptology* (Englewood Cliffs, NJ: Prentice Hall, 2001), pp. 280–81 (Theorem 4.3.2) is also readable but does not generalize quite as easily.

(Page 199) **works anyway**: It turns out that if a prime does divide both P and n, it has to divide P at least as many times as it divides n. Once again, I'm not going to try to prove it, but the references I gave in the endnote for page 197 might be useful.

(Page 199) **Pohlig and Hellman considered composite moduli**: Hellman, *Oral History Interview by Jeffrey R. Yost*, pp. 43–44.

Chapter 7 Public-Key Ciphers

(Page 201) **agree on the key**: And possibly the system, depending on how seriously they are taking Kerckhoffs' principle.

(Page 201) **"simple, but inefficient"**: Arnd Weber (ed.), "Secure communications over insecure channels (1974)" (January 16, 2002), http://www.itas.kit.edu/pub/m/2002/mewe02a.htm.

(Page 202) **project proposal**: Merkle's original project proposal is posted at "CS 244 project proposal" (Fall 1974), http://merkle.com/1974/CS244ProjectProposal.pdf.

(Page 202) **Merkle's computer security class**: Levy, *Crypto*, pp. 77–79.

(Page 202) **several versions**: Weber (ed.), "Secure communications."

(Page 202) **version that was finally published**: Ralph Merkle, "Secure communications over insecure channels," *Communications of the Association for Computing Machinery* 21:4 (1978). This version was published after three and a half years and much arguing with reviewers. Weber (ed.), "Secure communications"; Levy, *Crypto*, p. 81.

(Page 202) **"tedious, but quite possible"**: Merkle, "Secure communications over insecure channels," p. 296.

(Page 202) **cipher with a 128-bit key**: In particular, he suggested a version of Lucifer which Horst Feistel had published in 1973. (Feistel, "Cryptography and computer privacy.") A modern implementation might use AES.

(Page 203) **check number**: The check number is hardly necessary in our example, since all of the numbers are spelled out and it should be obvious to Bob when he solves the puzzle. However, if the numbers were encrypted in some other fashion, it might not be possible to tell for sure without the check number.

(Page 205) **250 decryptions**: This is not strictly true in our example, since there are much faster known-plaintext attacks that Eve could try on each puzzle instead. This is why Merkle suggested using a cipher with much stronger resistance to known-plaintext attacks and a large block size, but restricting the set of keys. I could have done that here, but it would have made the example much more complicated.

(Page 206) **key-agreement system**: Often this is called a key-exchange system, but that's not really accurate. The things that are exchanged can't be used as secret keys, but in the end Alice and Bob do agree on a secret key.

(Page 206) **Merkle recognized**: Levy, *Crypto*, pp. 82–83.

(Page 206) **one of them has to spend twice as long**: Or each of them has to spend roughly 1.4 times as long.

(Page 207) **Diffie's story**: Levy, *Crypto*, pp. 20–31.

(Page 207) **"two problems and a misunderstanding"**: Whitfield Diffie, "The first ten years of public-key cryptography," *Proceedings of the IEEE* 76:5 (1988).

(Page 207) **"digital signatures"**: Merkle also considered this question, but without much success. Merkle, *CS 244 Project Proposal*.

(Page 207) **"What good would it do . . . ."**: Diffie, "The first ten years of public-key cryptography," p. 560.

(Page 207) **three people**: And at least one more, as we see in Appendix A.

(Page 208) **privacy and self-reliance on the minds of Diffie and Hellman**: Levy, *Crypto*, for example, p. 34.

(Page 208) **Diffie and Hellman's paper**: Whitfield Diffie and Martin E. Hellman, "Multiuser cryptographic techniques," in Stanley Winkler (ed.), *Proceedings of the June 7–10, 1976, National Computer Conference and Exposition* (New York: ACM, 1976).

(Page 208) **draft copy**: Levy, *Crypto*, p. 81–82. Back before the Internet, it was customary for scientists in many fields to send copies of papers that had not yet been published to colleagues who might be interested. This was especially important in fast-moving fields like computer science, where a paper might become obsolete between the time it was written and was published. Today, these drafts are often posted on a Web site.

(Page 208) **Diffie, Hellman, and Merkle**: Levy, *Crypto*, pp. 76–83.

(Page 208) **one-way functions**: Diffie thinking about: Levy, *Crypto*, p. 28; Merkle thinking about: Ralph Merkle, "CS 244 project proposal" Fall 1974).

(Page 208) **Diffie-Hellman key agreement**: Levy, *Crypto*, p. 84.

(Page 209) **"We stand today . . ."**: Whitfield Diffie and Martin E. Hellman, "New directions in cryptography," *IEEE Transactions on Information Theory* 22:6 (1976).

(Page 209) **large prime number**: The Diffie-Hellman system, like the Pohlig-Hellman cipher, can also be done using finite field arithmetic modulo 2. Alice's and Bob's computations become quicker on a computer, but so do Eve's, so there isn't a lot of practical advantage in the end. Schneier, *Applied Cryptography*, p. 515.

(Page 209) **600 digits or more**: That is, 2048 bits. David Adrian et al., "Imperfect forward secrecy: How Diffie-Hellman fails in practice," in *22nd ACM Conference on Computer and Communications Security*, Association for Computing Machinery Special Interest Group on Security, Audit and Control (New York: ACM Press, 2015).

(Page 209) **generator modulo p**: Sometimes you will also see this called a **primitive root** modulo p.

(Page 209) **every prime has a generator**: First proved, once again, by Gauss. Gauss, *Disquisitiones arithmeticae*, Articles 54–55.

(Page 209) **fine to look them up**: But see Section 7.8 for an important caveat to this.

(Page 211) $p = 2819$: Of course, this isn't nearly large enough for real-life security. But this is just an illustration.

(Page 211) **94 and 305**: 94305 is the Stanford zip code.

(Page 213) **discrete logarithm modulo 232-digit prime**: Thorsten Kleinjung, "Discrete Logarithms in GF(p)—768 bits," Email sent to the NMBRTHRY mailing list, 2016, https://listserv.nodak.edu/cgi-bin/wa.exe?A2=NMBRTHRY;a0c66b63.1606.

(Page 213) **discrete logarithm record**: Larger computations have been done over finite fields. The record as of this writing is a computation over a field with 2^{9234} elements. The size of this field is a 2779-digit, or 9234-bit, number. Jens Zumbrägel, "Discrete logarithms in GF(2^9234)," E-mail sent to the NMBRTHRY mailing list, 2014, https://listserv.nodak.edu/cgi-bin/wa.exe?A2=NMBRTHRY;9aa2b043.1401.

(Page 213) **Diffie-Hellman in VPNs and IPv6**: The security system used by these networks is known as Internet Protocol Security, or IPsec. William Stallings, *Cryptography and Network Security: Principles and Practice*, 6th ed. (Boston: Pearson, 2014), Section 20.1. The cryptographic system used in IPsec is based on Diffie-Hellman, with additions to provide added security and authentication. Stallings, *Cryptology and Security*, Section 20.5.

(Page 213) **very difficult to find the decryption key from the encryption key**: Often the reverse is also true, but it will not be a requirement for the systems in this chapter.

(Page 214) **Diffie and Hellman's analogy**: Diffie and Hellman, "New directions in cryptography," p. 652.

(Page 216) **1976 paper**: Diffie and Hellman, "Multiuser cryptographic techniques."

(Page 216) **knapsack ciphers**: Simson Garfinkel, *PGP: Pretty Good Privacy* (Sebastopol, CA: O'Reilly Media, 1995), pp. 79–82.

(Page 216) **Rivest and Shamir excited; Adleman less so**: Levy, *Crypto*, pp. 92–95.

(Page 216) **Settled into a pattern**: Levy, *Crypto*, pp. 95–97.

(Page 217) **factoring as a one-way function**: Diffie and Hellman also briefly considered using factoring for their one-way function but didn't pursue it. Levy, *Crypto*, p. 83.

(Page 217) **Passover seder**: Levy, *Crypto*, p. 98.

(Page 217) **lay down on the couch**: According to one source this was a common practice when he was thinking about something. Levy, *Crypto*, p. 98. Other sources say that he lay down because he had a headache. Garfinkel, *PGP*, p. 74; Jim Gillogly and Paul Syverson, "Notes on Crypto '95 invited talks by Morris and Shamir," *Cipher: Electronic Newsletter of the Technical Committee on Security & Privacy, A Technical Committee of the Computer Society of the IEEE.* Electronic issue 9 (1995). It is not clear whether the wine was involved.

(Page 217) **exponentiation cipher:** There is no evidence that Rivest had actually seen Pohlig and Hellman's work on the exponentiation cipher at this point. He may very well have independently reinvented it.

(Page 217) **600 digits for n:** Again, 2048 bits. Benjamin Beurdouche et al., "A messy state of the union: Taming the composite state machines of TLS," in *2015 IEEE Symposium on Security and Privacy (SP)*, (Los Alamitos, CA: IEEE Computer Society, 2015).

(Page 217) $e = 17$: In fact, $e = 17$ is a fairly common choice even in the real world. It's small enough so that encryption is fast, but not so small that Eve can usually take advantage of it. It's prime, so the GCD of 17 and $\phi(n)$ is usually 1. And it's of the special form $17 = 2^4 + 1$, which makes it easy to do exponentiation using the most common computer technique.

(Page 218) "**Just the factors, ma'am**": See Barbara Mikkelson and David Mikkelson, "Just the facts," snopes.com, 2008, http://www.snopes.com/radiotv/tv/dragnet.asp.

(Page 219) **the morning of April 4 and the order of authors:** Levy, *Crypto*, pp. 100–1.

(Page 219) **RSA Technical Memo:** Ronald L. Rivest, et al., "A method for obtaining digital signatures and public-key cryptosystems," technical Memo number MIT-LCS-TM-082, MIT, April 4, 1977.

(Page 219) **paper describing RSA:** R. L. Rivest, et al., "A method for obtaining digital signatures and public-key cryptosystems," *Communications of the Association for Computing Machinery* 21:2 (1978).

(Page 219) **RSA patent:** Ronald L. Rivest et al., "Cryptographic communications system and method," United States patent: 4405829, 1983, http://www.google.com /patents?vid=4405829.

(Page 220) **Martin Gardner's column:** Martin Gardner, "Mathematical games: A new kind of cipher that would take millions of years to break," *Scientific American* 237:2 (1977).

(Page 220) **40 quadrillion years:** This estimate appears to have been a mistake; Rivest should have said that it would take 40 quadrillion operations. Garfinkel, *PGP*, p. 115. Levy, *Crypto*, p. 104, says it should have been "hundreds of millions of years"; my rough calculation gives 22,500 years based on Rivest, Shamir and Adleman, *Communications of the Association for Computing Machinery*. Your mileage may vary.

(Page 220) **over 3000 requests:** Garfinkel, *PGP*, p. 78.

(Page 220) **RSA in secure web servers:** Stallings, *Cryptology and Security*, Section 17.2.

(Page 221) **hybrid systems in web servers:** Stallings, *Cryptology and Security*, Section 17.2.

(Page 222) **tests have been known:** See, for example, Leonard Eugene Dickson, *Divisibility and Primality*, reprint of 1919 edition (Providence, RI: AMS Chelsea Publishing, 1966), p. 426.

(Page 222) **"The problem of distinguishing … "**: Gauss, *Disquisitiones arithmeticae*, Article 329.

(Page 223) **"the second is superior … "**: Gauss, *Disquisitiones arithmeticae*, Article 334.

(Page 223) **first pointed out**: R. Solovay and V. Strassen, "A fast Monte-Carlo test for primality," *SIAM Journal on Computing* 6:1 (1977); the paper was first received by the journal editors on June 12, 1974.

(Page 223) **probabilistic test**: Technically, a probabilistic procedure that is always fast but sometime wrong is called a **Monte Carlo algorithm**, whereas one that is always right but sometimes slow is called a **Las Vegas algorithm**. The Solovay-Strassen test is a Monte Carlo algorithm.

(Page 223) **liars and witnesses**: The standard terminology is for liar and witness to be opposites, even though lying witness and truthful witness might be more accurate. Notice that 1 is always going to be a liar for the Fermat test on a composite number. Can you see why?

(Page 225) **Rabin paper**: Michael O. Rabin, "Probabilistic algorithm for testing primality," *Journal of Number Theory* 12:1 (1980).

(Page 225) **Miller paper**: Gary L. Miller, "Riemann's hypothesis and tests for primality," in *Proceedings of Seventh Annual ACM Symposium on Theory of Computing*, Association for Computing Machinery Special Interest Group on Algorithms and Computation Theory (New York: ACM, 1975).

(Page 225) **Rabin-Miller test**: The Rabin-Miller test is actually not too hard to explain, but it would take us rather far off track. If you would like to check it out, Joseph H. Silverman, *A Friendly Introduction to Number Theory*, 3d ed. (Englewood Cliffs, NJ: Prentice Hall, 2005), pp. 130–31, has a succinct and readable description. Coutinho, *Mathematics of Ciphers*, pp. 100–4 (Sections 6.3–6.4) gives a few more details.

(Page 225) **Agrawal-Kayal-Saxena primality test**: The version that was finally published was Manindra Agrawal et al., "PRIMES is in P," *The Annals of Mathematics* 160:2 (2004). F. Bornemann, "PRIMES is in P: A breakthrough for 'everyman,'" *Notices of the AMS* 50:5 (2003) tells the story nicely, and you can ignore as much of the math as you want. (It's written for mathematicians who are not experts in the subject.) This discovery was taken as great encouragement by many young students! Kayal and Saxena had started their work as undergraduates and made their breakthrough during the first summer after graduation.

(Page 225) **creating a secure RSA key**: In practice, the most time-consuming part of the process should be generating unguessable random numbers to test for primality. Depending on how good a job your computer does of this, it could take up to a minute.

(Page 227) **factoring better than the obvious method**: For a good description of modern factoring techniques, see Carl Pomerance, "A tale of two sieves," *Notices of the American Mathematical Society* 43:12 (1996). There have been some improvements since that article was written, but the basic ideas there are still the state of the art 2016.

(Page 227) **129-digit challenge solved**: Garfinkel, *PGP*, p. 113–15; Derek Atkins et al., "The magic words are Squeamish Ossifrage," in Josef Pieprzyk and Reihanah

Safavi-Naini (eds.) *Advances in Cryptology—ASIACRYPT '94.* (Berlin/Heidelberg: Springer-Verlag, 1995).

(Page 228) **232-digit factorization**: Thorsten, Kleinjung Kazumaro Aoki, Jens Franke, Arjen Lenstra, Emmanuel Thomé, Joppe Bos, Pierrick Gaudry, et al., "Factorization of a 768-bit RSA modulus," cryptology ePrint Archive number 2010/006, 2010. Larger numbers have been factored but only if they have a special form.

(Page 228) **factoring n using a multiple of $\phi(n)$**: This algorithm is closely related to the Rabin-Miller primality test from Section 7.5. An early version of this algorithm is in Miller, "Riemann's hypothesis," but it relies on a widely believed but unproven conjecture. I don't know who came up with the modern version but you can find a description in Alfred J. Menezes et al., *Handbook of Applied Cryptography* (Boca Raton, FL: CRC, 1996), p. 287 (Section 8.2.2).

(Page 230) **chosen-ciphertext attack on RSA**: This also applies to the Pohlig-Hellman exponentiation cipher, by the way.

(Page 230) **Eve knows 243 and 3125**: She knows the other plaintext blocks too, since she encrypted them properly, but they aren't likely to help her.

(Page 231) **don't choose d too small**: You might wish you could do this in order to make decryption fast, the same way many people choose a small e in order to make encryption fast (see Section 7.4).

(Page 232) **$2214^6 \times 2019^{-1}$ modulo 3763**: Here 2019^{-1} is the same as the multiplicative inverse of 2019 modulo 3763.

(Page 233) **more details of attacks on RSA**: Schneier, *Applied Cryptography*, pp. 471–74 has a slightly more extensive summary with a few more attacks; some of them involve digital signatures. (See Section 8.4.) Dan Boneh, "Twenty years of attacks on the RSA cryptosystem," *Notices of the AMS* 46:2 (1999) has more details on many of the attacks.

(Page 233) **Diffie-Hellman-Merkle**: M. E. Hellman, "An overview of public key cryptography," *IEEE Communications Magazine* 40:5 (2002). The older terminology is probably too entrenched to be changed, however.

(Page 233) **patent**: Martin E. Hellman et al., "Cryptographic apparatus and method," United States Patent: 4200770, 1980, http://www.google.com/patents?vid=4200770.

(Page 233) **several internal NSA documents**: See Spiegel Staff, "Prying Eyes, Inside the NSA's war on Internet security," *Spiegel Online* (2014). and especially OTP VPN Exploitation Team, "Intro to the VPN exploitation process," http://www.spiegel.de/media/media-35515.pdf.

(Page 233) **"Logjam"**: For the name, see David Adrian et al., "The logjam attack," (May 20, 2015). https://weakdh.org/. For the technical description and the detailed rationale for believing that the NSA is using it, see Adrian et al. *Imperfect Forward Secrecy.*

(Page 234) **225 digits**: More accurately, 768 bits.

(Page 234) **150 digits**: More accurately, 512 bits.

(Page 234) **"FREAK"**: "Export" refers to the fact that small keys used to be required in software exported outside the United States. For more on the name, see Karthikeyan Bhargavan et al., "State Machine AttaCKs against TLS (SMACK TLS)," https://www.smacktls.com. For the technical description, see Beurdouche et al., *Messy State of the Union.*

(Page 235) **Ellis' story**: Levy, *Crypto*, pp. 313–19.

(Page 236) **Wasn't practical itself:** J. H. Ellis, "The history of non-secret encryption," *Cryptologia* 23:3 (1999).

(Page 236) **"It shows only ... ":** J. H. Ellis, "The possibility of secure non-secret digital encryption," UK Communications Electronics Security Group, January 1970.

(Page 236) **codebook:** Actually, Ellis was thinking not so much of a codebook as a block cipher taking, say, 100 bits of plaintext to 100 bits of ciphertext. Evidently his idea of a secure block size was similar to Feistel's. Such a block cipher is less vulnerable to frequency analysis, but I think a codebook is easier to visualize.

(Page 237) **Alice starts by asking Bob:** Remember that in the system that inspired Ellis, the recipient is responsible for the encryption.

(Page 238) **Some "process" could be found:** Ellis, "Possibility".

(Page 238) **"Because of the weakness ... :** Ellis, "History," p. 271.

(Page 238) **And this is how things stood:** Levy, *Crypto*, pp. 318–19.

(Page 238) **Cocks' story:** Levy, *Crypto*, pp. 319–22.

(Page 238) **exactly the kind of mathematics:** Number theory, the study of whole numbers and their properties.

(Page 238) **"I suppose it was actually also helpful ... ":** Levy, *Crypto*, p. 320.

(Page 238) **all essential ways:** One small difference was that Cocks, like Ellis, was still thinking about a system which started with Alice asking Bob for his public key. However, he did point out that once Alice had Bob's public key she could encrypt as many messages as she wanted using it.

(Page 239) **Cocks' paper:** C. C. Cocks, "A Note on non-secret encryption," UK Communications Electronics Security Group, November 20, 1973.

(Page 239) **Williamson's story:** Levy, *Crypto*, pp. 322–25. Williamson also lived in the same house as Cocks, but conversations about work, like writing about work, were forbidden while off of GCHQ grounds.

(Page 239) **Williamson's first paper:** M. J. Williamson, "Non-secret encryption using a finite field," UK Communications Electronics Security Group, January 21, 1974.

(Page 239) **Williamson's second paper:** Malcolm Williamson, "Thoughts on cheaper non-secret encryption," UK Communications Electronics Security Group, August 10, 1976.

(Page 240) **The fate of public-key encryption at GCHQ:** Levy, *Crypto*, pp. 324–29.

(Page 240) **"no further benefit ... ":** Ellis, "History."

(Page 240) **GCHQ posted five papers:** According to Williamson, the papers couldn't be made public "until a certain person retired." Levy, *Crypto*, p. 329.

CHAPTER 8 OTHER PUBLIC-KEY SYSTEMS

(Page 246) **"Tell me three times":** See Lewis Carroll, *The Hunting of the Snark: An Agony in Eight Fits* (London: Macmillan, 1876), Fit the First.

(Page 247) **Problems are about equally difficult:** There are a couple of catches: the three-pass protocol has more restrictions because the exponentiations have to be invertible, and you need to decide what happens if the case you are trying to solve doesn't have a valid solution. For those who know a little about the subject, the mathematical details are worked out in K. Sakurai and H. Shizuya, "A structural comparison of the computational difficulty of breaking discrete log cryptosystems," *Journal of Cryptology* 11:1 (1998).

(Page 247) **Technical report on mental poker**: Adi Shamir et al., "Mental poker," MIT, February 1, 1979.

(Page 247) **Collection dedicated to Martin Gardner**: A. Shamir et al., "Mental poker," in David A. Klarner (ed.), *The Mathematical Gardner* (Boston: Prindle, Weber & Schmidt; Belmont, CA: Wadsworth International, 1981). This is a very readable article intended for nonexperts. The three-pass protocol also appeared in Konheim, *Cryptography*, pp. 345–46, where it is described as "unpublished work" of Shamir's.

(Page 247) **Three-pass protocol reinvented by Omura**: J. L. Massey, "An introduction to contemporary cryptology," *Proceedings of the IEEE* 76:5 (1988).

(Page 247) **Major European conference**: J. Massey, "A new multiplicative algorithm over finite fields and its applicability in public-key cryptography," Presentation at EUROCRYPT '83 March 21–25, 1983.

(Page 247) **Massey and Omura's patent**: James L. Massey and Jimmy K. Omura, "Method and apparatus for maintaining the privacy of digital messages conveyed by public transmission," United States Patent: 4567600 January 28, 1986, http://www.google.com/patents?vid=4567600.

(Page 248) **Elgamal came up with an asymmetric-key system**: Taher ElGamal, "A public key cryptosystem and a signature scheme based on discrete logarithms," In George Robert Blakley and David Chaum (eds.), *Advances in Cryptology: Proceedings of CRYPTO '84* (Santa Barbara, CA: Springer-Verlag, 1985).

(Page 248) **Elgamal and ElGamal**: While the spelling "ElGamal" was used for the original papers and has become standard for this and other cryptographic systems, Taher Elgamal himself now prefers a lowercase *g*.

(Page 248) **using one that someone else is using**: But see the caveat in Section 7.8.

(Page 248) $p = 2819$: Just a reminder that in real life, p would be much larger than this.

(Page 248) **blind and hint**: This idea isn't entirely original to Elgamal. In fact, in a sense the idea of a random blind is the same idea as the one-time pad. The idea of sending a cryptographic hint with the blinded ciphertext seems to have originated in the early 1980s. Ronald L. Rivest and Alan T. Sherman, "Randomized Encryption Techniques," in David Chaum, Ronald L. Rivest, and Alan T. Sherman (eds.), *Advances in Cryptology: Proceedings of CRYPTO '82* (New York: Plenum Press, 1983) is a nice summary of the early history of probabilistic encryption, including blind and hint systems and the McEliece public-key system. The McEliece system uses a random blind but not a hint; instead it uses an error-correcting code to remove the blind. Rivest and Sherman attribute blind and hint systems similar to the ElGamal system to C. A. Asmuth and G. R. Blakley, "An efficient algorithm for constructing a cryptosystem which is harder to break than two other cryptosystems," *Computers & Mathematics with Applications* 7:6 (1981), where they use a related idea to construct the "join" of two encryption systems. As far as I know, however, Elgamal was the first to incorporate this into a public-key system.

(Page 250) **multiplicative cipher**: Elgamal pointed out that you could use other operations besides multiplication, but multiplication is convenient because we need to do it as part of the exponentiation anyway and it's reasonably fast compared to the exponentiation. ElGamal, "Public key cryptosystem".

(Page 251) **public-key options in PGP and GPG**: PGP: Jon Callas et al., "OpenPGP Message Format," IETF, November 2007; GPG: People of the GnuPG Project,

"GnuPG frequently asked questions," https://gnupg.org/faq/gnupg-faq.html. These are e-mail programs for which key agreement is not particularly well suited. Thus Diffie-Hellman is not a standard option, although ElGamal encryption is sometimes called "Diffie-Hellman encryption" in these programs. The PGP standard lists Diffie-Hellman as an option that "would be useful to use in an OpenPGP implementation, yet there are issues that prevent an implementer from actually implementing the algorithm."

(Page 251) **elliptic curve equations**: There is actually a more general form of the equation needed in some contexts, but this will do for our purposes.

(Page 258) **commutative, associative, identity, inverses**: The technical term for a set of objects with an operation that is associative and has an identity and inverses is a **group**. If it is also commutative, it is an **abelian group**. Numbers with addition, nonzero numbers with multiplication, numbers modulo n with addition, numbers relatively prime to n modulo n with multiplication, and elliptic curves are all examples of abelian groups. Permutations of length n with permutation products are also a group, but not abelian.

(Page 259) **need to find the inverse and can't**: Since the modulus is prime, the only numbers that don't have inverses are those that are the same as zero modulo that prime.

(Page 259) **elliptic curves modulo p**: It's also possible, and sometimes convenient, to consider elliptic curves where the coefficients and coordinates are elements of a finite field. The formulas are almost the same in that case, but not quite. We won't be worrying about it too much.

(Page 260) **modular exponentiation and elliptic curve discrete logarithm problems are hard**: And the factoring problem, but the factoring problem doesn't seem to have a good analog for elliptic curves.

(Page 260) **Neil Koblitz tells the story**: Neal Koblitz, *Random Curves: Journeys of a Mathematician* (Berlin/Heidelberg Springer-Verlag, 2008), pp. 298–310.

(Page 260) **Koblitz in the Soviet Union**: As a side note, Koblitz recalls that the first lecture that he ever gave on cryptography was in Moscow. He didn't talk about elliptic curve cryptography, but he did mention an application of public-key cryptography to nuclear test ban treaty verification.

(Page 261) **Miller's paper**: V. Miller, "Use of elliptic curves in cryptography," in Hugh C. Williams (ed.), *Advances in Cryptology–CRYPTO '85 Proceedings* (Berlin: Springer, 1986).

(Page 261) **"only" about 70 digits long**: That is, 224 to 255 bits; Elaine Barker et al., "Recommendation for key management—Part 1: General (Revision 3)," NIST, July 2012.

(Page 261) **looked up in a table**: The caveat in Section 7.8 may *not* apply here, since the precomputation technique mentioned there does not work on the elliptic curve discrete logarithm problem. See Section 8.5 for a different caveat, however.

(Page 261) **secret a and b**: It's convenient if these are less than the number of points generated by G but not vitally necessary.

(Page 261) **piece of secret information**: Note that this piece of shared secret information is actually a point, with an x- and a y-coordinate. It's most common to just use the x-coordinate, so you get a number modulo p.

(Page 261) **elliptic curve discrete logarithm record:** Joppe W. Bos et al., "Pollard rho on the PlayStation 3," in *SHARCS '09 Workshop Record*, Virtual Application and Implementation Research Lab within ECRYPT II European Network of Excellence in Cryptography Lausanne, Switzerland: 2009. The record for finite fields is a computation over a field with 2^{113} elements. The size of this field is a 113-bit number. Erich Wenger and Paul Wolfger, "Harder, better, faster, stronger: elliptic curve discrete logarithm computations on FPGAs," *Journal of Cryptographic Engineering* (September 3, 2015).

(Page 262) **Koblitz' paper:** Neal Koblitz, "Elliptic curve cryptosystems," *Mathematics of Computation* 48:177 (1987).

(Page 262) **elliptic curve ElGamal encryption:** You do have to find a way to represent your plaintext as a point on the elliptic curve, which is not completely trivial. Koblitz gives some ideas in Koblitz, "Elliptic curve cryptosystems," Section 3.

(Page 264) **look up f:** But see Section 8.5.

(Page 264) **Fast techniques to compute f:** Koblitz, "Elliptic curve cryptosystems" has more on this.

(Page 265) **message encrypted using that key:** Or a MAC calculated using that key.

(Page 265) **"a time and message dependent . . . ":** Diffie and Hellman, "Multiuser cryptographic techniques."

(Page 265) **need a couple of assumptions:** These are much less likely to be true with a probabilistic encryption system.

(Page 266) **"everywhere a sign":** Five Man Electrical Band, "Signs," Single. Lionel Records, 1971.

(Page 267) **genuine message:** The chance of Frank the forger being able to concoct a signature that gives a sensible English message when verified with v, without knowing σ, is extremely small. If the message is something other than text, this is one of those cases where Alice might want to send an unsigned copy of the message so Bob can compare.

(Page 268) **Alice signs and the encrypts:** There is some debate over whether one should sign first and then encrypt, as we have done here, or encrypt first and then sign. There are good arguments both ways. I choose to go with the "Horton principle": mean what you sign and sign what you mean—not just an encrypted version of what you mean. Ferguson et al., *Cryptography Engineering*, pp. 96–97 and 102–4; Dr. Seuss, *Horton Hatches the Egg* (Random House, 1940).

(Page 268) **certificates:** See Simson Garfinkel, *Web Security, Privacy and Commerce*, 2nd ed. (Sebastopol, CA: O'Reilly Media, 2002), pp. 160–93 for more on certificates and how they are used on the Internet.

(Page 268) **RSA digital signature certificates:** A 2013 scan of the Internet showed more than 99% of certificates were signed using RSA. Zakir Durumeric et al., "Analysis of the HTTPS certificate ecosystem," in *Proceedings of the 2013 Conference on Internet Measurement Conference*, Association for Computing Machinery Special Interest Groups on Data Communication and on Measurement and Evaluation (New York: ACM, 2013).

(Page 268) **RSA Data Security and Netscape:** Garfinkel, *Web Security*, pp. 175–76.

(Page 268) **VeriSign and Symantec:** The same 2013 scan showed that approximately 34% of certificates were issued by companies owned by Symantec. Approximately

10% of the total were issued by VeriSign itself. Durumeric et al., "Analysis of the HTTPS certificate ecosystem."

(Page 268) **Internet Explorer, Firefox, Chrome, and Safari**: To be exact, as of 2015 Internet Explorer and Firefox support RSA, Digital Signature Algorithm, and the Elliptic Curve Digital Signature Algorithm (ECDSA). Chrome and Safari seem to have skipped DSA and support only RSA and ECDSA. Qualys SSL Labs, "User agent capabilities," 2015. https://www.ssllabs.com/ssltest/clients.html. This Web site also gives you an option to test which algorithms your own browser supports.

(Page 269) **Bob may not see anything wrong**: If Alice and Bob are computers, then it is particularly likely that Bob won't see a problem. In that case the second sample message is a lot more likely than the first.

(Page 269) **synchronized clocks**: See Ferguson et al., *Cryptography Engineering*, Chapter 16, for lots more about the use and abuse of clocks in cryptography.

(Page 270) **short signatures**: This is often accomplished with the aid of a **hash function**, or **message digest function**. These functions are easy for anyone to compute without a key and take a message of arbitrary length to a value of fixed size, such as 512 bits. However, it should be hard to find a message with a given hash value or two messages with the same hash value. Hash functions are really beyond the scope of this book, but Barr, *Invitation to Cryptology*, Section 3.6 is a good introduction. Stallings, *Cryptology and Security*, Chapter 11 goes into more depth. It is also more up-to-date, including a section about the AES-style competition NIST recently held to choose a new hash function standard and its result. Ferguson et al., *Cryptography Engineering*, Chapter 5, has somewhat less detail about how hash functions work and more about how to use them.

(Page 270) **ElGamal signature scheme**: ElGamal, "Public key cryptosystem."

(Page 270) **DSA was controversial**: See Schneier, *Applied Cryptography*, Section 20.1 for early reactions to the DSA.

(Page 271) **Sony was using the same nonce**: The group calls itself "fail0verflow": bushing, marcan and sven, "Console hacking 2010: PS3 epic fail," slides from lecture presented at 27th Chaos Communication Congress, 2010, https://events.ccc.de/congress/2010/Fahrplan/events/4087.en.html.

(Page 271) **another hacker**: The hacker who published the key is George Hotz, a.k.a. "GeoHot." Jonathan Fildes, "iPhone hacker publishes secret Sony PlayStation 3 key," BBC News Web site, 2011. http://www.bbc.co.uk/news/technology-12116051.

(Page 271) **Sony's lawsuit**: David Kravets, "Sony settles PlayStation hacking lawsuit," *Wired Magazine* Web site, http://www.wired.com/2011/04/sony-settles-ps3-lawsuit. The legal documents may be found at Corynne McSherry, "Sony v. Hotz ends with a whimper, I mean a gag order," Electronic Frontier Foundation Deeplinks Blog, 2011, https://www.eff.org/deeplinks/2011/04/sony-v-hotz-ends-whimper-i-mean-gag-order. Hotz agreed not to share any more confidential information about Sony products as well as to refrain from hacking them.

(Page 272) **adaptive chosen-ciphertext attack**: In the original version of ElGamal encryption, if Eve has the ciphertext R and C and she can trick Bob into deciphering (for example) R and $2C$, then Bob's result will be $2P$, from which Eve can easily get P. She does not get the private key, however.

(Page 272) **DHIES and ECIES**: DHIES and ECIES were first described in Mihir, Bellare and Phillip Rogaway, "Minimizing the use of random oracles in authenticated encryption schemes," in Yongfei Han, Tatsuaki Okamoto, and Sihan Quing (eds.), *Proceedings of the First International Conference on Information and Communication Security* (Berlin/Heidelberg: Springer-Verlag, 1997) under the name DLAES, although you have to read very closely to find the mention of elliptic curves. The scheme has also been known as DHES and DHIES, and the modular exponentiation discrete logarithm version is sometimes called DLIES. As it says in Michel Abdalla et al., "The oracle Diffie-Hellman assumptions and an analysis of DHIES," in David Naccache (ed.), *Topics in Cryptology-CT-RSA 2001* (Berlin/Heidelberg: Springer-Verlag, 2001): "It is all the same scheme."

(Page 272) **hyperelliptic curves**: For more on hyperelliptic curves, see Hoffstein et al., *Introduction to Mathematical Cryptography*, Section 8.10.

(Page 273) **pairing function**: See Trappe and Washington, *Introduction to Cryptography*, Section 16.6, for an overview of pairing functions, and Hoffstein et al., *Introduction to Mathematical Cryptography*, Sections 6.8–6.10, for the gory details.

(Page 273) **tripartite Diffie-Hellman**: See Hoffstein et al., *Introduction to Mathematical Cryptography*, Sections 6.10.1 and the references there.

(Page 273) **identity-based encryption**: See Trappe and Washington, *Introduction to Cryptography*, Section 16.6, or Hoffstein et al., *Introduction to Mathematical Cryptography*, Section 6.10.2 for the details.

(Page 273) **Suite B**: NSA/CSS, "Cryptography Today," NSA/CSS Web site, https://www.nsa.gov/ia/programs/suitteb_cryptography/index.shtml. There apparently is also a "Suite A" for "especially sensitive information"; the very algorithms used are classified and not available to the public. NSA/CSS, "Fact sheet NSA Suite B cryptography," NSA/CSS Web site, http://wayback.archive.org/web /20051125141648/http://www.nsa.gov/ia/industry/crypto_suite_b.cfm. The reader may wish to consider this decision in light of Kerckhoffs' principle.

(Page 273) **original Suite B**: NSA/CSS, *"Fact Sheet NSA Suite B Cryptography".* The second algorithm for key agreement, elliptic curve MQV, was removed from the suite in 2008.

(Page 273) **algorithm for helping create short signatures**: That is, a hash function.

(Page 273) **commercial and government standards**: At the time, AES and the hash function were the only two algorithms in their classes fully endorsed by NIST, unlike the case for the key agreement and digital signature categories. AES is still the only fully endorsed symmetric encryption algorithm, although another endorsed hash function has been added.

(Page 273) **NSA particularly mentioned**: NSA/CSS, "The case for elliptic curve cryptography," NSA/CSS Web site, http://wayback.archive.org/web/ 20131209051540/http://www.nsa.gov/business/programs/elliptic_curve.shtml.

(Page 274) **Dual EC DRBG**: Bruce Schneier, "Did NSA put a secret backdoor in new encryption standard?" *Wired Magazine* Web site, http://archive.wired.com /politics/security/commentary/securitymatters/2007/11/securitymatters_1115.

(Page 274) **two researchers from Microsoft**: Dan Shumow and Niels Ferguson, "On the possibility of a back door in the NIST SP800-90 Dual EC PRNG," Slides from presentation at Rump Session of CRYPTO 2007, http://rump2007.cr.yp.to/15

-shumow.pdf. Apparently the existence of such a back door was suspected as early as 2005. Matthew Green, "A few more notes on NSA random number generators," A Few Thoughts on Cryptographic Engineering Blog, http://blog.cryptography engineering.com/2013/12/a-few-more-notes-on-nsa-random-number.html.

(Page 274) **Snowden documents on Dual EC DRBG**: Nicole Perlroth, "Government announces steps to restore confidence on encryption standards," New York Times Web site, http://bits.blogs.nytimes.com/2013/09/10/government-announces-steps-to -restore-confidence-on-encryption-standards/.

(Page 274) **NIST removed the system**: "NIST removes cryptography algorithm from random number generator recommendations," NIST Tech Beat Blog, http://www .nist.gov/itl/csd/sp800-90-042114.cfm.

(Page 275) **"constants that the NSA influences"**: Bruce Schneier, "NSA surveillance: A guide to staying secure," *The Guardian* (2013). In particular, if you do use elliptic curves, this might be a reason to compute the curves and generators yourself instead of looking them up in a table which might have been influenced by someone malicious.

(Page 275) **new set of algorithms**: NSA/CSS, *"Cryptography Today."*

(Page 275) **NIST report on quantum-resistant cryptography**: Lily Chen et al., Report on Post-Quantum Cryptography, NIST, April 2016.

CHAPTER 9 THE FUTURE OF CRYPTOGRAPHY

(Page 276) **Automatic food dispenser**: Schrödinger originally phrased the question somewhat differently, but I just can't deal with discussing dead cats. Even hypothetical ones. Sorry.

(Page 280) **Deutch's algorithm**: The problem and the algorithm were first described in D. Deutsch, "Quantum theory, the Church-Turing principle and the universal quantum computer," *Proceedings of the Royal Society of London. Series A, Mathematical and Physical Sciences* 400:1818 (1985).

(Page 280) **Shor's algorithm**: Shor's algorithm was first published in P. W. Shor, "Algorithms for quantum computation: Discrete logarithms and factoring," in *Proceedings, 35th Annual Symposium on Foundations of Computer Science*, IEEE Computer Society Technical Committee on Mathematical Foundations of Computing (Los Alamitos, CA: IEEE, 1994). A very nice nontechnical explanation of the ideas involved is Scott Aaronson, "Shor, I'll do it," in Reed Cartwright and Bora Zivkovic (eds.), *The Open Laboratory: The Best Science Writing on Blogs 2007* (Lulu.com, 2008).

(Page 281) **smallest number for Shor's algorithm**: Shor's algorithm doesn't work on even numbers, which are easy to factor anyway, or numbers like 9, which are a perfect power of a prime. Those can also be factored relatively quickly using special techniques.

(Page 281) **factorization of 15**: Lieven M. K. Vandersypen et al., "Experimental realization of Shor's quantum factoring algorithm using nuclear magnetic resonance," *Nature* 414:6866 (2001).

(Page 281) **factorization of 21**: Enrique Martin-Lopez et al., "Experimental realisation of Shor's quantum factoring algorithm using qubit recycling," *Nature Photonics* 6:11 (2012).

(Page 281) **factorization of 143**: Nanyang Xu et al., "Quantum factorization of 143 on a dipolar-coupling nuclear magnetic resonance system," *Physical Review Letters* 108:13 (2012). It is not clear whether or not this algorithm, called adiabatic quantum computation, is as fast as Shor's algorithm.

(Page 281) **factorization of 56153**: Nikesh S. Dattani and Nathaniel Bryans, "Quantum factorization of 56153 with only 4 qubits," arXiv number 1411.6758, November 27, 2014. As the authors point out, in general "this reduction will not allow us to crack big RSA codes [sic]."

(Page 281) **Grover's algorithm**: Grover's algorithm was first published in Lov K. Grover, "A fast quantum mechanical algorithm for database search," in *Proceedings of the Twenty-eighth Annual ACM Symposium on Theory of Computing*, Association for Computing Machinery Special Interest Group on Algorithms and Computation Theory (New York: ACM, 1996). Graham P Collins, "Exhaustive searching is less tiring with a bit of quantum magic," *Physics Today* 50:10 (1997), is a pretty readable summary of the technique.

(Page 281) **256-bit AES keys**: NSA/CSS, *Cryptography Today*.

(Page 281) **postquantum cryptography**: For a good, although somewhat technical, overview of postquantum cryptography, see Daniel J. Bernstein, "Introduction to post-quantum cryptography," in Daniel J. Bernstein, Johannes Buchmann, and Erik Dahmen (eds.), *Post-Quantum Cryptography* (Springer Berlin Heidelberg, 2009).

(Page 281) **Not known to be easily solvable**: Although, like most things in public-key cryptography, they are not definitively known to be difficult either.

(Page 284) **500 dimensions or more**: Hoffstein et al., *Introduction to Mathematical Cryptography*, Section 7.11.2. Note that the number of dimensions of the lattice is $2N$ for the values of N given in that section.

(Page 284) **example cryptographic system**: The first published cryptographic system explicitly based on lattices appears to be the one invented by **Miklós Ajtai** and **Cynthia Dwork** in 1997. (Miklós Ajtai and Cynthia Dwork, "A Public-key cryptosystem with worst-case/average-case equivalence," in *Proceedings of the Twenty-ninth Annual ACM Symposium on Theory of Computing*, Association for Computing Machinery Special Interest Group on Algorithms and Computation Theory (New York; ACM, 1997).) The Ajtai-Dwork system was based on a variant of the shortest vector problem and is currently considered to be secure but impractical. The system I describe here was invented at about the same time and is currently considered practical but insecure.

(Page 284) **Babai's algorithm**: L Babai, "On Lovász' lattice reduction and the nearest lattice point problem," *Combinatorica* 6:1 (1986).

(Page 287) **both a "good" set and a "bad" set**: I'm skipping over the details of how Bob would find the generators. The short answer is that he finds a set of points with angles close to right angles, makes that the good set, and uses it to calculate a bad set. For more details, see the references for the GGH cryptosystem (page 291).

(Page 289) **a very small amount of information**: We will see that Eve can often recover numbers somewhat near the original plaintext, even if she can't recover the actual plaintext. When there is less information in each number, it is harder for Eve to guess the plaintext from the "somewhat near" information. Our example cipher would be even more secure if we encoded each letter using binary bits and took each bit separately. Furthermore, we should really add in some extra random bits to avoid frequency attacks. All this makes the messages very long, unfortunately. This effect is called **message expansion**.

(Page 289) **"lattice now"**: See James Agee and Walker Evans, *Let Us Now Praise Famous Men* (Boston: Houghton Mifflin, 1941).

(Page 289) **almost certain**: Unlike most cryptographic systems we have looked at, there is a small chance that Bob's decryption will not correctly match the original message. If so, it probably will not make any sense so it is usually easy to tell. This is similar to the situation with primality testing in Section 7.5. As long as the chance of an accidental error is very small, the system is good enough.

(Page 291) **GGH cryptosystem**: Oded Goldreich et al., "Public-key cryptosystems from lattice reduction problems," in Burton S. Kaliski Jr. (ed.), *Advances in Cryptology—CRYPTO '97* (Springer Berlin Heidelberg, 1997). Other sources for more details of the system are Hoffstein et al., *Introduction to Mathematical Cryptography*, Section 7.8 and Daniele Micciancio and Oded Regev, "Lattice-based cryptography," in Daniel J. Bernstein, Johannes Buchmann, and Erik Dahmen (eds.), *Post-Quantum Cryptography* (Springer Berlin Heidelberg, 2009), Section 5.

(Page 292) **GGH is insecure**: Phong Nguyen, "Cryptanalysis of the Goldreich-Goldwasser-Halevi cryptosystem from CRYPTO '97," in Michael Wiener (ed.), *Advances in Cryptology—CRYPTO '99*, (Springer Berlin Heidelberg, 1999).

(Page 292) **other systems similar to GGH**: See, for example, Micciancio and Regev, "Lattice-based cryptography," Section 5.

(Page 292) **most promising**: Ray A. Perlner and David A. Cooper, "Quantum resistant public key cryptography: A survey," in Kent Seamons, Neal McBurnett, and Tim Polk, (eds.) *Proceedings of the 8th Symposium on Identity and Trust on the Internet* (New York: ACM Press, 2009).

(Page 292) **NTRU**: NTRU was originally described in Jeffrey Hoffstein et al., "Public key cryptosystem method and apparatus," United States Patent: 6081597, 2000http://www.google.com/patents/US6081597 and Jeffrey Hoffstein et al., "NTRU: A ring-based public key cryptosystem," in Joe P. Buhler (ed.), *Algorithmic Number Theory* (Berlin/Heidelberg: Springer, 1998). For the lattice description and other information, see Hoffstein et al., *Introduction to Mathematical Cryptography*, Section 17.10, Micciancio and Regev, "Lattice-based cryptography," Section 5.2, or Trappe and Washington, *Introduction to Cryptography*, Section 17.4.

(Page 292) **rumors suggest**: Trappe and Washington, *Introduction to Cryptography*, Section 17.4.

(Page 292) **Jeff Hoffstein replied**: Personal communication, June 22, 1998. I was at a conference at Reed College in a van with Carl Pomerance, headed toward the conference banquet. Hoffstein was walking down the street when Pomerance yelled the question at him out the window.

(Page 292) **digital signature systems**: GGH digital signatures, like GGH encryption, have been shown to be insecure. (Phong Q. Nguyen, and Oded Regev, "Learning a parallelepiped: Cryptanalysis of GGH and NTRU signatures," *Journal of Cryptology* 22:2 (2008).) Early versions of NTRU digital signatures were also shown to be insecure; the latest version was proposed in 2014 and has not been broken so far. The inventors point out that it "will require years of scrutiny before it can be deemed secure." (Hoffstein et al., *Introduction to Mathematical Cryptography*, Section 17.12.5)

(Page 292) **Wiesner's ideas**: See Levy, *Crypto*, pp. 332–38 for Wiesner's story. The paper was eventually published as Stephen Wiesner, "Conjugate coding," *SIGACT News* 15:1 (1983).

(Page 293) **Bennett and Brassard**: See Levy, *Crypto*, pp. 338–39, for Bennett's background and G. Brassard, "Brief history of quantum cryptography: A personal perspective," in *IEEE Information Theory Workshop on Theory and Practice in Information-Theoretic Security, 2005,* Piscataway, NJ: IEEE Information Theory Society in cooperation with the International Association for Cryptologic Research (IACR) for Bennett and Brassard's meeting.

(Page 293) **BB84**: C. H. Bennett and G. Brassard, "Quantum cryptography: Public key distribution and coin tossing," in *Proceedings of the IEEE International Conference on Computers, Systems, and Signal Processing,* IEEE Computer Society, IEEE Circuits and Systems Society, Indian Institute of Science (Bangalore, India, 1984).

(Page 293) **as you look at it**: Not that a human can generally see a single photon.

(Page 296) **keep about half the bits**: In this example, they did a little better.

(Page 297) **What if Eve is listening?**: We will see shortly that Eve actually has an extra problem here, but let's ignore it for the moment.

(Page 298) **then Eve is listening**: Or possibly it's just noise on the line, but there are ways that Alice and Bob can account for that, too. In some cases they can proceed even if Eve has discovered some of the bits. Samuel J. Lomonaco Jr., "A talk on quantum cryptography, or how Alice outwits Eve," in David Joyner (ed.), *Coding Theory and Cryptography: From Enigma and Geheimschreiber to Quantum Theory* (Berlin/Heidelberg; New York: Springer, 2000), has a good introduction to BB84, BB84 with noise on the line, and several other protocols. It also explains some of the possibly strange-looking notation used in this subject, although it helps to know some linear algebra. Samuel J. Lomonaco Jr., "A quick glance at quantum cryptography," *Cryptologia* 23:1 (1999), is an earlier version with somewhat more depth and more references but less introductory material.

(Page 298) **Needed to build a prototype**: C. H. Bennett and G. Brassard, "The dawn of a new era for quantum cryptography: The experimental prototype is working!" *ACM SIGACT News* 20:4 (1989); Brassard, "Brief history."

(Page 298) **First-ever key agreement by quantum cryptography**: Bennett and Brassard, "Dawn of a new era"; C. H. Bennett et al., "Experimental quantum cryptography," *Journal of Cryptology* 5:1 (1992); Brassard, "Brief history."

(Page 298) **quantum key distribution over fiber optics**: Boris Korzh et al., "Provably secure and practical quantum key distribution over 307 km of optical fibre," *Nature Photonics* 9:3 (2015). This system used the "coherent one-way protocol" (COW) instead of BB84. Nicolas Gisin et al., "Towards practical and fast quantum cryptography," arXiv number quant-ph/0411022, November 3, 2004. One of the problems in transmitting quantum particles is that the communications channel currently has to be a single link. Any attempt to boost or redirect the signal will destroy the quantum characteristics that the system depends on, although researchers are working on getting around that.

(Page 299) **BB84 through the open air**: Tobias Schmitt-Manderbach et al., "Experimental demonstration of free-space decoy-state quantum key distribution over 144 km," *Physical Review Letters* 98:1 (2007).

(Page 299) **First bank transfer protected by quantum cryptography**: A. Poppe et al., "Practical quantum key distribution with polarization entangled photons," *Optics Express* 12:16 (2004). This system did not use BB84, but rather a somewhat related protocol known as E91, which was first published in Artur K. Ekert, "Quantum cryptography based on Bell's theorem," *Physical Review Letters* 67:6 (1991). For a less technical description of E91, see Artur Ekert, "Cracking codes, part II," *Plus Magazine* No. 35 (2005).

(Page 299) **Quantum-cryptographic equipment for sale**: See, for example, Andrew Shields and Zhiliang Yuan, "Key to the quantum industry," *Physics World* 20:3 (2007).

(Page 299) **Computer networks protected by quantum cryptography**: United States: Shields and Yuan, "Key to the quantum industry," Richard J. Hughes et al., "Network-centric quantum communications with application to critical infrastructure protection," (May 1, 2013); Austria: Roland Pease, "'Unbreakable' encryption unveiled," BBC News Web site, http://news.bbc.co.uk/2/hi/science /nature/7661311.stm; Switzerland: D. Stucki et al., "Long-term performance of the SwissQuantum quantum key distribution network in a field environment," *New Journal of Physics* 13:12 (2011); Japan: M. Sasaki et al., "Field test of quantum key distribution in the Tokyo QKD Network," *Optics Express* 19:11 (2011); China: Jian-Yu Wang et al., "Direct and full-scale experimental verifications towards ground-satellite quantum key distribution," *Nature Photonics* 7:5 (2013).

(Page 299) **"I don't know . . . "**: Clay Dillow, "Unbreakable encryption comes to the U.S.," fortune.com, http://fortune.com/2013/10/14/unbreakable-encryption -comes-to-the-u-s/, quoting Don Hayford of Battelle Memorial Institute.

(Page 299) **pure cryptanalysis**: These techniques are usually the most interesting mathematically, which is the reason I have focused on them.

(Page 300) **Photon number-splitting attack**: This attack was named in Gilles Brassard et al., "Limitations on practical quantum cryptography," *Physical Review Letters* 85:6 (2000); it is described there as a modification of an earlier idea.

(Page 300) **keep her captured photons**: Note that she can't use the storage trick in the single-photon case, since she would have to both store the photon *and* send it on to Bob.

(Page 301) **Modifications to BB84**: The most well-known of these is called SARG04; it was first published in Valerio Scarani et al., "Quantum cryptography protocols robust against photon number splitting attacks for weak laser pulse implementations," *Physical Review Letters* 92:5 (2004).

(Page 301) **Decoy-pulse method**: Won-Young Hwang, "Quantum key distribution with high loss: Toward global secure communication," *Physical Review Letters* 91:5 (2003). Some of the links in the Japanese quantum network, among others, used this method (Sasaki et al., "Field test of quantum key distribution").

(Page 301) **Bright illumination attack**: Lars Lydersen et al., "Hacking commercial quantum cryptography systems by tailored bright illumination," *Nature Photonics* 4:10 (2010). Other active attacks include the **time-shift attack**, which takes advantage of the possibility that some detectors might be more likely to fail to register bits that are 1 as opposed to 0, or vice versa (Yi Zhao et al., "Quantum hacking: Experimental demonstration of time-shift attack against practical quantum-key-distribution systems," *Physical Review A* 78:4 (2008)), and the **phase-remapping attack**, which attacks systems in which Alice's equipment can receive photons as well as send them (Feihu Xu et al., "Experimental demonstration of phase-remapping attack in a practical quantum key distribution system," *New Journal of Physics* 12:11 (2010)).

(Page 301) **"It may be roundly asserted . . . "**: Edgar Allen Poe, "A few words on secret writing," *Graham's Magazine* 19:1 (1841).

■ ■ ■ SUGGESTIONS FOR FURTHER READING ■ ■ ■

As I said in the Preface, this book is about a particular aspect of cryptography, and there are lots of other ways that you can study the subject. Here are some suggestions for how to start; the detailed information about each source is in the bibliography.

If you would like to pursue an academic approach to cryptography, there are lots of great textbooks out there. I can't mention all of them here, but I'll tell you a few of my favorites. For something at about the same mathematical level as this book, I like *Invitation to Cryptology*, by Thomas Barr. It's a little out of date, but I've had good luck using it with students and it has lots of great exercises. For something a little more challenging, I use *Introduction to Cryptography with Coding Theory*, by Wade Trappe and Lawrence Washington, in my courses for upper-level math and computer science majors. It also has great exercises and covers some topics that this book hasn't, such as error-correcting codes. If you really want to push yourself mathematically, try *An Introduction to Mathematical Cryptography*, by Jeffrey Hoffstein, Hill Pipher, and Joseph Silverman. It's written for advanced undergraduates and beginning graduate students and focuses on public-key cryptography and digital signatures.

If you are interested in the practical side of actually using modern cryptography, there are lots of good textbooks there, too. I've had good luck with *Cryptography and Network Security*, by William Stallings. It covers cryptography and mathematics, with a focus on what's used in modern computers, and then goes on to talk about the specific hardware and software that is used to keep those computers safe. Even if you don't need a textbook and are just curious about how your computer works with the cryptographic systems we've discussed, I recommend it. But for a handbook of exactly what to do and what not to do, you can't beat *Cryptography Engineering*, by Niels Ferguson, Bruce Schneier, and Tadayoshi Kohno. They describe the book as narrow and focused: "we don't give you dozens of choices; we give you one option and tell you how to implement it correctly" (p. xxviii).

A less technical overview of the practical side is *Secrets and Lies: Digital Security in a Networked World*, by Bruce Schneier. Schneier is both a noted cryptographer and one of my favorite authors on cryptography, writing on everything from the technical aspects of ciphers to the practical to the social. I'll point out a couple of his books by name, but I recommend everything he's written. *Secrets and Lies* is aimed at businesspeople who want to understand how digital security impacts their business, but it's extremely readable, and I recommend it to anyone who wants to understand practical cryptography without wading through too much jargon.

If you want to be a professional cryptographer, that is, someone whose job it is to invent and/or break the systems that keep secrets safe, you should read *Applied*

Cryptography, by Bruce Schneier. *Applied Cryptography* is somewhat dated now, but it covers the mathematical (and many other) details of *every* important modern cipher known up to 1996. It was an invaluable reference while writing this book. Follow that with *Handbook of Applied Cryptography*, by Alfred Menezes, Paul van Oorschot, and Scott Vanstone, which has slightly different coverage and is slightly more up to date. Then read *The Design of Rijndael: AES—The Advanced Encryption Standard*, by Joan Daemen and Vincent Rijmen. As the winners of the AES competition, they arguably know as much as anyone about how to design a cipher. Even better, they explain both the details and the motivations behind the cipher in a surprisingly clear way for those who can keep up with the math.

If you are interested in the history of cryptography, the must-read book is *The Codebreakers*, by David Kahn. The first edition was published in 1967, and it's *the* definitive work on the history of cryptography as it stood up until that point. Two things have happened since then, however. First, a lot of previously classified historical material about cryptography has been released, particularly about cryptography in World War II. Secondly, the use of cryptography in computers has exploded, leading to the development of lots of new ciphers with interesting backgrounds. The second edition of *The Codebreakers*, from 1996, has a short chapter on these developments, but you might want more. There are lots of good books now on World War II cryptography; I've mentioned a few in the bibliography. I don't have a particular favorite. For the development of computer cryptography up until 2001, I like *Crypto* by Steven Levy. Unfortunately, that book came out just before the announcement of the winner of the AES competition. In my opinion, the first really good history of cryptography at the beginning of the twenty-first century has yet to be written. While we are waiting, I recommend the historical vignettes in the later chapters of *Secret History: The Story of Cryptology*, by Craig Bauer. Bauer's book is a mixture of history and mathematics written by an expert in the history of cryptography. It can be used as a textbook, a reference, or just something to pick up and read a few entertaining pages at random.

One thing I haven't said much about in this book is the social implications of cryptography, particularly its role in the protection of personal privacy. A good introduction to digital technology and privacy for nonexperts is *Blown to Bits: Your Life, Liberty, and Happiness After the Digital Explosion*, by Hal Abelson, Ken Ledeen, and Harry Lewis. This covers many aspects of modern privacy, including cryptography. *Privacy on the Line: The Politics of Wiretapping and Encryption*, by Whitfield Diffie and Susan Landau, focuses more specifically on communications technology and goes into much more scholarly detail. Landau's *Surveillance or Security? The Risks Posed by New Wiretapping Technologies* covers many of the same topics but brings them further up to date. As I'm writing this, Bruce Schneier has just published *Data and Goliath: The Hidden Battles to Collect Your Data and Control Your World*. I confess I haven't read it yet, but I'm really looking forward to it.

I've mentioned more than once that modern cryptography is a fast-moving field. Unsurprisingly, cryptographers use the web heavily for both disseminating and obtaining the latest news. Many of them write blogs, and I'll mention a few of my favorites. Bruce Schneier posts almost every day to *Schneier on Security*, https://www.schneier .com. Like the rest of his writing, it ranges from cryptography to computer security to wider issues of security and privacy. Many of the entries are short snippets of news articles, with links. When Schneier posts one of his own essays, it's especially worth reading.

Matthew Green posts about once a month to *A Few Thoughts on Cryptographic Engineering*, http://blog.cryptographyengineering.com. Many of the posts are about technical topics but written in a very readable way. They often start with a nontechnical summary before attempting to explain the details. Matt Blaze wrote a similar blog until 2013, *Matt Blaze's Exhaustive Search*, http://www.crypto.com/blog. The blog doesn't seem to be currently active, but the page has archives and links, including Blaze's Twitter feed, which is active. Steve Bellovin posts about once a month to *SMBlog: Pseudo-Random Thoughts on Computers, Society, and Security*, https://www.cs.columbia.edu /~smb/blog. I would describe these as technically informed opinion essays, often inspired by the latest news but not just reporting on it. Bellovin's page also has links to several other blogs that are less related to cryptography but that readers interested in cryptography might find interesting.

I will also be maintaining a blog devoted to updating the material in this book with new developments in cryptography and new historical discoveries. Many of these will be pulled from the sources I've already mentioned, but I will also post recommendations for new sources you might want to read. The blog will be accessible through the web page for this book at http://press.princeton.edu/titles/10826.html.

Finally, if you really want to see the latest research in cryptography in all its technical glory, there are two main places on line where preprints of technical papers are posted for free download. The more general one is *arXiv*, http://arxiv.org, which has sections for physics (including quantum physics), mathematics, computer science, and a few other fields. The *Cryptology ePrint Archive*, http://eprint.iacr.org/, is more restricted.

■ ■ ■ BIBLIOGRAPHY ■ ■ ■

Aaronson, Scott. "Shor, I'll do it." In Reed Cartwright and Bora Zivkovic (eds.), *The Open Laboratory: The Best Science Writing on Blogs 2007.* Lulu.com, January 23, 2008, 197–202. Originally published on the blog "Shtetl-Optimized," http://scottaaronson.com/blog/?p=208, February 24, 2007.

ABC. "The Muppet Show: Sex and Violence." Television. March 19, 1975.

Abdalla, Michel, Mihir Bellare and Phillip Rogaway. "The oracle Diffie-Hellman assumptions and an analysis of DHIES." In David Naccache (ed.), *Topics in Cryptology-CT-RSA 2001.* Berlin/Heidelberg: Springer-Verlag, 2001, 143–58.

Abelson, Hal, Ken Ledeen, and Harry Lewis. *Blown to Bits: Your Life, Liberty, and Happiness After the Digital Explosion.* Upper Saddle River, NJ: Addison-Wesley Professional, 2008. Also available as a free download from http://www.bitsbook.com.

Abu Nuwas, Al-Hasan ibn Hani al-Hakami. "Don't cry for Layla." Princeton Online Arabic Poetry Project. https://www.princeton.edu/~arabic/poetry/layla.swf.

Adrian, David, Karthikeyan Bhargavan, Zakir Durumeric, Pierrick Gaudry, Matthew Green, J. Alex Halderman, Nadia Heninger, et al. "Imperfect forward secrecy: How Diffie-Hellman fails in practice." In *22nd ACM Conference on Computer and Communications Security.* Association for Computing Machinery Special Interest Group on Security, Audit and Control. New York: ACM Press, October 2015, 5–17.

———. "The Logjam Attack." (May 20, 2015). https://weakdh.org/.

Agee, James, and Walker Evans. *Let Us Now Praise Famous Men.* Boston: Houghton Mifflin, 1941.

Agrawal, Manindra, Neeraj Kayal, and Nitin Saxena. "PRIMES is in P." *The Annals of Mathematics* 160:2 (September 2004), 781–793. http://www.jstor.org/stable/3597229.

Ajtai, Miklós, and Cynthia Dwork. "A Public-key cryptosystem with worst-case/average-case equivalence." In *Proceedings of the Twenty-ninth Annual ACM Symposium on Theory of Computing.* Association for Computing Machinery Special Interest Group on Algorithms and Computation Theory. New York: ACM, 1997, 284–93.

Al-Kadi, Ibrahim A. "Origins of cryptology: The Arab contributions." *Cryptologia* 16 (1992), 97–126.

al Mutanabbi, Abu at-Tayyib Ahmad ibn al-Husayn. "al-Mutanabbi to Sayf al-Dawla." Princeton Online Arabic Poetry Project. http://www.princeton.edu/~arabic/poetry/al_mu_to_sayf.html.

Alvarez, Gonzalo, Dolores De La Guía, Fausto Montoya, and Alberto Peinado. "Akelarre: A new block cipher algorithm." In Stafford Tavares and Henk Meijer (eds.), *Proceedings of the SAC '96 Workshop.* Kingston, ON: Queen's University, August 1996, 1–14.

Anderson, Ross. "A5 (Was: HACKING DIGITAL PHONES)." Posted in uk.telecom (Usenet group). June 17, 1994. http://groups.google.com/group/uk.telecom/msg /ba76615fef32ba32.

———. "On Fibonacci keystream generators." In *Fast Software Encryption: Second International Workshop Leuven, Belgium, December 14–16, 1994 Proceedings*. Berlin/Heidelberg: Springer, 1995, 346–52. http://dx.doi.org/10.1007/3-540 -60590-8_26.

André, Frédéric. "Hagelin C-36." http://fredandre.fr/c36.php?lang=en.

Asmuth, C. A., and G. R. Blakley. "An efficient algorithm for constructing a cryptosystem which is harder to break than two other cryptosystems." *Computers & Mathematics with Applications* 7:6 (1981), 447–50. http://www.sciencedirect.com /science/article/B6TYJ-45DHSNX-17/2/8877c15616bb560298d056788b59aff6.

Atkins, Derek, Michael Graff, Arjen K. Lenstra, and Paul C. Leyland. "The magic words are Squeamish Ossifrage." In *Advances in Cryptology—ASIACRYPT '94*. Josef Pieprzyk and Reihanah Safavi-Naini (eds.). Berlin/Heidelberg: Springer-Verlag, 1995, 261–77.

Babai, L. "On Lovász' lattice reduction and the nearest lattice point problem." *Combinatorica* 6:1 (March 1986), 1–13.

Bacon, Francis. *Of The Advancement And Proficience Of Learning or the Partitions Of Sciences IX Bookes Written in Latin by the Most Eminent Illustrations & Famous Lord Francis Bacon Baron of Verulam Vicont St. Alban*. Oxford: Printed by Leon Lichfield, printer to the University, for Rob Young and Ed Forrest, 1640. Translated by Gilbert Watts from the Latin *De augmentis scientarium*, which is an enlargement, translated into Latin, of the *Proficience and Advancement of Learning* of 1605.

Barkan, Elad, and Eli Biham. "Conditional estimators: An effective attack on A5/1." In *Selected Areas in Cryptography*. Berlin/Heidelberg: Springer, 2006, 1–19. http://dx .doi.org/10.1007/11693383_1.

Barkan, Elad, Eli Biham, and Nathan Keller. "Instant ciphertext-only cryptanalysis of GSM encrypted communication." In *Advances in Cryptology—CRYPTO 2003*. Berlin/Heidelberg: Springer, 2003, 600–16. http://dx.doi.org/10.1007/978-3-540 -45146-4_35.

Barker, Elaine, William Barker, William Burr, William Polk, and Miles Smid. "Recommendation for key management—Part 1: General (Revision 3)." NIST Special Publications Number 800-57, Part 1. NIST, July 2012. http://csrc.nist.gov/ publications/nistpubs/800-57/sp800-57_part1_rev3_general.pdf.

Barker, Wayne G. *Cryptanalysis of the Hagelin Cryptograph*. Laguna Hills, CA: Aegean Park Press, June 1981.

Barr, Thomas H. *Invitation to Cryptology*. Englewood Cliffs, NJ: Prentice Hall, 2001.

Bauer, Craig P. *Secret History: The Story of Cryptology*. Boca Raton, FL: CRC Press, 2013.

Bauer, Friedrich. *Decrypted Secrets: Methods and Maxims of Cryptology*. 3rd, rev., updated ed. Berlin [u.a.]: Springer, 2002.

Bauer, Friedrich L. "An error in the history of rotor encryption devices." *Cryptologia* 23:3 (1999), 206–10. http://www.informaworld.com/10.1080/0161-119991887847.

Baum, L. Frank. *The Wonderful Wizard of Oz*. Chicago: George M. Hill, 1900. http:// www.gutenberg.org/ebooks/55.

Beker, Henry, and Fred Piper. *Cipher Systems: The Protection of Communications.* New York: Wiley, 1982.

Bellare, Mihir, and Phillip Rogaway. "Minimizing the use of random oracles in authenticated encryption schemes." In Yongfei Han, Tatsuaki Okamoto, and Sihan Quing, (eds.), *Proceedings of the First International Conference on Information and Communication Security.* Berlin/Heidelberg: Springer, 1997, 1–16.

Bellovin, Steven M. "Frank Miller: Inventor of the one-time pad." *Cryptologia* 35:3 (July 2011), 203–22. http://www.tandfonline.com/doi/abs/10.1080/01611194.2011.583711.

———. "Vernam, Mauborgne, and Friedman: The one-time pad and the index of coincidence." Columbia University Computer Science Technical Reports Number CUCS-014-14. Department of Computer Science, Columbia University. May 2014. http://dx.doi.org/10.7916/D8Z0369C.

Bennett, C. H., F. Bessette, G. Brassard, L. Salvail, and J. Smolin. "Experimental quantum cryptography." *Journal of Cryptology* 5:1 (1992), 3–28.

Bennett, C. H., and G. Brassard. "Quantum cryptography: Public key distribution and coin tossing." In *Proceedings of the IEEE International Conference on Computers, Systems, and Signal Processing.* IEEE Computer Society, IEEE Circuits and Systems Society, Indian Institute of Science. Bangalore, India, December 1984, 175–79.

———. "The dawn of a new era for quantum cryptography: The experimental prototype is working!" *ACM SIGACT News* 20:4 (1989), 78–80. http://portal.acm.org/citation.cfm?id=74087.

Bernstein, Daniel J. "Introduction to post-quantum cryptography." In Daniel J. Bernstein, Johannes Buchmann, and Erik Dahmen (eds.), *Post-Quantum Cryptography.* Berlin Heidelberg: Springer, 2009, 1–14. http://link.springer.com/chapter/10.1007/978-3-540-88702-7_1. Also available from http://pqcrypto.org/.

Beurdouche, Benjamin, Karthikeyan Bhargavan, Antoine Delignat-Lavaud, Cedric Fournet, Markulf Kohlweiss, Alfredo Pironti, Pierre-Yves Strub, and Jean-Karim Zinzindohou. "A messy state of the union: Taming the composite state machines of TLS." In *2015 IEEE Symposium on Security and Privacy (SP).* Los Alamitos, CA: IEEE Computer Society, May 2015, 535–52.

Bhargavan, Karthikeyan, Antoine Delignat-Lavaud, Cédric Fournet, Markulf Kohlweiss, Alfredo Pironti, Pierre-Yves Strub, Santiago Zanella-Béguelin, Jean-Karim Zinzindohoué, and Benjamin Beurdouche. "State Machine AttaCKs against TLS (SMACK TLS)." https://www.smacktls.com.

Biham, Eli. "How to make a difference: Early history of differential cryptanalysis." Slides from invited talk presented at Fast Software Encryption, 13th International Workshop. March 2006. http://www.cs.technion.ac.il/$\sim$biham/Reports/Slides/fse2006-history-dc.pdf.

Biham, Eli, and Adi Shamir. *Differential Cryptanalysis of the Data Encryption Standard.* New York: Springer, 1993.

Biryukov, Alex, and Eyal Kushilevitz. "From differential cryptanalysis to ciphertext-only attacks." In Hugo Krawczyk (ed.), *Advances in Cryptology—CRYPTO '98.* Berlin/Heidelberg: Springer, January 1998, 72–88.

Biryukov, Alex, Adi Shamir, and David Wagner. "Real time cryptanalysis of A5/1 on a PC." In *Fast Software Encryption: 7th International Workshop, FSE 2000 New York,*

NY, USA, April 10–12, 2000 Proceedings. Berlin/Heidelberg: Springer, 2001, 37–44. http://dx.doi.org/10.1007/3-540-44706-7_1.

Boak, David G. "A history of U.S. communications security (Volume I)." National Security Agency. July 1973. http://www.nsa.gov/public_info/_files/cryptologic _histories/history_comsec.pdf.

Bogdanov, Andrey, Dmitry Khovratovich, and Christian Rechberger. "Biclique cryptanalysis of the full AES." In Dong Hoon Lee and Xiaoyun Wang (eds.), *Advances in Cryptology—ASIACRYPT 2011.* Berlin Heidelberg: Springer 2011, 344–71. http://link.springer.com/chapter/10.1007/978-3-642-25385-0_19.

Boneh, Dan. "Twenty years of attacks on the RSA cryptosystem." *Notices of the AMS* 46:2 (February 1999), 203–13. http://www.ams.org/notices/199902/boneh.pdf.

Bornemann, F. "PRIMES is in P: A breakthrough for 'everyman.'" *Notices of the AMS* 50:5 (May 2003), 545–52. http://www.ams.org/notices/200305/fea-bornemann.pdf.

Bos, Joppe W., Marcelo E. Kaihara, and Peter L. Montgomery. "Pollard rho on the PlayStation 3." In *SHARCS '09 Workshop Record.* Virtual Application and Implementation Research Lab within ECRYPT II European Network of Excellence in Cryptography, Lausanne, Switzerland: 2009, 35–50.

Brassard, G. "Brief history of quantum cryptography: A personal perspective." In *IEEE Information Theory Workshop on Theory and Practice in Information-Theoretic Security, 2005.* Piscataway, NJ: IEEE Information Theory Society in cooperation with the International Association for Cryptologic Research (IACR), 19–23.

Brassard, Gilles, Norbert Lütkenhaus, Tal Mor, and Barry C. Sanders. "Limitations on practical quantum cryptography." *Physical Review Letters* 85:6 (2000), 1330. http://link.aps.org/doi/10.1103/PhysRevLett.85.1330.

Brown, Dan. *The Da Vinci Code.* 1st ed.. New York: Doubleday, 2003.

Buonafalce, Augusto. "Bellaso's reciprocal ciphers." *Cryptologia* 30:1 (2006), 39. http://www.informaworld.com/10.1080/01611190500383581.

bushing, marcan, and sven. "Console hacking 2010: PS3 epic fail." Slides from lecture presented at 27th Chaos Communication Congress, December 29, 2010. https://events.ccc.de/congress/2010/Fahrplan/events/4087.en.html.

Callas, Jon, Lutz Donnerhacke, Hal Finney, David Shaw, and Rodney Thayer. OpenPGP Message Format. Request for Comments Number 4880. IETF. November 2007. https://tools.ietf.org/html/rfc4880 (accessed July 28, 2015).

Cannière, Christophe De, and Bart Preneel. "Trivium." In Matthew Robshaw and Olivier Billet (eds.), *New Stream Cipher Designs.* Berlin, New York: Springer, 2008, 244–266.

Carroll, Lewis. *Alice's Adventures in Wonderland.* London: Macmillan, 1865. http://www.gutenberg.org/ebooks/11.

———. *Through the Looking-Glass, and What Alice Found There.* London: Macmillan, 1871. http://www.gutenberg.org/ebooks/12.

———. *The Hunting of the Snark: An Agony in Eight Fits.* London: Macmillan, 1876. http://www.gutenberg.org/ebooks/13.

Chambers, W. "On random mappings and random permutations." In *Fast Software Encryption: Second International Workshop Leuven, Belgium, December 14–16, 1994 Proceedings.* Berlin/Heidelberg: Springer, 1995, 22–28. http://dx.doi.org/10.1007 /3-540-60590-8_3.

Chambers, W. G. and S. J. Shepherd. "Mutually clock-controlled cipher keystream generators." *Electronics Letters* 33:12 (1997), 1020–21.

Chen, Lily, Stephen Jordan, Yi-Kai Liu, Dustin Moody, Rene Peralta, Ray Perlner, and Daniel Smith-Tone. Report on Post-Quantum Cryptography. National Institute of Standards and Technology Internal Report Number 8105. NIST, April 2016. http://nvlpubs.nist.gov/nistpubs/ir/2016/NIST.IR.8105.pdf.

Cid, Carlos, and Ralf-Philipp Weinmann. "Block ciphers: Algebraic cryptanalysis and Gröbner bases." In Massimiliano Sala, Shojiro Sakata, Teo Mora, Carlo Traverso, and Ludovic Perret (eds.), *Gröbner Bases, Coding, and Cryptography*. Berlin/Heidelberg: Springer, 2009, 307–27. http://link.springer.com/chapter/10.1007/978-3-540-93806-4_17.

Clark, Ronald William. *The Man Who Broke Purple: The Life of Colonel William F. Friedman, Who Deciphered the Japanese Code in World War II*. Boston: Little Brown, 1977.

Cocks, C. C. "A Note on non-secret encryption." UK Communications Electronics Security Group. November 20, 1973. http://www.cesg.gov.uk/publications/media/notense.pdf.

Collins, Graham P. "Exhaustive searching is less tiring with a bit of quantum magic." *Physics Today* 50:10 (1997), 19–21. http://dx.doi.org/10.1063/1.881969.

Coppersmith, D. "The Data Encryption Standard (DES) and its strength against attacks." *IBM Journal of Research and Development* 38:3 (1994), 243–50. http://portal.acm.org/citation.cfm?id=185915.

Coutinho, S. C. *The Mathematics of Ciphers: Number Theory and RSA Cryptography*. Natick, MA: AK Peters, Ltd., 1998.

Daemen, Joan, and Vincent Rijmen. "AES proposal: Rijndael." NIST. September 1999. http://csrc.nist.gov/archive/aes/rijndael/Rijndael-ammended.pdf. Series AES proposals, Document version 2.

———. *The Design of Rijndael: AES—The Advanced Encryption Standard*, 1st ed. Berlin/Heidelberg; New York: Springer, 2002.

Dattani, Nikesh S., and Nathaniel Bryans. "Quantum factorization of 56153 with only 4 qubits." ArXiv Number 1411.6758. November 27, 2014. http://arxiv.org/abs/1411.6758.

de Leeuw, Karl. "The Dutch invention of the rotor machine, 1915–1923." *Cryptologia* 27:1 (2003). 73. http://www.informaworld.com/10.1080/0161-110391891775.

Dettman, Alex, Wilhelm Fenner, Wilhelm Flicke, Kurt Friederichsohn, and Adolf Paschke. *Russian Cryptology During World War II*. Laguna Hills, CA: Aegean Park Press, 1999.

Deutsch, D. "Quantum theory, the Church-Turing principle and the universal quantum computer." *Proceedings of the Royal Society of London. Series A, Mathematical and Physical Sciences* 400:1818 (July 1985), 97–117. http://www.jstor.org/stable/2397601.

Dickson, Leonard Eugene. *Divisibility and Primality*. Reprint of 1919 edition. Volume 1 of *History of the Theory of Numbers*. Providence, RI: AMS Chelsea Publishing, 1966.

Diffie, Whitfield. "The first ten years of public-key cryptography." *Proceedings of the IEEE* 76:5 (1988), 560–77.

Diffie, Whitfield, and Martin E. Hellman. "Multiuser cryptographic techniques." In Stanley Winkler (ed.), *Proceedings of the June 7–10, 1976, National Computer Conference and Exposition*. New York: ACM, 1976, 109–12.

Diffie, Whitfield, and Martin E. Hellman. "New directions in cryptography." *IEEE Transactions on Information Theory* 22:6 (1976), 644–54.

Diffie, Whitfield, and Susan Landau. *Privacy on the Line: The Politics of Wiretapping and Encryption,* updated and expanded edition. Cambridge, MA: MIT Press, 2010.

Dillow, Clay. "Unbreakable encryption comes to the U.S." fortune.com. October 14, 2013. http://fortune.com/2013/10/14/unbreakable-encryption-comes-to-the-u-s/.

Durumeric, Zakir, James Kasten, Michael Baily, and J. Alex Halderman. "Analysis of the HTTPS certificate ecosystem." In *Proceedings of the 2013 Internet Measurement Conference.* Association for Computing Machinery Special Interest Groups on Data Communication and on Measurement and Evaluation. New York: ACM, October 2013, 291–304.

Dworkin, Morris. "Recommendation for block cipher modes of operation: Methods for format-preserving encryption." NIST Special Publications Number 800-38G Draft. NIST, July 2013. http://csrc.nist.gov/publications/drafts/800-38g/sp800_38g_draft.pdf.

Dworkin, Morris, and Ray Perlner. Analysis of VAES3 (FF2). Cryptology ePrint Archive Number 2015/306. 2015. http://eprint.iacr.org/2015/306. A slightly abridged version is at http://csrc.nist.gov/groups/ST/toolkit/BCM/documents/comments/800-38_Series-Drafts/FPE/analysis-of-VAES3.pdf.

ECRYPT Network of Excellence. "eSTREAM: The eSTREAM stream cipher project." http://www.ecrypt.eu.org/stream/index.html.

———. "Call for stream cipher primitives, version 1.3" (April 12, 2005). http://www.ecrypt.eu.org/stream/call.

Ekert, Artur. "Cracking codes, part II." *Plus Magazine* No. 35 (May 2005). http://plus.maths.org/issue35/features/ekert/index.html.

Ekert, Artur K. "Quantum cryptography based on Bell's theorem." *Physical Review Letters* 67:6 (1991), 661–63. http://link.aps.org/doi/10.1103/PhysRevLett.67.661.

Electronic Frontier Foundation. "Frequently asked questions (FAQ) about the Electronic Frontier Foundation's 'DES cracker' machine." http://w2.eff.org/Privacy/Crypto/Crypto_misc/DESCracker/HTML/19980716_eff_des_faq.html.

ElGamal, Taher. "A public key cryptosystem and a signature scheme based on discrete logarithms." In George Robert Blakley and David Chaum (eds.), *Advances in Cryptology: Proceedings of CRYPTO '84.* Santa Barbara, CA: Springer-Verlag, 1985)., 10–18.

Ellis, J. H. "The possibility of secure non-secret digital encryption." UK Communications Electronics Security Group. January 1970. http://web.archive.org/web/20061013203932/www.cesg.gov.uk/site/publications/media/possnse.pdf.

———. "The history of non-secret encryption." *Cryptologia* 23:3 (1999), 267–73. http://www.informaworld.com/10.1080/0161-119991887919.

Ernst, Thomas. "The numerical-astrological ciphers in the third book of Trithemius's Steganographia," *Cryptologia* 22:4 (1998), 318. http://www.informaworld.com/10.1080/0161-119891886957.

Euler, Leonhard. "Theoremata Arithmetica Nova Methodo Demonstrata," *Novi Commentarii Academiae Scientiarum Petropolitanae* 8 (1763), 74–104. http://www.math.dartmouth.edu/~euler/pages/E271.html.

Falconer, John (J. F.). *Rules for Explaining and Decyphering All Manner of Secret Writing, Plain and Demonstrative with Exact Methods for Understanding Intimations by Signs, Gestures, or Speech...* 2nd ed. London: Printed for Dan. Brown... and Sam. Manship ..., 1692.

Feistel, Horst. "Cryptography and computer privacy," *Scientific American* 228:5 (May 1973), 15–23. http://www.apprendre-en-ligne.net/crypto/bibliotheque/feistel/index.html.

Ferguson, Niels, and Bruce Schneier. "Cryptanalysis of Akelarre." In Carlisle Adams and Mike Just (eds.), *Proceedings of the SAC '97 Workshop.* Ottawa, ON: Carleton University, 1997, 201–12.

Ferguson, Niels, Bruce Schneier and Tadayoshi Kohno. *Cryptography Engineering: Design Principles and Practical Applications.* New York: Wiley, 2010. This is a revised edition of *Practical Cryptography*, by Ferguson and Schneier.

Fildes, Jonathan. "iPhone hacker publishes secret Sony PlayStation 3 Key." BBC News Web site, January 6, 2011. http://www.bbc.co.uk/news/technology-12116051.

Five Man Electrical Band. "Signs." Single. Lionel Records. May 1971.

Franksen, Ole Immanuel. "Babbage and cryptography. Or, the mystery of Admiral Beaufort's cipher." *Mathematics and Computers in Simulation* 35:4 (October 1993), 327–67. http://sciencedirect.com/science/article/B6V0T-45GMGDR-34/2/ba2cfbe86bd5e3c8f912778454feb549.

Friedman, William. *Advanced Military Cryptography.* Laguana Hills, CA: Aegean Park Press, 1976.

_____. *Military Cryptanalysis. Part II, Simpler Varieties of Polyalphabetic Substitution Systems.* Cryptographic Series Number 40. Laguna Hills, CA: Aegean Park Press, 1984. http://www.nsagov/public_info/_files/military_cryptanalysis/mil_crypt_II.pdf. Originally published in 1938.

_____. *Military Cryptanalysis. Part III, Simpler Varieties of Aperiodic Substitution Systems.* Cryptographic Series Number 60. Laguna Hills, CA: Aegean Park Press, 1992. http://www.nsa.gov/public_info/_files/military_cryptanalysis/mil_crypt_III.pdf. Reprint of a US military text, originally published in 1939. Declassified December 1992.

_____. *Military Cryptanalysis. Part IV, Transposition and Fractionating Systems.* Cryptographic Series Number 61. Laguna Hills, CA: Aegean Park Press, 1992. http://www.nsa.gov/public_info/_files/military_cryptanalysis/mil_crypt_IV.pdf. Reprint of a US military text, originally published in 1941. Declassified December 1992.

Gaddy, David W. "The first U.S. Government Manual on Cryptography." *Cryptologic Quarterly* 11:4 (1992). https://www.nsa.gov/public_info/_files/cryptologic_quarterly/manual_on_cryptography.pdf.

_____. "Internal struggle: The Civil War." In *Masked Dispatches: Cryptograms and Cryptology in American History, 1775–1900*, 3rd ed. Fort George G. Meade, MD: National Security Agency Center for Cryptologic History, 2013, 88–103.

Gardner, Martin. "Mathematical games: A new kind of cipher that would take millions of years to break," *Scientific American* 237:2 (August 1977), 120–24.

Garfinkel, Simson. *PGP: Pretty Good Privacy.* Sebastopol, CA: O'Reilly Media, 1995.

_____. *Web Security, Privacy and Commerce*, 2nd ed. Sebastopol, CA: O'Reilly Media, 2002. With Gene Spafford.

Garis, Howard Roger. *Uncle Wiggily's Adventures.* New York: A. L. Burt, 1912. http://www.gutenberg.org/ebooks/15281.

Garliński, Józef. *The Enigma War: The Inside Story of the German Enigma Codes and How the Allies Broke Them*, hardcover 1st American ed. New York: Charles Scribners, 1980. Appendix by Tadeusz Lisicki.

Gauss, Carl Friedrich. *Disquisitiones arithmeticae*. New Haven and London: Yale University Press, 1966. Translated by Arthur A. Clarke, S.J.

Gentry, Craig. "Fully homomorphic encryption using ideal lattices," in *Proceedings of the Forty-first Annual ACM Symposium on Theory of Computing*. Association for Computing Machinery Special Interest Group on Algorithms and Computation Theory. New York: ACM, 2009, 169–178.

———. "Computing arbitrary functions of encrypted data," *Communications of the ACM* 53:3 (March 2010), 97.

Gillogly, Jim, and Paul Syverson. "Notes on Crypto '95 invited talks by Morris and Shamir," *Cipher: Electronic Newsletter of the Technical Committe on Security & Privacy, A Technical Committee of the Computer Society of the IEEE*. Electronic issue 9 (September 18, 1995). http://www.ieee-security.org/Cipher/ConfReports/conf-rep-Crypto95.html.

Gisin, Nicolas, et al. "Towards practical and fast quantum cryptography," ArXiv Number quant-ph/0411022. November 3, 2004. http://arxiv.org/abs/quant-ph/0411022.

Givierge, M. *Cours de cryptographie*. Paris: Berger-Levrault, 1925.

Goldreich, Oded, Shafi Goldwasser, and Shai Halevi. "Public-key cryptosystems from lattice reduction problems." In Burton S. Kaliski Jr. (ed.), *Advances in Cryptology—CRYPTO '97*. Berlin/Heidelberg: Springer, 1997, 112–31.

Golic, Jovan Dj. "Cryptanalysis of alleged A5 stream cipher." In Walter Fumy (ed.), *Advances in Cryptology—EUROCRYPT '97: Proceedings of the 16th Annual International Conference on the Theory and Application of Cryptographic Techniques*. Konstanz, Germany: Springer-Verlag, 1997, 239–55.

Golomb, Solomon. *Shift Register Sequences*, rev. ed. Laguna Hills, CA: Aegean Park Press, 1982.

Green, Matthew. "A few more notes on NSA random number generators," A Few Thoughts on Cryptographic Engineering Blog. December 28, 2013. http://blog.cryptographyengineering.com/2013/12/a-few-more-notes-on-nsa-random-number.html.

Grover, Lov K. "A fast quantum mechanical algorithm for database search." In *Proceedings of the Twenty-eighth Annual ACM Symposium on Theory of Computing*. Association for Computing Machinery Special Interest Group on Algorithms and Computation Theory. New York: ACM, 1996, 212–19.

GSM Association. "GSMA statement on media reports relating to the breaking of GSM encryption." Press release (December 30, 2009). http://gsmworld.com/newsroom/press-releases/2009/4490.htm.

Hall, W. J. "The Gromark cipher (Part 1)." *The Cryptogram* 35:2 (April 1969), 25.

Hamer, David H., Geoff Sullivan, and Frode Weierud. "Enigma variations: An extended family of machines." *Cryptologia* 22:3 (1998), 211–29.

Hawking, Stephen W. *A Brief History of Time: From the Big Bang to Black Holes*. Toronto; New York: Bantam, 1988.

Hellman, M. E. "An overview of public key cryptography," *IEEE Communications Magazine* 40:5 (2002), 42–49. http://ieeexplore.ieee.org/xpls/abs_all.jsp?arnumber =1006971.

Hellman, Martin. "Oral history interview by Jeffrey R. Yost." Number OH 375. Charles Babbage Institute, University of Minnesota, Minneapolis (November 22, 2004). http://purl.umn.edu/107353.

Hellman, Martin E., Bailey W. Diffie, and Ralph C. Merkle. "Cryptographic apparatus and method." United States Patent: 4200770 (April 29, 1980). http://www.google .com/patents?vid=4200770.

Hellman, M. E. and S. C. Pohlig. "Exponentiation cryptographic apparatus and method." United States Patent: 4424414 (January 1984). http://www.google.com /patents?vid=4424414.

Hill, Lester S. "Cryptography in an algebraic alphabet." *The American Mathematical Monthly* 36:6 (1929), 306–12. http://www.jstor.org/stable/2298294.

Hitt, Parker. *Manual for the Solution of Military Ciphers.* Fort Leavenworth, KS: Press of the Army Service Schools, 1916.

Hoffstein, Jeffrey, Jill Pipher, and Joseph H. Silverman. "NTRU: A ring-based public key cryptosystem." In Joe P. Buhler (ed.), *Algorithmic Number Theory.* Berlin Heidelberg: Springer, June 1998, 267–88.

———. "Public key cryptosystem method and apparatus." United States Patent: 6081597 (June 27, 2000). http://www.google.com/patents/US6081597. Priority date August 19, 1996.

———. *An Introduction to Mathematical Cryptography,* 2nd ed. New York: Springer, 2014. http://dx.doi.org/10.1007/978-1-4939-1711-2.

Hughes, Richard J., Jane E. Nordholt, Kevin P. McCabe, Raymond T. Newell, Charles G. Peterson, and Rolando D. Somma. "Network-centric quantum communications with application to critical infrastructure protection." ArXiv Number 1305.0305 (May 1, 2013). http://arxiv.org/abs/1305.0305.

Hwang, Won-Young. "Quantum key distribution with high loss: Toward global secure communication." *Physical Review Letters* 91:5 (2003), 057901. http://link.aps.org/doi /10.1103/PhysRevLett.91.057901.

Ivory, James. "Demonstration of a theorem respecting prime numbers." *New Series of The Mathematical Respository* Volume I, Part II (1806), 6–8.

Johnson, Thomas R. *American Cryptology During the Cold War, 1945–1989; Book I: The Struggle for Centralization, 1945–1960.* Volume 5 of *United States Cryptologic History Series VI, The NSA Period, 1952–Present.* Fort George G. Meade, MD: Center for Cryptologic History, National Security Agency, 1995. http://www.nsa.gov/public _info/_files/cryptologic_histories/cold_war_i.pdf.

———. *American Cryptology During the Cold War, 1945–1989; Book III: Retrenchment and Reform, 1972–1980.* Volume 5 of *United States Cryptologic History Series VI, The NSA Period, 1952–Present.* Fort George G Meade, MD: Center for Cryptologic History, National Security Agency, 1995. http://www.nsa.gov/public_info/_files /cryptologic_histories/cold_war_iii.pdf. A differently redacted version is available at http://www.cryptome.org/0001/nsa-meyer.htm.

Kahn, David. "In memoriam: Georges-Jean Painvin." *Cryptologia* 6:2 (1982), 120. http://www.informaworld.com/10.1080/0161-118291856939.

———. "Two Soviet spy ciphers." In *Kahn on Codes: Secrets of the New Cryptology.* New York: Macmillan, 1984, 146–64. Originally presented at the annual convention of the American Cryptogram Association, September 3, 1960, and published that year as a monograph. Later published in a Central Intelligence Agency journal.

———. *Seizing the Enigma: The Race to Break the German U-Boats Codes, 1939-1943,* 1st ed. Boston: Houghton Mifflin, March 1991.

———. *The Codebreakers: The Story of Secret Writing,* rev. ed. New York: Scribner, 1996.

Kaliski, B. S., and Yiqun Lisa Yin. "On the security of the RC5 encryption algorithm." Technical Report Number TR-602, Version 1.0. RSA Laboratories (September 1998). ftp://ftp.rsasecurity.com/pub/rsalabs/rc5/rc5-report.pdf.

Kelly, Thomas. "The myth of the skytale." *Cryptologia* 22 (1998), 244–60.

Kerckhoffs, Auguste. "La cryptographie militaire, I." *Journal des sciences militaires* IX (1883), 5–38.

Kim, Kwangjo, Tsutomu Matsumoto, and Hideki Imai. "A recursive construction method of S-boxes satisfying strict avalanche criterion," in Alfred Menezes and Scott A. Vanstone (eds.), *CRYPTO '90: Proceedings of the 10th Annual International Cryptology Conference on Advances in Cryptology.* Berlin/Heidelberg, New York: Springer, 1991, 564–74.

Kipling, Rudyard. *The Jungle Book.* 1894. http://www.gutenberg.org/ebooks/236.

———. *Just So Stories.* 1902. http://www.gutenberg.org/ebooks/2781.

Klein, Melville. *Securing Record Communications: The TSEC/KW-26.* Fort George G. Meade, MD: Center for Cryptologic History, National Security Agency, 2003. http://www.nsa.gov/about/_files/cryptologic_heritage/publications/misc/tsec_kw26.pdf.

Kleinjung, Thorsten. "Discrete Logarithms in GF(p)—768 bits." Email sent to the NMBRTHRY mailing list. June 16, 2016. https://listserv.nodak.edu/cgi-bin/wa.exe?A2=NMBRTHRY;a0c66b63.1606.

Kleinjung, Thorsten, Kazumaro Aoki, Jens Franke, Arjen Lenstra, Emmanuel Thomé, Joppe Bos, Pierrick Gaudry, et al. "Factorization of a 768-bit RSA modulus." Cryptology ePrint Archive Number 2010/006. 2010. http://eprint.iacr.org/2010/006.

Knudsen, Lars R., and Vincent Rijmen. "Ciphertext-only attack on Akelarre." *Cryptologia* 24:2 (2000), 135–47. http://www.tandfonline.com/doi/abs/10.1080/01611190008984238.

Koblitz, Neal. "Elliptic curve cryptosystems." *Mathematics of Computation* 48:177 (1987), 203–9. http://www.ams.org/journals/mcom/1987-48-177/S0025-5718-1987-0866109-5/.

———. *Random Curves: Journeys of a Mathematician.* Berlin/Heidelberg: Springer, 2008.

Konheim, Alan G. *Cryptography, A Primer.* New York: Wiley, 1981.

———. *Computer Security and Cryptography.* Hoboken, NJ: Wiley-Interscience, 2007.

Korzh, Boris, Charles Ci Wen Lim, Raphael Houlmann, Nicolas Gisin, Ming Jun Li, Daniel Nolan, Bruno Sanguinetti, Rob Thew, and Hugo Zbinden. "Provably secure and practical quantum key distribution over 307 km of optical fibre." *Nature Photonics* 9:3 (March 2015), 163–68.

Kotel'nikova, Natal'ya V. "Vladimir Aleksandrovich Kotel'nikov: The life's journey of a scientist." *Physics-Uspekhi* 49:7 (2006), 727–36. http://www.iop.org/EJ/abstract /1063-7869/49/7/A05.

Kravets, David. "Sony settles PlayStation hacking lawsuit." Wired Magazine Web site, April 11, 2011. http://www.wired.com/2011/04/sony-settles-ps3-lawsuit/.

Kruh, Louis, and C. A. Deavours. "The Typex Cryptograph." *Cryptologia* 7:2 (1983), 145. http://www.informaworld.com/10.1080/0161-118391857874.

Kullback, Solomon. *Statistical Methods in Cryptanalysis.* Laguna Hills, CA: Aegean Park Press, 1976. Originally published in 1938.

Landau, Susan. "Communications security for the twenty-first century: The Advanced Encryption Standard." *Notices of the AMS* 47:4 (April 2000), 450–59.

———. "Standing the test of time: The Data Encryption Standard." *Notices of the AMS* 47:3 (March 2000), 341–49.

———. *Surveillance or Security? The Risks Posed by New Wiretapping Technologies.* Cambridge, MA: MIT Press, 2011.

Lange, André, and Émile-Arthur Soudart. *Treatise on Cryptography* (Washington, DC: U.S. Government Printing Office, 1940). Laguna Hills, CA: Aegean Park Press Reprint.

Levy, Steven. *Crypto: How The Code Rebels Beat The Government—Saving Privacy In The Digital Age.* 1st paperback ed. New York: Penguin (Non-Classics), January 2002.

Lewand, Robert Edward. *Cryptological Mathematics.* Washington, DC: The Mathematical Association of America (December 2000).

Lidl, R., and H. Niederreiter. *Introduction to Finite Fields.* Cambridge, UK: Cambridge University Press, 1986.

Lomonaco, Samuel J. Jr. "A quick glance at quantum cryptography." *Cryptologia* 23:1 (1999), 1–41. http://www.informaworld.com/10.1080/0161-119991887739.

———. "A talk on quantum cryptography, or how Alice outwits Eve." In David Joyner (ed.), *Coding Theory and Cryptography: From Enigma and Geheimschreiber to Quantum Theory.* Berlin/Heidelberg; New York: Springer, January 2000, 144–74. A revised version is available from http://arxiv.org/abs/quant-ph/0102016.

Lydersen, Lars, Carlos Wiechers, Christoffer Wittmann, Dominique Elser, Johannes Skaar, and Vadim Makarov. "Hacking commercial quantum cryptography systems by tailored bright illumination," *Nature Photonics* 4:10 (October 2010), 686–689. http://dx.doi.org/10.1038/nphoton.2010.214.

Madryga, W. E. "A high performance encryption algorithm." In James H. Finch and E. Graham Dougall (eds.), *Proceedings of 2nd IFIP International Conference on Computer Security: a Global Challenge.* Amsterdam: North-Holland, 1984, 557–69.

Mahoney, Michael. *The Mathematical Career of Pierre de Fermat (1601–1665).* Princeton, NJ: Princeton University Press, 1973.

Marks, Leo. *Between Silk and Cyanide: A Codemaker's War, 1941–1945.* 1st US ed. New York: Free Press, June 1999.

Martin-Lopez, Enrique, Anthony Laing, Thomas Lawson, Roberto Alvarez, Xiao-Qi Zhou, and Jeremy L. O'Brien. "Experimental realisation of Shor's quantum factoring algorithm using qubit recycling." *Nature Photonics* 6:11 (November 2012), 773–76. http://www.nature.com/nphoton/journal/v6/n11/full/nphoton.2012.259.html.

Massey, J. "A new multiplicative algorithm over finite fields and its applicability in public-key cryptography." Presentation at EUROCRYPT '83 (March 21–25, 1983).

Massey, J. L. "An introduction to contemporary cryptology." *Proceedings of the IEEE* 76:5 (1988), 533–49.

Massey, James L., and Jimmy K. Omura. "Method and apparatus for maintaining the privacy of digital messages conveyed by public transmission." United States Patent: 4567600 (January 28, 1986). http://www.google.com/patents?vid=4567600.

McSherry, Corynne. "Sony v. Hotz ends with a whimper, I mean a gag order." Electronic Frontier Foundation Deeplinks Blog (April 12, 2011). https://www.eff.org /deeplinks/2011/04/sony-v-hotz-ends-whimper-i-mean-gag-order.

Mendelsohn, C. J. "Blaise de Vigenère and the 'Chiffre Carré.'" *Proceedings of the American Philosophical Society* 82:2 (1940), 103–29.

Menezes, Alfred J., Paul C. van Oorschot, and Scott A. Vanstone. *Handbook of Applied Cryptography*. Boca Raton, FL: CRC, October 1996. The full text is available online at http://www.cacr.math.uwaterloo.ca/hac/.

Merkle, Ralph. "CS 244 project proposal" (Fall 1974). http://merkle.com/1974/CS244 ProjectProposal.pdf.

———. "Secure communications over insecure channels." *Communications of the Association for Computing Machinery* 21:4 (April 1978), 294–99.

Micciancio, Daniele, and Oded Regev. "Lattice-based cryptography." In Daniel J. Bernstein, Johannes Buchmann, and Erik Dahmen (eds.), *Post-Quantum Cryptography* Springer Berlin Heidelberg, 2009, 147–91. http://link.springer.com /chapter/10.1007/978-3-540-88702-7_5.

Mikkelson, Barbara, and David Mikkelson. "Just the facts." snopes.com (December 13, 2008). http://www.snopes.com/radiotv/tv/dragnet.asp.

Miller, Gary L. "Riemann's hypothesis and tests for primality." In *Proceedings of Seventh Annual ACM Symposium on Theory of Computing*. Association for Computing Machinery Special Interest Group on Algorithms and Computation Theory. New York: ACM, 1975, 234–39.

Miller, V. "Use of elliptic curves in cryptography." In Hugh C. Williams (ed.), *Advances in Cryptology–CRYPTO '85 Proceedings*. Berlin: Springer, 1986, 417–26.

Milne, A. A. *Winnie-the-Pooh*. Reissue ed. New York: Puffin, August 1992.

Molotkov, Sergei N. "Quantum cryptography and V A Kotel'nikov's one-time key and sampling theorems." *Physics-Uspekhi* 49:7 (2006), 750–61. http://www.iop.org/EJ /abstract/1063-7869/49/7/A09.

Monty Python. "Decomposing composers." *Monty Python's Contractual Obligation Album*. Charisma Records. 1980.

Morris, Robert. "The Hagelin cipher machine (M-209): Reconstruction of the internal settings." *Cryptologia* 2:3 (1978), 267. http://www.informaworld.com/10.1080/0161 -117891853126.

NBS. "Guidelines for implementing and using the NBS Data Encryption Standard." Federal Information Processing Standards Number 74. NBS. April 1981. https:// www.thc.org/root/docs/cryptography/fips74.html.

Neal, Dave. "AES encryption is cracked." *The Inquirer* (August 17, 2011). http://www .theinquirer.net/inquirer/news/2102435/aes-encryption-cracked.

Nechvatal, James, Elaine Barker, Lawrence Bassham, William Burr, Morris Dworkin, James Foti, and Edward Roback. Report on the development of the Advanced

Encryption Standard (AES). NIST (October 2000). http://csrc.nist.gov/archive/aes /round2/r2report.pdf.

Nguyen, Phong. "Cryptanalysis of the Goldreich-Goldwasser-Halevi cryptosystem from Crypto'97." In Michael Wiener (ed.), *Advances in Cryptology—CRYPTO '99*, Berlin Heidelberg: Springer, August 1999, 288–304.

Nguyen, Phong Q, and Oded Regev. "Learning a parallelepiped: Cryptanalysis of GGH and NTRU signatures." *Journal of Cryptology* 22:2 (November 2008), 139–60.

NIST. "Computer data authentication." Federal Information Processing Standards Number 113 (May 1985). http://csrc.nist.gov/publications/fips/fips113/fips113.html.

_____. "Announcing request for candidate algorithm nominations for the Advanced Encryption Standard (AES)." *Federal Register* 62:177 (September 1997), 48051–58. http://csrc.nist.gov/archive/aes/pre-round1/aes_9709.htm.

_____. "Announcing the Advanced Encryption Standard (AES)." Federal Information Processing Standards Number 197 (November 2001). http://csrc.nist.gov/publications /fips/fips197/fips-197.pdf.

_____. "NIST removes cryptography algorithm from random number generator recommendations." NIST Tech Beat Blog (April 21, 2014). http://www.nist.gov/itl /csd/sp800-90-042114.cfm.

NIST Computer Security Division. "Computer Security Resource Center: Current modes." http://csrc.nist.gov/groups/ST/toolkit/BCM/current_modes.html.

NSA. "GSM classification guide" (September 20, 2006). https://s3.amazonaws.com/s3 .documentcloud.org/documents/888710/gsm-classification-guide-20-sept-2006.pdf.

_____. "Summer mathematics, R21, and the Director's Summer Program." *The EDGE: National Information Assurance Research Laboratory (NIARL) Science, Technology, and Personnel Highlights.* September 2008. http://www.spiegel.de/media/media -35550.pdf.

NSA/CSS. "Fact sheet NSA Suite B cryptography." NSA/CSS Web site. http://wayback. archive.org/web/20051125141648/http://www.nsa.gov/ia/industry/crypto_suite_b .cfm. Archived by the *Internet Archive* from http://www.nsa.gov/ia/industry/ crypto_suite_b.cfm on November 25, 2005.

_____. "The case for elliptic curve cryptography." NSA/CSS Web site (January 15, 2009). http://wayback.archive.org/web/20131209051540/http://www.nsa.gov /business/programs/elliptic_curve.shtml. Archived by the *Internet Archive* from http://www.nsa.gov/business/programs/elliptic_curve.shtml on December 9, 2013.

_____. "Cryptography today." NSA/CSS Web site (August 19, 2015). https://www.nsa .gov/ia/programs/suiteb_cryptography/index.shtml.

NSA Research Directorate Staff. "Securing the cloud with homomorphic encryption." *The Next Wave* 20:3 (2014). https://www.nsa.gov/research/tnw/tnw203/articles/pdfs /TNW203_article5.pdf.

OTP VPN Exploitation Team. "Intro to the VPN exploitation process." (September 13, 2010). http://www.spiegel.de/media/media-35515.pdf.

Paget, Chris, and Karsten Nohl. "GSM: SRSLY?" Slides from lecture presented at 26th Chaos Communication Congress (December 27, 2009). http://events.ccc.de/congress /2009/Fahrplan/events/3654.en.html.

Pease, Roland. "'Unbreakable' encryption unveiled." BBC News Web site (October 9, 2008). http://news.bbc.co.uk/2/hi/science/nature/7661311.stm.

People of the GnuPG Project. "GnuPG frequently asked questions." October 23, 2014. https://gnupg.org/faq/gnupg-faq.html.

Perlner, Ray A., and David A. Cooper. "Quantum resistant public key cryptography: A survey." In Kent Seamons, Neal McBurnett, and Tim Polk (eds.), *Proceedings of the 8th Symposium on Identity and Trust on the Internet.* New York: ACM Press, 2009, 85–93.

Perlroth, Nicole. "Government announces steps to restore confidence on encryption standards." New York Times Web site (September 10, 2013). http://bits.blogs .nytimes.com/2013/09/10/government-announces-steps-to-restore -confidence-on-encryption-standards/.

Plutarch. *Plutarch's Lives.* London; New York: W. Heinemann; Macmillan, 1914. http:// penelope.uchicago.edu/Thayer/E/Roman/Texts/Plutarch/Lives. Translated by Bernadotte Perrin.

Poe, Edgar Allen. "A few words on secret writing." *Graham's Magazine* 19:1 (July 1841), 33–38.

Pohlig, S. and M. Hellman. "An improved algorithm for computing logarithms over GF(p) and its cryptographic significance (corresp.)" *IEEE Transactions on Information Theory* 24 (1978), 106–10.

Polybius. *The Histories.* Cambridge, MA: Harvard University Press, 1922–1927. http://penelope.uchicago.edu/Thayer/E/Roman/Texts/Polybius. Translated by W. R. Paton.

Pomerance, Carl. "A tale of two sieves." *Notices of the American Mathematical Society.* 43:12 (December 1996), 1473–85. http://www.ams.org/notices/199612/pomerance.pdf.

Poppe, A., A. Fedrizzi, R. Ursin, H. Böhm, T. Lorünser, O. Maurhardt, M. Peev, et al. "Practical quantum key distribution with polarization entangled photons." *Optics Express* 12:16 (2004), 3865–71. http://www.opticsexpress.org/abstract.cfm? URI=oe-12-16-3865.

Proc, Jerry. "Hagelin C-362." http://www.jproc.ca/crypto/c362.html.

Qualys SSL Labs. "User agent capabilities." 2015. https://www.ssllabs.com/ssltest /clients.html.

Rabin, Michael O. "Probabilistic algorithm for testing primality." *Journal of Number Theory* 12:1 (February 1980), 128–38. http://dx.doi.org/10.1016/0022-314X(80)90084-0.

Reeds, Jim. "Solved: The ciphers in Book III of Trithemius's Steganographia." *Cryptologia* 22:4 (1998), 291. http://www.informaworld.com/10.1080 /0161-119891886948.

Reinke, Edgar C. "Classical cryptography." *The Classical Journal* 58:3 (December 1962), 113–21. http://www.jstor.org/stable/3295135.

Reuvers, Paul, and Marc Simons. "Fialka" (May 26, 2015). http://www.cryptomuseum .com/crypto/fialka/.

Rijmen, Vincent. "The Rijndael page." http://www.ktana.eu/html/theRijndaelPage.htm. Formerly at http://www.esat.kuleuven.ac.be/~rijmen/rijndael.

Rivest, R. L., A. Shamir and L. Adleman. "A method for obtaining digital signatures and public-key cryptosystems." *Communications of the Association for Computing Machinery* 21:2 (1978), 120–26.

Rivest, Ronald L. "The RC5 Encryption Algorithm." In Bart Preneel (ed.), *Fast Software Encryption.* Berlin/Heidelberg: Springer, January 1995, 86–96.

Rivest, Ronald L., Len Adleman, and Michael L. Dertouzos. "On data banks and privacy homomorphisms." In Richard A. DeMillo, David P. Dobkin, Anita K. Jones, and Richard J. Lipton (eds.), *Foundations of Secure Computation.* New York: Academic Press, 1978, 165–79. https://people.csail.mit.edu/rivest/pubs/RAD78.pdf.

Rivest, Ronald L., M.J.B. Robshaw, Ray Sidney, and Yigun Lisa Yin. "The RC6™ block cipher." NIST. August 1998. ftp://cs.usu.edu.ru/crypto/RC6/rc6v11.pdf, series AES Proposals. Version 1.1.

Rivest, Ronald L., Adi Shamir, and Leonard M. Adleman. "Cryptographic communications system and method." United States patent: 4405829. September 20, 1983. http://www.google.com/patents?vid=4405829.

_____. "A method for obtaining digital signatures and public-key cryptosystems." Technical Memo Number MIT-LCS-TM-082. MIT. April 4, 1977. http://publications .csail.mit.edu/lcs/specpub.php?id=81.

Rivest, Ronald L., and Alan T. Sherman. "Randomized Encryption Techniques." In David Chaum, Ronald L. Rivest, and Alan T. Sherman (eds.), *Advances in Cryptology: Proceedings of CRYPTO '82.* New York: Plenum Press, 1983, 145–63.

Robshaw, Matthew, and Olivier Billet (eds.). *New Stream Cipher Designs: The eSTREAM Finalists.* Berlin; New York: Springer, 2008.

Sachkov, Vladimir N. "V A Kotel'nikov and encrypted communications in our country." *Physics-Uspekhi* 49:7 (2006), 748–50. http://www.iop.org/EJ/abstract/1063-7869/49/7 /A08.

Sakurai, K., and H. Shizuya. "A structural comparison of the computational difficulty of breaking discrete log cryptosystems." *Journal of Cryptology* 11:1 (1998), 29–43. http://www.springerlink.com/content/ykxnr0e24p80h9x3/.

Sasaki, M., M. Fujiwara, H. Ishizuka, W. Klaus, K. Wakui, M. Takeoka, S. Miki, et al. "Field test of quantum key distribution in the Tokyo QKD Network." *Optics Express* 19:11 (May 2011), 10387. https://www.opticsinfobase.org/oe/fulltext.cfm? uri=oe-19-11-10387&id=213840.

Scarani, Valerio, Antonio Acín, Grégoire Ribordy, and Nicolas Gisin. "Quantum cryptography protocols robust against photon number splitting attacks for weak laser pulse implementations." *Physical Review Letters* 92:5 (February 6, 2004), 057901. http://link.aps.org/doi/10.1103/PhysRevLett.92.057901.

Schmitt-Manderbach, Tobias, Henning Weier, Martin Fürst, Rupert Ursin, Felix Tiefenbacher, Thomas Scheidl, Josep Perdigues, et al. "Experimental demonstration of free-space decoy-state quantum key distribution over 144 km." *Physical Review Letters* 98:1 (January 2007), 010504. http://link.aps.org/doi/10.1103/PhysRevLett.98 .010504.

Schneier, Bruce. *Applied Cryptography: Protocols, Algorithms and Source Code in C.* 2d ed. New York: Wiley, 1996.

_____. *Data and Goliath: The Hidden Battles to Collect Your Data and Control Your World.* New York: Norton, 2015.

_____. "Did NSA put a secret backdoor in new encryption standard?" *Wired Magazine* Web site. November 15, 2007. http://archive.wired.com/politics/security/ commentary/securitymatters/2007/11/securitymatters_1115.

———. "NSA surveillance: A guide to staying secure." *The Guardian* (September 9, 2013). http://www.theguardian.com/world/2013/sep/05/nsa-how-to-remain-secure-surveillance.

———. *Secrets and Lies: Digital Security in a Networked World.* New York: Wiley, 2011.

Seuss, Dr. *Horton Hatches the Egg.* New York: Random House, 1940.

Shakespeare, William. *Julius Caesar.* 1599. http://www.gutenberg.org/ebooks/2263.

Shamir, A., R. L. Rivest, and L. M. Adleman. "Mental poker." In David A. Klarner (ed.), *The Mathematical Gardner.* Boston: Prindle, Weber & Schmidt; Belmont, CA: Wadsworth International, 1981, 37–43.

Shamir, Adi, Ronald L. Rivest and Leonard M. Adelman. "Mental poker." Technical Memo Number MIT-LCS-TM-125. MIT. February 1, 1979. http://publications.csail.mit.edu/lcs/specpub.php?id=124.

Shannon, C. E. "Communication theory of secrecy systems." *Bell System Technical Journal* 28:4 (1949), 656–715.

Shields, Andrew, and Zhiliang Yuan. "Key to the quantum industry." *Physics World* 20:3 (March 1, 2007), 24–29. http://physicsworld.com/cws/article/print/27161.

Shor, P. W. "Algorithms for quantum computation: Discrete logarithms and factoring." In *Proceedings, 35th Annual Symposium on Foundations of Computer Science.* IEEE Computer Society Technical Committee on Mathematical Foundations of Computing. Los Alamitos, CA: IEEE, 1994, 124–134.

Shumow, Dan, and Niels Ferguson. "On the possibility of a back door in the NIST SP800-90 Dual EC PRNG." Slides from presentation at Rump Session of CRYPTO 2007. August 21, 2007. http://rump2007.cr.yp.to/15-shumow.pdf.

Silverman, Joseph H. *A Friendly Introduction to Number Theory.* 3d ed. Englewood Cliffs, NJ: Prentice Hall, 2005.

Solovay, R., and V. Strassen. "A fast Monte-Carlo test for primality." *SIAM Journal on Computing* 6:1 (March 1977), 84–85. http://link.aip.org/link/?SMJ/6/84/1.

Soltani, Ashkan, and Craig Timberg. "T-Mobile quietly hardens part of its U.S. cellular network against snooping." *The Washington Post* (October 22, 2014). https://www.washingtonpost.com/blogs/the-switch/wp/2014/10/22/t-mobile-quietly-hardens-part-of-its-u-s-cellular-network-against-snooping/.

Spiegel Staff. "Prying eyes: Inside the NSA's war on Internet security." *Spiegel Online* (December 28 2014). http://www.spiegel.de/international/germany/inside-the-nsa-s-war-on-internet-security-a-1010361.html. Translated from the German edition of *Der Speigel.*

Stallings, William. *Cryptography and Network Security: Principles and Practice.* 6th ed. Boston: Pearson, 2014.

Stevenson, Frank A. "[A51] Cracks beginning to show in A5/1…" Email sent to the A51 mailing list. May 1, 2010. http://lists.lists.reflextor.com/pipermail/a51/2010-May/000605.html.

Stevenson, Robert Louis. *Treasure Island.* London: Cassell, 1883. http://www.gutenberg.org/ebooks/120.

Strachey, Edward. "The soldier's duty." *The Contemporary Review* XVI (February 1871), 480–85.

Stucki, D., M. Legré, F. Buntschu, B. Clausen, N. Felber, N. Gisin, L. Henzen, et al. "Long-term performance of the SwissQuantum quantum key distribution network in a field environment." *New Journal of Physics* 13:12 (December 2011), 123001. http:// iopscience.iop.org/1367-2630/13/12/123001.

Suetonius. *The Divine Augustus*. New York: R. Worthington, 1883. http://www .fordham.edu/halsall/ancient/suetonius-augustus.html. Translated by Alexander Thomson.

――――. *De Vita Caesarum, Divus Iulius (The Lives of the Caesars, The Deified Julius)*. Cambridge, MA: Harvard University Press, 1920. http://www.fordham.edu/halsall /ancient/suetonius-julius.html. Translated by J. C. Rolfe.

Timberg, Craig, and Ashkan Soltani. "By cracking cellphone code, NSA has ability to decode private conversations." *The Washington Post* (December 13, 2013). http:// www.washingtonpost.com/business/technology/by-cracking-cellphone-code -nsa-has-capacity-for-decoding-private-conversations/2013/12/13/e119b598-612f -11e3-bf45-61f69f54fc5f_story.html.

Trappe, Wade, and Lawrence C. Washington. *Introduction to Cryptography with Coding Theory*. 2nd ed. Upper Saddle River, NJ: Prentice Hall, 2005.

Twain, Mark. *The Adventures of Tom Sawyer*. 1876. http://www.gutenberg.org /ebooks/74.

van der Meulen, Michael. "The road to German diplomatic ciphers—1919 to 1945." *Cryptologia* 22:2 (1998), 141–66. http://www.informaworld.com/10.1080/0161 -119891886858.

Vandersypen, Lieven M. K., Matthias Steffen, Gregory Breyta, Costantino S. Yannoni, Mark H. Sherwood and Isaac L. Chuang. "Experimental realization of Shor's quantum factoring algorithm using nuclear magnetic resonance." *Nature* 414:6866 (December 20, 2001), 883–87. http://dx.doi.org/10.1038/414883a.

Vansize, William V. "A new page-printing telegraph." *Transactions of the American Institute of Electrical Engineers* 18 (1902), 7–44.

Vernam, Gilbert. "Secret signaling system." United States Patent: 1310719. July 1919. http://www.google.com/patents?vid=1310719.

Vigenère, Blaise de. *Traicté des Chiffres, ou Secrètes Manières d'Escrire (Treatise on Ciphers, or Secret Methods of Writing)*. Paris: A. L'Angelier, 1586. http://gallica.bnf .fr/ark:/12148/bpt6k1040608n.

Wang, Jian-Yu, Bin Yang, Sheng-Kai Liao, Liang Zhang, Qi Shen, Xiao-Fang Hu, Jin-Cai Wu, et al. "Direct and full-scale experimental verifications towards ground-satellite quantum key distribution." *Nature Photonics* 7:5 (April 21, 2013), 387–93.

Weber, Arnd (ed.). "Secure communications over insecure channels (1974), by Ralph Merkle, with an interview from the year 1995." (January 16, 2002), http://www.itas. kit.edu/pub/m/2002/mewe02a.htm.

Weisner, Louis, and Lester Hill. "Message protector." United States Patent: 1845947. February 16, 1932. http://www.google.com/patents?vid=1845947.

Wenger, Erich, and Paul Wolfger. "Harder, better, faster, stronger: elliptic curve discrete logarithm computations on FPGAs." *Journal of Cryptographic Engineering* (September 3, 2015), 1–11.

Wiesner, Stephen. "Conjugate coding." *SIGACT News* 15:1 (1983), 78–88. http://portal .acm.org/citation.cfm?id=1008920.

Williamson, M. J. "Non-secret encryption using a finite field." UK Communications Electronics Security Group. January 21, 1974. http://www.cesg.gov.uk/publications /media/secenc.pdf (accessed 2011-01-02).

Williamson, Malcolm. "Thoughts on cheaper non-secret encryption." UK Communications Electronics Security Group. August 10, 1976. http://web.archive. org/web/20070107090748/http://www.cesg.gov.uk/site/publications/media/cheapnse .pdf.

Xu, Feihu, Bing Qi, and Hoi-Kwong Lo. "Experimental demonstration of phase-remapping attack in a practical quantum key distribution system." *New Journal of Physics* 12:11 (2010), 113026. http://iopscience.iop.org/1367-2630/12/11 /113026.

Xu, Nanyang, Jing Zhu, Dawei Lu, Xianyi Zhou, Xinhua Peng, and Jiangfeng Du. "Quantum factorization of 143 on a dipolar-coupling nuclear magnetic resonance system." *Physical Review Letters* 108:13 (March 30, 2012), 130501. http://link.aps.org /doi/10.1103/PhysRevLett.108.130501.

Zhao, Yi, Chi-Hang Fred Fung, Bing Qi, Christine Chen, and Hoi-Kwong Lo. "Quantum hacking: Experimental demonstration of time-shift attack against practical quantum-key-distribution systems." *Physical Review A* 78:4 (October 2008), 042333. http://link.aps.org/doi/10.1103/PhysRevA.78.042333.

Zumbrägel, Jens. "Discrete logarithms in GF($2^{\wedge}9234$)." E-mail sent to the NMBRTHRY mailing list. January 31, 2014. https://listserv.nodak.edu/cgi-bin/wa.exe? A2=NMBRTHRY;9aa2b043.1401.

■ ■ ■ INDEX ■ ■ ■